The Pathogenic Spirochetes: Strategies for Evasion of Host Immunity and Persistence

Monica E. Embers
Editor

The Pathogenic Spirochetes: Strategies for Evasion of Host Immunity and Persistence

Springer

Editor
Monica E. Embers
Covington, LA, USA

ISBN 978-1-4614-5403-8 ISBN 978-1-4614-5404-5 (eBook)
DOI 10.1007/978-1-4614-5404-5
Springer New York Heidelberg Dordrecht London

Library of Congress Control Number: 2012948596

Printed on acid-free paper

Springer is part of Springer Science+Business Media (www.springer.com)

"He who knew syphilis knew medicine"
-Sir William Osler

There was a young man from Back Bay
Who thought syphilis just went away
He believed that a chancre
Was only a canker
That healed in a week and a day.

But now he has 'acne vulgaris'
Or whatever they call it in Paris;
On his skin it has spread
From his feet to his head,
And his friends want to know where his hair is.

There's more to his terrible plight:
His pupils won't close in the light
His heart is cavorting,
His wife is aborting,
And he squints through his gun-barrel sight.

Arthralgia cuts into his slumber;
His aorta is in need of a plumber;
But now he has tabes,
And saber-shinned babies,
While of gummas he has quite a number.

He's been treated in every known way,
But his spirochetes grow day by day;
He's developed paresis,
Has long talks with Jesus,
And thinks he's the Queen of the May.

Anonymous

Introduction

Long-term survival by an infectious microbe within the host requires the ability to fend off the immune response or become undetectable. Pathogens may utilize varied strategies for this purpose, but the spirochetes are uniquely adapted to combat host immunity on several fronts to ensure persistence for long periods of time. The symptoms and disease course of these systemic pathogens reflect their ability to spread through the host without generating acute responses from immune system recognition. The pathogenic spirochetes are indeed masters of immune evasion and persistence.

The Order Spirochaetales contains three families, encompassing nine Genuses, three of which are associated with disease in humans: Borrelia, Leptospira, and Treponema. Most research has been focused on the species from which the more common diseases result: Lyme disease, Syphilis, Leptospirosis, relapsing fever, and periodontal disease. In general, these organisms are genetically complex and utilize a unique variety of virulence factors, many known and likely many yet to be discovered. Each, however, has distinctly become adapted to different environments for sustenance and transmission.

For the tick-borne *Borreliae*, immune evasion essentially begins as the spirochetes traverse the salivary glands of the tick into the mammalian host. We know that factors in tick saliva inhibit the immune response both generally, and in specific assistance to the *Borrelia* spirochetes. From that point, dissemination, tissue invasion, and seclusion, both immunological and physical, ensues. The spirochete bacteria are well adapted for tissue invasion, with their small-diameter, long-width, and periplasmic flagella. Other commonalities amongst the spirochetes are their fastidious growth in microaerophilic environments and the systemic nature of disease caused by them.

In this book, we explore the many mechanisms by which the most prevalent Spirochetal pathogens persist in a healthy immune-competent host. Among them are the direct and indirect suppression of host immune signals, phase and antigenic variation, escaping recognition by host complement proteins, and seclusion into immune-privileged sites. We also explore antibiotic therapy for control of infection. A baffling topic which warrants further research.

Much of the emphasis for this compendium on the ability of the spirochetal pathogens to evade immune responses in the host and persist falls on the Lyme disease

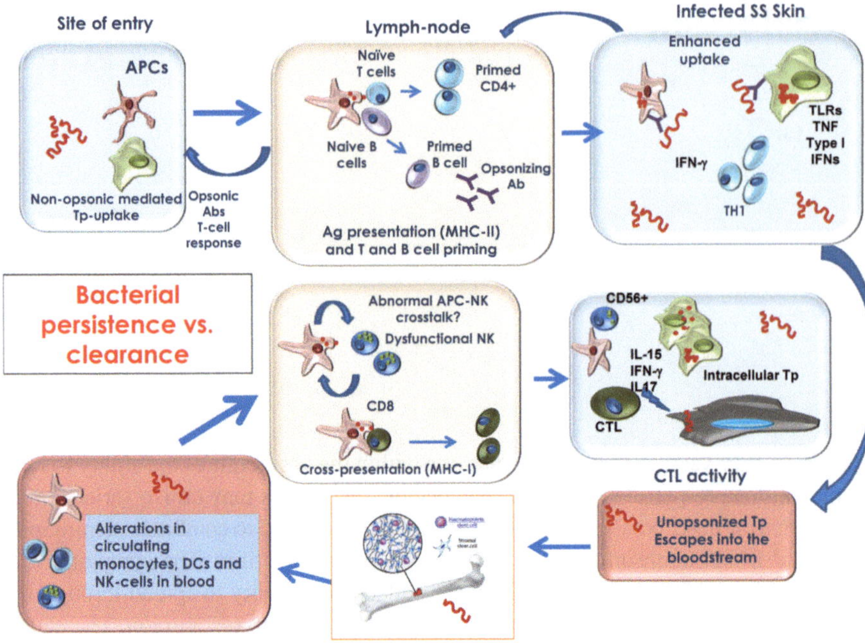

Fig. 1 Model of syphilis pathogenesis that integrates innate and adaptive immune responses to the bacterium (kindly provided by J. Salazar)

spirochete, *Borrelia burgdorferi*. This is largely because it is the most studied of the spirochetes and in part due to impact of disease caused by this organism in the USA and Europe.

The Treponemes

Spirochetes of the Treponema genus cause diseases of current and historically high prevalence worldwide, such as syphilis (*T. pallidum* subspecies pallidum) and periodontitis (*T. denticola*). Less prevalent are T. pallidum ssp. responsible for Bejel and Yaws, and *T. carateum*, responsible for the disease named Pinta in Central and South America. Research over the years on *T. pallidum* has revealed multiple mechanisms of immune evasion, affecting the host response at many levels.

The progression of syphilis follows several different phases (primary, secondary, and tertiary) with variation amongst individuals. At each level of infection, a shift in the balance of immune response versus immune evasion by *T. pallidum* can lead to either clearance or persistence. Using past and current research findings, Dr. Juan C. Salazar has proposed a model of immunologic events in the early stages of syphilis. As shown in Fig. 1, phagocytosis by antigen-presenting cells (APCs) is central to the immune response against *T. pallidum*. After primary infection, spirochetes replicate at the site, elude the innate response, and rapidly disseminate to the skin and other tissues. Initial uptake by phagocytic cells is inefficient due to absence of opsonic antibodies, but the small number that have phagocytosed spirochetes migrate to the draining lymph nodes where they present antigen to naïve T and B cells. Newly sensitized T helper cells return to the primary lesion (chancre), where they recognize *T. pallidum* antigens on APCs and secrete interferon-γ (IFN-γ), which activates tissue macrophages. Concomitant increases in opsonic antibodies facilitate phagocytosis and clearance, resulting in chancre resolution. Meanwhile, treponemal spirochete densities at other sites in the skin begin to increase and trigger local inflammatory responses, resulting in the skin lesions characteristic of secondary syphilis. While primed $CD4^+$ and $CD8^+$ T cells, along with opsonic antibodies may be present, the outer membrane of the spirochete provides few antigenic targets and some treponemal subpopulations may become resistant to opsonophagocytosis. The predominance of $CD8^+$ T cells in secondary syphilis lesions may indicate a misdirected cellular immune response resulting from inefficient $CD4^+$ T cell priming in the lymph nodes. Lack of effective opsonization will, thus, prevent uptake by tissue-based macrophages, leading to further dissemination through the bloodstream to other end organs and tissues, including the bone marrow. In this compartment, *T. pallidum* may affect development of monocytes, dendritic cells, and natural killer (NK) cells. Over time, the generation of more antigen-specific T cells, IFN-γ-producing $CD56^+$ NK cells, and treponeme-specific opsonic antibodies may then lead to control of spirochete infection.

Four chapters of this book relate to the Treponema species. Two chapters are dedicated to research findings on the mechanisms of invasion and establishment of infection by the agents of Syphilis (S. Houston and C. Cameron) and periodontal disease (J. McDowell and R. Marconi). The subject of molecular genetic variation of a specific antigen, TprK, by the *T. pallidum* spirochete is included in a subsequent chapter by L. Giacani and A. Centurion-Lara. Finally, the potential for syphilis spirochetes to resist antibiotic treatment is explored in a chapter authored by L. Stamm.

The Leptospires

Leptospirosis is the most common zoonosis worldwide, but is relatively rare in humans, with 100–200 cases in the USA (>50% in Hawaii); however, this is probably an underestimate. The disease is caused by the spirochete *Leptospira interrogans* (multiple serotypes), which infects via invasion of intact mucous membranes or abraded skin. Leptospirosis was first described by Adolf Weil in 1886 as an "acute infectious disease with enlargement of spleen, jaundice and nephritis." Interestingly, leptospirosis was thought to be a primary cause of death in Native Americans in 1620 (present-day Massachusetts) before the arrival of Pilgrims. A comprehensive and elegantly written chapter by S. Faisal, S. McDonough, and Yung-Fu Chang describes the disease and pathogenic mechanisms utilized by *L. interrogans*.

The Borreliae

The Lyme disease spirochete. While pathogenic spirochetes are found worldwide, those associated with Lyme disease are prevalent in Europe and the USA, with emergence in the Asian continent. Even within the USA, the disease appears to be slowly expanding from northern States to the middle and southern USA. New species have also been isolated and identified, such as *Borrelia lonestari* (Fig. 2) and *Borrelia bissettii* from the southern USA. However, their actual pathogenicity and prevalence remain to be elucidated.

Since the discovery of the cause–effect relationship between *B. burgdorferi* and Lyme disease, only a few studies have probed into the acquired immune response to this pathogen specifically. Very recent studies have uncovered not only multiple mechanisms utilized by the pathogen to escape immune recognition but also a disconcerted response to infection of mice, specifically with respect to the B cell response.

It has become clear that protection afforded by experimental animal models against *B. burgdorferi* relies on the humoral (antibody) response (Barthold and Bockenstedt 1993; Barthold 1999; Fikrig et al. 1994; McKisic and Barthold 2000; Belperron et al. 2007). In several of these studies, passive transfer of immune serum either prevented infection or controlled immune-mediated disease pathology. However, while *B. burgdorferi* antigens generate antibodies that can be protective, the adaptation and selection of spirochetes throughout the course of infection preclude clearance by the host antibody response. This is due in large part to molecular adaptation by these spirochetes. A gene array analysis of the mRNA expression of 137 putative lipoprotein genes revealed two phases of expression (Liang et al. 2002). At primary infection, prior to the generation of significant antibody titers, most (116 of 137) lipoprotein genes were transcribed. In the second phase, at the advent of the humoral response, between 17 and 30 days post-infection, transcription was downregulated for most of these genes. This result suggests that immune selection or signaling may drive the molecular adaptation. More recent results from the Liang lab indicate that *Borrelia burgdorferi* lipoproteins may serve to protect the spirochetes from innate responses. In fact, the exact identity of the lipoprotein constitution may not be as important as the simple presence of the lipoprotein cloak itself (Xu et al. 2008).

As the story of immune evasion by *B. burgdorferi* unfolds, it appears that this spirochete has somehow developed the means to delay a proper immune response, allowing its molecular adaptation and dissemination to occur, unrecognized. The absence of immunity to reinfection by *B. burgdorferi* in immune-competent hosts (Piesman et al. 1997; Nowakowski et al. 2003) prompted inquiry into the apparent inability of these hosts to generate an effective memory response. Early infection with *B. burgdorferi* was shown to elicit hypercellularity in the draining lymph nodes of infected mice (Fig. 3), seen as lymphadenopathy (Tunev et al. 2011). This was the result of abundant B cell proliferation, which occurred without concomitant T cell proliferation. A testament to the uncanny immune evasive tactics of this pathogen was their accumulation within the lymph node during this intense response. In terms of the specificity of antigen and isotype, abnormally high proportions of antibodies were IgM and they targeted select few dominant *B. burgdorferi* antigens.

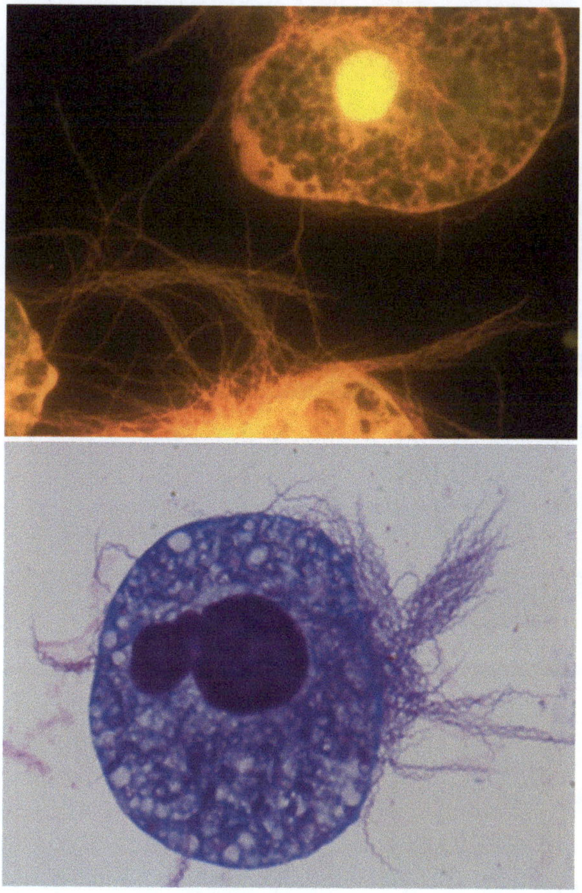

Fig. 2 Bundles of Borrelia lonestari spirochetes grown with AAE2 (Amblyomma americanum embryonic) tick cells. Preparations were stained with acridine orange (top) or Diff-Quick (bottom). Photos courtesy of Dr. Andrea Varela-Stokes, Mississippi State University

The authors of this work have probed further into the phenomenological ineffective B cell response to *B. burgdorferi*. Recent studies by the N. Baumgarth lab, in collaboration with S. Barthold have again found ample B cell accumulation in mice in response to infection with *B. burgdorferi,* much of which occurs without a strong induction of germinal centers. Looking at the response over time, the authors found that T-cell dependent B cell responses in germinal centers do occur, but that they are slow to develop and decline rapidly in comparison to what is seen with other infections, despite the continued presence of *B. burgdorferi* antigens (Hastey et al. 2012). So, not only is there a predilection towards the generation of low-affinity IgM responses, but the T-dependent response may be delayed to a point that host adaptation by the spirochete likely supersedes the generation of antibodies that would be effective at clearing the pathogen.

Relapsing Fever (RF) spirochetes. The relapsing fever spirochetes can be transmitted by either the body louse (epidemic RF) or by soft ticks (endemic RF). For

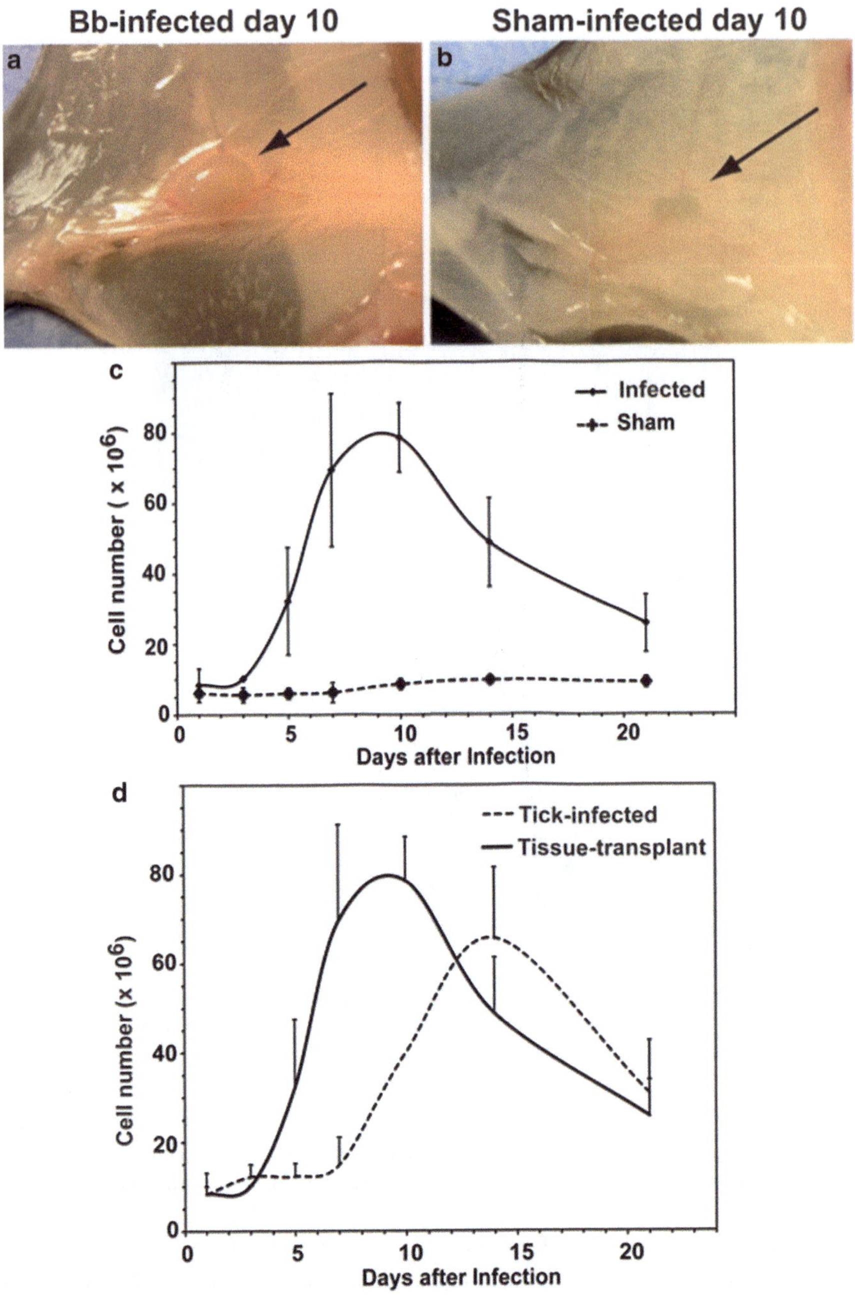

Fig. 3 Infection with host-adapted *B. burgdorferi* induces lymphadenopathy near the site of infection. Lymphadenopathy of the right inguinal lymph node of C57BL/6 mice 10 days after infection via subcutaneous transplantation of small pieces of ear from *B. burgdorferi*-infected (**a**) or noninfected (**b**) congenic SCID mice into their right dorsal tarsal region. (**c**) Shown are mean cell numbers ± SD of right inguinal lymph nodes from groups of four mice per time point. (**d**) Comparison of lymph node cellularity (mean ± SD) obtained from draining lymph nodes in (**c**) and from the axillary lymph nodes of tick-infected mice. Reproduced from (Tunev et al. 2011) with permission from PLoS Pathogens

one of the more common agents of RF in the USA, *B. hermsii,* a New World relapsing fever species, approximately 30 serotypes have been derived from a single bacterial cell. The serotype of a Borrelia cell depends on its major surface antigen. The 30 or so antigens are divided approximately equally between two families: Variable Large Proteins (Vlp) of approximately 36 kDa and Variable Small Proteins (Vsp) of approximately 20 kDa. In Chapter 9, we (Embers and Lopez) discuss RF prevalence and disease, along with host immune responses and mechanisms for antigenic variation by the spirochetes, including gene conversion/full or partial, rearrangement, and hypermutation.

Seven chapters of this book are dedicated primarily to the Borreliae. Covering seclusion by *B. burgdorferi* into immune-privileged sites and the establishment of chronic (untreated) infection is a chapter by Robert Gilmore Jr. With regard to active evasion of host immunity, the role of complement-binding proteins (Peter Kraiczy and Reinhard Wallich), antigenic variation (Troy Bankhead) and modulation of immune mediators (Dennis, Gautam and Dixit) are discussed. Finally, we (Embers and Barthold) examine the topic of persistence by the Lyme disease spirochete following antibiotic therapy.

References

Barthold SW, Bockenstedt LK (1993) Passive immunizing activity of sera from mice infected with Borrelia burgdorferi. Infect Immun 61(11):4696–4702

Barthold SW (1999) Specificity of infection-induced immunity among Borrelia burgdorferi sensu lato species. Infect Immun 67(1):36–42

Fikrig E, Bockenstedt LK, Barthold SW, Chen M, Tao H, Ali-Salaam P, Telford SR, Flavell RA (1994) Sera from patients with chronic Lyme disease protect mice from Lyme borreliosis. J Infect Dis 169(3):568–574

McKisic MD, Barthold SW (2000) T-cell-independent responses to *Borrelia burgdorferi* are critical for protective immunity and resolution of lyme disease. Infect Immun 68(9):5190–5197

Belperron AA, Dailey CM, Booth CJ, Bockenstedt LK (2007) Marginal zone B-cell depletion impairs murine host defense against *Borr,elia burgdorferi* infection. Infect Immun 75(7):3354–3360

Liang FT, Nelson FK, Fikrig E (2002) Molecular adaptation of Borrelia burgdorferi in the murine host. J Exp Med 196(2):275–280

Xu Q, McShan K, Liang FT (2008) Essential protective role attributed to the surface lipoproteins of Borrelia burgdorferi against innate defences. Mol Microbiol 69(1):15–29

Piesman J, Dolan MC, Happ CM, Luft BJ, Rooney SE, Mather TN, Golde WT (1997) Duration of immunity to reinfection with tick-transmitted Borrelia burgdorferi in naturally infected mice. Infect Immun 65(10):4043–4047

Nowakowski J, Nadelman RB, Sell R, McKenna D, Cavaliere LF, Holmgren D, Gaidici A, Wormser GP (2003) Long-term follow-up of patients with culture-confirmed Lyme disease. Am J Med 115(2):91–96

Tunev SS, Hastey CJ, Hodzic E, Feng S, Barthold SW, Baumgarth N (2011) Lymphoadenopathy during Lyme Borreliosis Is Caused by Spirochete Migration-Induced Specific B Cell Activation. PLoS Pathog **7**(5):e1002066

Hastey C, Elsner, RA, Barthold, SW and N. Baumgarth (2012) Delays and diversions mark the development of B cell responses to Borrelial burgdorferi infection. J Immun, 188(11):5612–5622

Contents

Part I
Invasion and Dissemination

Chapter 1
Treponema pallidum Dissemination; Facilitating Immune Evasion and Bacterial Persistence

Simon Houston and Caroline E. Cameron

1.1 Dissemination Potential of *Treponema pallidum*

1.1.1 T. pallidum, the Causative Agent of Syphilis

Syphilis is a chronic multistage disease, usually transmitted sexually or in utero, which affects more than 12 million people globally each year and is caused by the bacterium *T. pallidum* subsp. *pallidum*. Although developing nations account for the majority of new syphilis infections (Gerbase et al. 1998), rapidly increasing syphilis infection rates have also been observed in Europe (Righarts et al. 2004), Australia (Jin et al. 2005), and North America (Kent and Romanelli 2008) over the course of the past decade. Two additional ways in which syphilis infections impact health are congenital syphilis, which remains a major health concern in developing nations and results in spontaneous abortion, stillbirth, postpartum death or significant newborn malformations (Walker and Walker 2007), and the well-documented capacity of *T. pallidum* infection to significantly increase the risk of HIV transmission and acquisition (Nusbaum et al. 2004).

T. pallidum is a spirochete bacterium with a small genome (~1.14 Mbp) encoding approximately 1,040 proteins (Fraser et al. 1998; Matejkova et al. 2008). The bacterium is inherently fragile due to its unusual envelope arrangement in which the peptidoglycan layer is found within a cytoplasmic-membrane-proximal location rather than occupying the typical outer-membrane-proximal location as in conventional Gram-negative bacteria (Izard et al. 2009; Liu et al. 2010). This unusual ultrastructure and associated extreme fragility make laboratory manipulation of this bacterium exceedingly difficult. Further, the bacterium is an obligate human pathogen which

S. Houston, Ph.D • C.E. Cameron, Ph.D. (✉)
Department of Biochemistry and Microbiology, University of Victoria, Victoria, British Columbia, Canada
e-mail: caroc@uvic.ca

M.E. Embers (ed.), *The Pathogenic Spirochetes: Strategies for Evasion of Host Immunity and Persistence*, DOI 10.1007/978-1-4614-5404-5_1,

to date has not been successfully cultured continuously in vitro. The combination of experimental limitations associated with research on this pathogen and lack of conserved virulence factors that can be easily identified via bioinformatic analyses of the *T. pallidum* genome has hindered the identification of virulence factors that contribute to the infection process of this pathogen.

1.1.2 The Highly Invasive Nature of *T. pallidum*

The highly invasive nature of *T. pallidum* has been well established through numerous in vitro and in vivo laboratory studies (Cumberland and Turner 1949; Raiziss and Severac 1937; Riviere et al. 1989; Thomas et al. 1988, 1989). The widespread clinical manifestations associated with syphilis infection further emphasize the highly invasive nature of *T. pallidum*. Establishment of the initial localized infection is usually made evident by the appearance of a painless lesion, known as a chancre. It has also been demonstrated in vivo that the bacteria are capable of invading the tissue barrier and undergoing rapid and widespread dissemination via the circulatory system within hours of exposure (Cumberland and Turner 1949; Raiziss and Severac 1937). The highly invasive nature of *T. pallidum* is further underpinned by the fact that the pathogen can be detected in several sites distal to the primary infection site, including the kidneys, liver and central nervous system (CNS), during the early stages of syphilis. *T. pallidum* is also one of only a few highly invasive pathogens capable of traversing the vascular basement membrane of the brain (the blood–brain barrier; BBB), which is composed primarily of collagen type IV and laminin, but also contains collagen type I, entactin, fibronectin, perlecan, and vitronectin (Carbonell et al. 2009).

1.1.3 Spread of *T. pallidum* During Each of the Stages of Infection

There are four stages of syphilis: primary, secondary, latent, and tertiary. During primary syphilis, *T. pallidum* replicates at the site of infection and forms a single chancre which spontaneously heals within 3–8 weeks (Baughn and Musher 2005; LaFond and Lukehart 2006). Despite this formation of a characteristic, localized lesion, in vivo studies have demonstrated that *T. pallidum* does not remain localized at the site of infection but instead is able to enter the bloodstream minutes after experimental inoculation (Cumberland and Turner 1949; Raiziss and Severac 1937) followed by widespread dissemination within hours (Collart et al. 1971). Consistent with these observations, by the secondary stage of infection *T. pallidum* can be found to affect multiple organs and tissue sites, including the skin, oral cavity and tongue (Mindel et al. 1989), stomach (Greenstein et al. 1994), kidney (Bansal et al. 1978; Tourville et al. 1976), liver (Hira et al. 1987; Lee et al. 1971; Mindel et al. 1989), spleen (Hira et al. 1987; Mindel et al. 1989), and CNS (Chapel 1980; Lukehart et al.

1988; Rolfs et al. 1997). Latent syphilis is divided into two phases: (1) early latent syphilis, which comprises the year following secondary syphilis infection and which may involve a relapse of secondary symptoms, and (2) late latent syphilis, an asymptomatic period of infection lasting for at least 1 year (Kent and Romanelli 2008; LaFond and Lukehart 2006). The tertiary stage of syphilis, which occurs in approximately 30 % of untreated cases, further emphasizes the highly invasive nature of the pathogen, as the disease produces widespread clinical manifestations, including cardiovascular syphilis, neurosyphilis, and gumma formation which most commonly affects the skin, bone, and liver (Singh and Romanowski 1999). Furthermore, several studies have successfully demonstrated the presence of *T. pallidum* (or *T. pallidum* DNA) in cerebrospinal fluid (CSF) of patients with neurosyphilis (Lukehart et al. 1988; Marra et al. 2004), tertiary stage gummas (Handsfield et al. 1983; Zoechling et al. 1997), and tertiary stage syphilitic aortitis (O'Regan et al. 2002).

1.2 Spread of *T. pallidum* to Immunologically Privileged Sites

One mechanism by which *T. pallidum* may evade the immune system and promote chronic infection is through dissemination to, and invasion of, immunologically privileged sites. These are body sites in which there is a reduced or absent innate immune response, and include the brain, eyes, placenta, fetus, ovaries and testicles. It is believed that immune privilege is critical for the protection of specific vital organs where an inflammatory immune response would result in organ destruction and failure (Ferguson et al. 2002). In vivo animal studies have demonstrated that *T. pallidum* is capable of infecting known immunologically privileged sites, including the brain (Collart et al. 1971; Rosahn et al. 1948), testes (Sell et al. 1980), and eyes (Marra et al. 1991). PCR analyses have also confirmed the presence of *T. pallidum* in the eye (Muller et al. 2007; Rajan et al. 2006) and CSF (Burstain et al. 1991; Hay et al. 1990; Marra et al. 1996; Noordhoek et al. 1991) of syphilis patients. This capability may allow the pathogen to take advantage of the reduced immune surveillance and subsequently aid bacterial survival for extended periods of time in widely distributed anatomical sites. Bacteria residing in these immunologically privileged sites may explain, at least in part, the capacity of *T. pallidum* to persist during latency despite the induction of a vigorous immune response within non-immunologically privileged host sites, and would provide a reservoir of replicating bacteria that could contribute to the recurrences of spirochetemia that are observed during early latency.

1.3 Mechanisms of Dissemination

Numerous studies have demonstrated that *T. pallidum* is capable of invading the tissue barrier and undergoing rapid widespread dissemination via the circulatory system (Cumberland and Turner 1949; Mahoney and Bryant 1934; Raiziss and Severac 1937; Riviere et al. 1989; Sell et al. 1980; Thomas et al. 1988, 1989).

Although the virulence factors involved in *T. pallidum* dissemination remain to be definitively identified, several mechanisms have been proposed to be responsible for this striking aspect of treponemal pathogenesis.

1.3.1 Attachment Mechanisms

Bacterial attachment to host components is an essential early step in dissemination and invasion of many pathogenic bacteria, including *T. pallidum*. One common mechanism by which pathogenic bacteria interact with their host is through the expression of surface-exposed adherence factors or adhesins. Adhesins that have been implicated or proven to be involved in host component interactions include fimbriae, surface polysaccharides (including bacterial capsules) and, as may be the case with *T. pallidum*, surface exposed outer-membrane proteins.

1.3.1.1 Adhesins and Host Attachment

T. pallidum has been shown to bind to several mammalian cell types (Baseman and Hayes 1980; Fitzgerald et al. 1975, 1977; Hayes et al. 1977; Thomas et al. 1988, 1989) as well as isolated components of the extracellular matrix, including collagen, laminin, and fibronectin (Cameron 2003; Fitzgerald and Repesh 1985; Fitzgerald et al. 1984; Thomas et al. 1985). Data regarding the molecular identity and contribution of *T. pallidum* adhesins to host–pathogen interactions and dissemination have only recently begun to emerge. Palzkill and colleagues (Brinkman et al. 2008) identified the treponemal protein Tp0136 as a fibronectin-binding adhesin that elicits a strong immune response during human and experimental animal infection and exhibits extensive sequence polymorphism among *T. pallidum* subspecies and strains. An additional study used in silico analysis of the sequenced *T. pallidum* genome (Fraser et al. 1998) to identify 27 potential outer membrane proteins (Cameron 2003), 10 of which were investigated further based on their predicted high likelihood of surface exposure and possible role as adhesins (Cameron 2003; Cameron et al. 2004). Two of the proteins, Tp0155 and Tp0483, when purified in their recombinant forms, were found to bind in a dose-dependent manner to matrix fibronectin and matrix and soluble fibronectin, respectively. Furthermore, the two proteins inhibited attachment of *T. pallidum* to fibronectin-coated slides (Cameron et al. 2004). An additional *T. pallidum* adhesin, Tp0751, was identified and shown to bind specifically to a variety of laminin isoforms (Cameron 2003), abundant glycoprotein components of the basement membrane that underlies endothelial cells. Heterologous expression of Tp0751 on the surface of the non-pathogenic spirochete *Treponema phagedenis* was also shown to confer laminin binding on this non-adherent spirochete (Cameron et al. 2008). More recently it has been shown that Tp0751 is capable of binding human fibrinogen (Houston et al. 2010), a key structural protein that is integral to the process of blood coagulation. Under normal physiological conditions within the host, fibrin clot formation is induced upon conversion of soluble fibrinogen to insoluble fibrin via activation of thrombin by the coagulation cascade (Levi et al. 2004). It has been reported that this process is likely to play an important role in bacterial containment and inhibition of bacterial dissemination (Rivera et al. 2007). Similar to other invasive bacteria (Chang et al. 2005; Liu et al. 2007), the interaction of *T. pallidum* with coagulation proteins such as fibrinogen may contribute to the rapid widespread dissemination of this pathogen by interfering with this host defence mechanism to limit the spread of infectious agents. Collectively these findings suggest that *T. pallidum* is capable of attaching to a broad range of host components that have a widespread anatomical distribution, a necessity for such a highly invasive bacterium.

1.3.2 Motility

For many flagellated pathogenic bacteria, motility is an important virulence factor which aids in bacterial colonization, nutrient acquisition, and dissemination of infection (Ottemann and Miller 1997). Similar to other spirochetes, *T. pallidum* exhibits a corkscrew-like motility through rotation around its longitudinal axis. Unlike the surface-exposed flagella found in other motile bacterial species, *T. pallidum* cellular propulsion is accomplished through a motility system that is unique to spirochetes whereby the flagellar filaments are located in the periplasmic space between the inner and outer membranes (LaFond and Lukehart 2006). The *T. pallidum* flagellar filaments are composed of several proteins, including the sheath protein, FlaA, which overlies the core proteins FlaB1, FlaB2, and FlaB3 (Cockayne et al. 1987; Norris et al. 1988; Radolf et al. 1986). This unusual structural arrangement facilitates treponemal immune evasion through sequestering of these highly immunogenic motility proteins within a subsurface cellular locale. In addition, the corkscrew-like motility promotes movement through more viscous material compared to externally flagellated bacteria (Charon and Goldstein 2002). In turn, this virulence attribute contributes to enhanced tissue penetration, dissemination of infection, and invasion of a wide variety of distant host sites (Berg and Turner 1979; Liu et al. 2010; Norris 1993).

1.3.3 Ultrastructure

T. pallidum is characterized by several unusual ultrastructural features which may aid in dissemination of infection, evasion of the immune response and bacterial persistence.

1.3.3.1 Low Protein Content in Outer Membrane

It has been demonstrated that the outer membrane of *T. pallidum* is relatively immunologically inert compared to other pathogenic bacteria, a characteristic

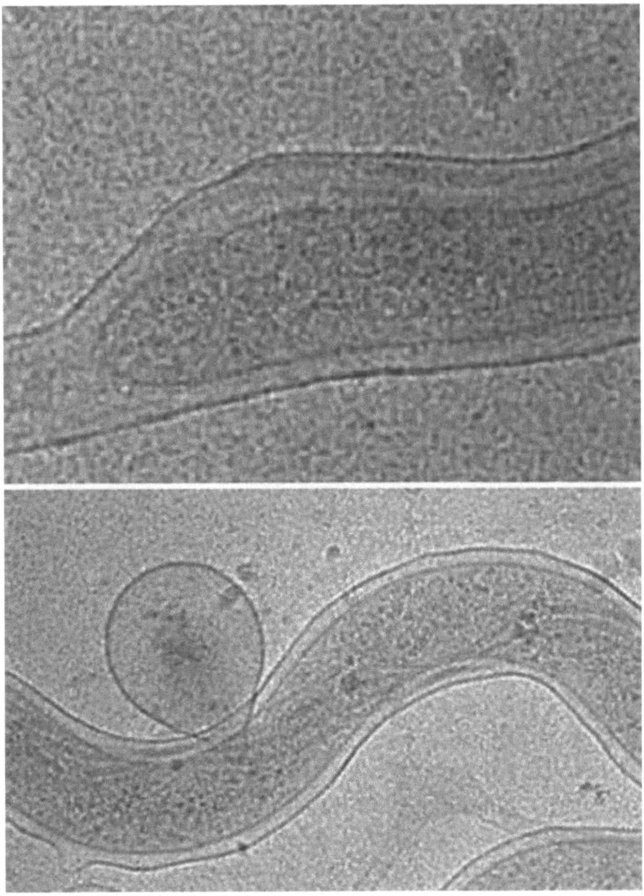

Fig. 1.1 Cryo-electron micrographs of *T. pallidum* show outer membrane (OM) blebs (*bottom*), and the bulge in the OM over a basal body (*top*). Provided by Justin Radolf

which is thought to be responsible for the ability of this "stealth" pathogen to establish long-term latency (Cameron 2005; Radolf 1995). The underlying mechanism responsible for the lack of immunological reactivity observed with the *T. pallidum* outer membrane (Fig. 1.1) has been investigated using several elegant microscopy techniques, including freeze-fracture electron microscopy and cryo-electron tomography (Fig. 1.2) (Cox et al. 1992, 2010; Izard et al. 2009; Liu et al. 2010; Radolf et al. 1989; Walker et al. 1989). These studies have confirmed that there is a paucity of integral outer membrane proteins on the surface of this bacterium, with freeze-fracture investigations providing evidence for approximately 100-fold fewer integral membrane proteins present on the surface of *T. pallidum* compared to *Escherichia coli* (Walker et al. 1989). The use of in vitro *T. pallidum* immobilization assays combined with freeze-fracture investigations has shown that aggregation of outer membrane proteins in the presence of immune rabbit

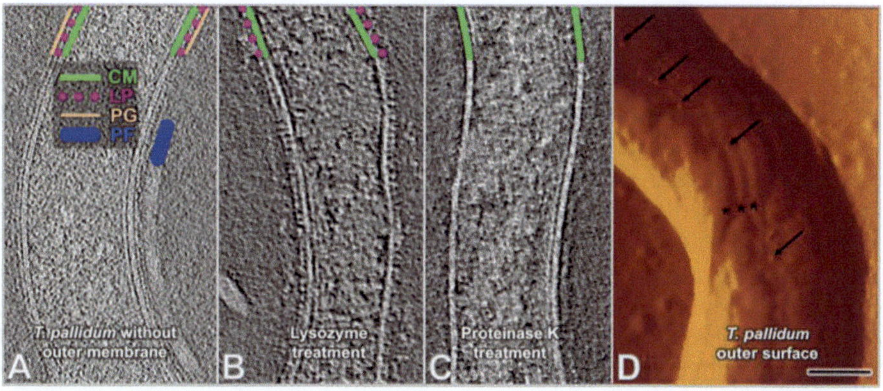

Fig. 1.2 Confirmation of *T. pallidum* peptidoglycan layer, cytoplasmic membrane surface proteins, and rare outer surface proteins. Cells in all *panels* were previously treated with distilled water to remove the outer membrane and expose subsurface structures. (**a**) Cells without subsequent enzyme treatment. Locations of the cytoplasmic membrane (*green line*), surface proteins (*purple circles*), peptidoglycan (*orange line*), and a periplasmic flagellum (*blue line*) are shown. (**b**) With lysozyme treatment. The peptidoglycan layer was removed, whereas the putative lipoproteins (*purple circles*) are still visible on the outer surface of the cytoplasmic membrane. (**c**) With proteinase K treatment, the layer of putative lipoproteins was removed. (**d**) Presence of intramembranous particles on the outer surface of *T. pallidum*, as revealed by scanning probe microscopy (SPM). Freshly prepared, Percoll-purified *T. pallidum* were applied to a mica surface, air-dried, and visualized by SPM. A portion of one *T. pallidum* cell is shown using the amplitude mode. Small protrusions are visible on the outer surface (*arrows*) and often are located on the bulge in the outer membrane created by the underlying periplasmic flagella (*asterisks*). Scale bar is 100 nm. Reproduced from (Liu et al. 2010) with permission from the Journal of Molecular Biology

serum requires an extended period of 16 h, which is a considerably longer timeframe than observed for similar aggregation within other bacteria. (Blanco et al. 1990). This emphasizes the rare nature of outer membrane proteins within *T. pallidum*. Given that surface-exposed proteins represent the first antigens to be targeted by the immune system following infection, these findings help explain the immunologically inert nature of the *T. pallidum* outer membrane. The presentation of this unusual bacterial surface to the host would directly aid evasion of the immune response and, accordingly, would facilitate bacterial dissemination and establishment of persistent infection.

1.3.3.2 Lipoproteins, Lipopolysaccharide and the Proinflammatory Response

One way in which proteins are often found attached to the bacterial surface, including within the related spirochetes *Leptospira* and *Borrelia*, is through outer membrane anchorage via an N-terminal lipid-modified cysteine residue (Haake 2000). Whole genome sequencing of *T. pallidum* originally indicated the presence of only

22 putative lipoproteins compared to 105 in *Borrelia burgdorferi* (Fraser et al. 1998). More recently, computational analyses and in silico studies indicate that there are 46 potential lipoproteins in the *T. pallidum* genome, whereas the genome of *B. burgdorferi* was shown to contain 127 putative lipoprotein genes (Setubal et al. 2006). Recent studies using high resolution cryo-electron tomography suggest that, unlike the situation within the related spirochetes *Leptospira* (Haake and Matsunaga 2010) and *Borrelia* (Liu et al. 2009), the majority of lipoproteins found within *T. pallidum* are localized to the outer leaflet of the cytoplasmic membrane (Izard et al. 2009; Liu et al. 2010). To date only two predicted lipoproteins have been suggested to be surface exposed within *T. pallidum*. The first, Tp0136, is a fibronectin-binding protein that was suggested to be surface-exposed through a combination of immunofluorescence and immunoelectron microscopy (Brinkman et al. 2008), although an independent investigation reported discordant results for surface exposure of this protein within *T. pallidum* (Cox et al. 2010). The second predicted lipoprotein, Tp0751, is suggested to be surface-exposed through the functional capacity of this protein to bind to host components laminin (Cameron 2003) and fibrinogen (Houston et al. 2010), and through the observation that heterologous expression of Tp0751 on the surface of the related spirochete *T. phagedenis* induces lipidation of this protein (Houston et al. 2010). Although the definitive answer of whether these predicted lipoproteins are surface-exposed within *T. pallidum* awaits further investigation, it is apparent that lipoproteins, if present on the treponemal surface, mirror the rare occurrence observed for integral membrane proteins. Since it has been shown that *T. pallidum* lipoproteins function as proinflammatory agonists (Akins et al. 1993; Radolf et al. 1995; Salazar et al. 2002) and elicit strong antibody responses (McGill et al. 2010), suggesting a propensity for these molecules to elicit a vigorous immune response during infection, the predominant location of *T. pallidum* lipoproteins within a subsurface locale undoubtedly contributes to the immune evasion capacity of this pathogen and its ability to successfully disseminate and establish chronic infection. In addition, *T. pallidum* does not possess lipopolysaccharide (LPS) (Belisle et al. 1994; Fraser et al. 1998; Hardy and Levin 1983), a major inflammatory-inducing endotoxin which activates the innate immune system via Toll-like receptor 4 (TLR4) and which is associated with the outer membrane of Gram-negative bacteria (Ulevitch and Tobias 1999). The unusual absence of this bacterial surface component would cause a reduction in the proinflammatory response induced by *T. pallidum* during infection, thereby directly enhancing its chance of survival within the host environment.

1.3.4 Chemotaxis

Evidence indicates that *T. pallidum* may utilize chemotaxis to further promote survival and dissemination during infection. Thirteen chemotaxis genes, including four methyl-accepting chemotaxis protein (MCP)-encoding genes, which potentially encode for proteins with specificity for amino acids and carbohydrates, were identified from the whole genome sequence (Fraser et al. 1998). A putative *T. pallidum* MCP gene, *mcp1*, has also been cloned and shown to encode a 64 kDa protein which is highly homologous to at least 69 other chemotaxis proteins. Furthermore, the authors were able to demonstrate the existence of a 64 kDa methylated *T. pallidum* protein using intrinsic radiolabelling with L-[methyl-3H] methionine (Hagman et al. 1997). Two operons encoding putative chemotaxis (Che) response regulators were also identified in the *T. pallidum* genome (Fraser et al. 1998) and four of these chemotaxis proteins (CheA, CheW, CheX, and CheY) were expressed in vitro (Greene et al. 1997). Each of the *che* genes was found to be co-transcribed using reverse transcriptase-PCR and shown to exhibit significant homology to known chemotaxis genes (Greene et al. 1997). The presence of so many potential chemotaxis genes in the relatively small *T. pallidum* genome further suggests that the pathogen uses chemotaxis in response to nutrient gradients and hostile host environments, processes which likely contribute to *T. pallidum* dissemination.

1.3.5 Antigenic Variation of the Tpr Protein Family

Evidence suggests that *T. pallidum* also utilizes antigenic variation as a mechanism to facilitate immune evasion. The *T. pallidum* repeat (*tpr*) gene family is comprised of 12 paralogous genes (*tprA-L*) divided into three families (subfamily I–III) (Centurion-Lara et al. 1999a). Several of these proteins are predicted to be located in the outer membrane and most have been shown to elicit an immune response during experimental syphilis infection (Leader et al. 2003; Morgan et al. 2002; Sun et al. 2004). Recently, it has been shown that members of this protein family undergo both antigenic (Centurion-Lara et al. 1999b, 2004; LaFond et al. 2003) and phase variation (Giacani et al. 2007, 2009). In particular, *tprK* has been shown to undergo strain-specific antigenic variation at seven discrete variable regions, whereas *tprC* and *tprD* exhibit strain-independent variation (Centurion-Lara et al. 2004; LaFond et al. 2006; LaFond and Lukehart 2006). Together with antigenic variation, it has been proposed that "off/on" gene regulation, or phase variation, of Tpr proteins for which an immune response has, or has not, been generated, respectively, may aid *T. pallidum* in escaping detection by the host immune response (LaFond and Lukehart 2006). Antigenic variation by *T. pallidum* is discussed in more detail in Chap. 2.

1.3.6 Proteolytic Degradation of Host Components

Pathogenic bacteria often express surface-exposed and extracellular proteases as a means to facilitate host component and tissue destruction, an important mechanism that can contribute to dissemination of infection and tissue invasion. Potential host-interacting proteases were not specifically identified following *T. pallidum* complete genome sequencing (Fraser et al. 1998). However, the highly invasive nature of the pathogen supports their existence. Five potential haemolysins with sequence similarity to *B. burgdorferi* proteins were identified (Fraser et al. 1998), although direct haemolytic activity was not demonstrated. One study did report that *T. pallidum* is capable of inducing interstitial collagenase (matrix metalloprotease-1, MMP-1) expression when added to human dermal fibroblast cultures (Chung et al. 2002). Its target, collagen-1, is the major component of connective tissues and bones and is also found in all organs. However, no further studies have been published regarding this mechanism, so it is difficult to ascertain the exact contribution of host protease up-regulation by *T. pallidum* in bacterial dissemination.

By using heterologous protein expression and in vitro protease assays, it has been demonstrated that soluble recombinant forms of the Tp0751 adhesin produced in both *E. coli* and insect expression systems were capable of degrading human fibrinogen and laminin. The authors also demonstrated that the observed protease activity was abolished by metalloprotease inhibitors, including the zinc-chelator 1,10 phenanthroline, but not by serine or cysteine protease inhibitors. Inductively coupled plasma-mass spectrometry revealed that the protease binds zinc and calcium (Houston et al. 2010). These results indicate that Tp0751, which has been designated pallilysin, is a zinc-dependent host component-degrading protease. The ability of pathogenic bacteria to degrade fibrinogen is recognized as an important virulence factor by which the pathogen prevents or inhibits coagulation-based containment. This mechanism may be of particular importance at the initial site of infection and during dissemination via the circulatory system. Likewise, laminin is a major component of basement membranes, obstacles which *T. pallidum* must traverse during the course of infection. This study provides the first description of a *T. pallidum* protein that is capable of degrading human proteins and thus provides novel insight into a potentially important mechanism of *T. pallidum* dissemination.

1.4 Effective Vaccine Design

The limited effectiveness of current worldwide public health programs in controlling syphilis highlights the need for the development of an effective vaccine. In light of increasing global *T. pallidum* infection rates, the extreme morbidity associated with the tertiary stage of syphilis, and the fact that syphilis infection significantly increases the risk of HIV acquisition and transmission, development of an effective syphilis vaccine would have a significant impact on global public health by preventing sexually and congenitally transmitted syphilis infections and by reducing transmission and acquisition of HIV. In order for a vaccine to be effective, it would have to not only prevent dissemination of infection, but also prevent establishment of later stages of the disease. It is probable that the identification of rare surface-exposed outer membrane proteins will be essential in syphilis vaccine development, as these antigens are the first targets of the immune system (LaFond and Lukehart 2006) and are also most likely to induce protective immunity since clearance of *T. pallidum* is dependent upon opsonophagocytosis and subsequent killing by macrophages (Baker-Zander and Lukehart 1992; Baker-Zander et al. 1993; Cullen and Cameron 2006; Lukehart and Miller 1978; Sell et al. 1982). In addition to the difficulties associated with the definitive identification of rare outer membrane proteins in *T. pallidum*, it must also be noted that a successful vaccine will likely rely on the production of recombinant vaccine candidate targets that closely resemble the conformation of the native treponemal antigens.

1.5 Future Directions and Areas of Needed Research

The ability to disseminate so widely and rapidly is an important mechanism of *T. pallidum* pathogenesis. Although the highly disseminating nature of *T. pallidum* is most likely a multi-factorial process, no factors have been definitively identified that explain this striking pathogenic characteristic. Host-interacting proteases are common virulence factors in many pathogenic bacteria. In particular, proteases cause host tissue damage and destruction thereby promoting bacterial dissemination via the degradation of structural tissue components and plasma proteins. Although this mechanism of pathogenesis is common to many disseminating and invasive pathogens, due to the fact that genome sequencing failed to identify potential host-interacting proteases (Fraser et al. 1998) this potentially important mechanism has yet to be fully explored in *T. pallidum*. However, recent studies have begun to identify such potential virulence factors, including "pallilysin" (Tp0751), a protein that has been shown to function as an adhesin and a host-degrading protease (Cameron 2003; Cameron et al. 2005, 2008; Houston et al. 2010). Although these interactions are hypothesized to be important for *T. pallidum* tissue invasion and dissemination (Cameron 2003; Houston et al. 2010), the exact role of pallilysin remains to be determined. It is possible that several host-interacting proteins, including pallilysin, contribute to the virulence of *T. pallidum* by promoting the attachment to and degradation of a wide variety of host molecules, thereby facilitating rapid widespread dissemination of *T. pallidum*. Identification of additional *T. pallidum* host-interacting proteases and detailed investigation of their binding and proteolytic capabilities will provide an indication of the importance of *T. pallidum*-mediated host component degradation to the process of treponemal dissemination. Ultimately, this knowledge may be utilized for the identification of suitable targets for vaccine design.

References

Akins DR, Purcell BK, Mitra MM, Norgard MV, Radolf JD (1993) Lipid modification of the 17-kilodalton membrane immunogen of *Treponema pallidum* determines macrophage activation as well as amphiphilicity. Infect Immun 61:1202–1210

Baker-Zander SA, Lukehart SA (1992) Macrophage-mediated killing of opsonized *Treponema pallidum*. J Infect Dis 165:69–74

Baker-Zander SA, Shaffer JM, Lukehart SA (1993) Characterization of the serum requirement for macrophage-mediated killing of *Treponema pallidum* ssp. *pallidum*: relationship to the development of opsonizing antibodies. FEMS Immunol Med Microbiol 6:273–279

Bansal RC, Cohn H, Fani K, Lynfield YL (1978) Nephrotic syndrome and granulomatous hepatitis in secondary syphilis. Arch Dermatol 114:1228–1229

Baseman JB, Hayes EC (1980) Molecular characterization of receptor binding proteins and immunogens of virulent *Treponema pallidum*. J Exp Med 151:573–586

Baughn RE, Musher DM (2005) Secondary syphilitic lesions. Clin Microbiol Rev 18:205–216

Belisle JT, Brandt ME, Radolf JD, Norgard MV (1994) Fatty acids of *Treponema pallidum* and *Borrelia burgdorferi* lipoproteins. J Bacteriol 176:2151–2157

Berg HC, Turner L (1979) Movement of microorganisms in viscous environments. Nature 278:349–351

Blanco DR, Walker EM, Haake DA, Champion CI, Miller JN, Lovett MA (1990) Complement activation limits the rate of *in vitro* treponemicidal activity and correlates with antibody-mediated aggregation of *Treponema pallidum* rare outer membrane protein. J Immunol 144: 1914–1921

Brinkman MB, McGill MA, Pettersson J, Rogers A, Matejkova P, Smajs D, Weinstock GM, Norris SJ, Palzkill T (2008) A novel *Treponema pallidum* antigen, TP0136, is an outer membrane protein that binds human fibronectin. Infect Immun 76:1848–1857

Burstain JM, Grimprel E, Lukehart SA, Norgard MV, Radolf JD (1991) Sensitive detection of *Treponema pallidum* by using the polymerase chain reaction. J Clin Microbiol 29:62–69

Cameron CE (2003) Identification of a *Treponema pallidum* laminin-binding protein. Infect Immun 71:2525–2533

Cameron CE (2005) *T. pallidum* outer membrane and outer membrane proteins. In: Radolf JD, Lukehart SA (eds) Pathogenic *Treponema*. Molecular and cellular biology. Caister Academic Press, Norfolk, England, pp 237–266

Cameron CE, Brown EL, Kuroiwa JMY, Schnapp LM, Brouwer NL (2004) *Treponema pallidum* fibronectin-binding proteins. J Bacteriol 186:7019–7022

Cameron CE, Brouwer NL, Tisch LM, Kuroiwa JMY (2005) Defining the interaction of the *Treponema pallidum* adhesin Tp0751 with laminin. Infect Immun 73:7485–7494

Cameron CE, Kuroiwa JM, Yamada M, Francescutti T, Chi B, Kuramitsu HK (2008) Heterologous expression of the *Treponema pallidum* laminin-binding adhesin Tp0751 in the culturable spirochete *Treponema phagedenis*. J Bacteriol 190:2565–2571

Carbonell WS, Ansorge O, Sibson N, Muschel R (2009) The vascular basement membrane as "soil" in brain metastasis. PLoS One 4:e5857

Centurion-Lara A, Castro C, Barrett L, Cameron C, Mostowfi M, Van Voorhis WC, Lukehart SA (1999a) *Treponema pallidum* major sheath protein homologue Tpr K is a target of opsonic antibody and the protective immune response. J Exp Med 189:647–656

Centurion-Lara A, Gordones C, Castro C, Van Voorhis WC, Lukehart SA (1999b) The *tprK* gene is heterogeneous among *Treponema pallidum* strains and has multiple alleles. Infect Immun 68:824–831

Centurion-Lara A, LaFond RE, Hevner K, Godornes C, Molini BJ, Van Voorhis WC, Lukehart SA (2004) Gene conversion: a mechanism for generation of heterogeneity in the *tprK* gene of *Treponema pallidum* during infection. Mol Microbiol 52:1579–1596

Chang AK, Kim HY, Park JE, Acharya P, Park IS, Yoon SM, You HJ, Hahm KS, Park JK, Lee JS (2005) *Vibrio vulnificus* secretes a broad-specificity metalloprotease capable of interfering with blood homeostasis through prothrombin activation and fibrinolysis. J Bacteriol 187: 6909–6916

Chapel TA (1980) The signs and symptoms of secondary syphilis. Sex Transm Dis 7:161–164

Charon NW, Goldstein SF (2002) Genetics of motility and chemotaxis of a fascinating group of bacteria: the spirochetes. Annu Rev Genet 36:47–73

Chung KY, Kim KS, Lee MG, Chang NS, Lee JB (2002) *Treponema pallidum* induces up-regulation of interstitial collagenase in human dermal fibroblasts. Acta Derm Venereol 82:174–178

Cockayne A, Bailey MJ, Penn CW (1987) Analysis of sheath and core structures of the axial filament of *Treponema pallidum*. J Gen Microbiol 133:1397–1407

Collart P, Franceschini P, Durel P (1971) Experimental rabbit syphilis. Br J Vener Dis 47:389–400

Cox DL, Chang P, McDowall AW, Radolf JD (1992) The outer membrane, not a coat of host proteins, limits antigenicity of virulent *Treponema pallidum*. Infect Immun 60:1076–1083

Cox DL, Luthra A, Dunham-Ems S, Desrosiers DC, Salazar JC, Caimano MJ, Radolf JD (2010) Surface immunolabeling and consensus computational framework to identify candidate rare outer membrane proteins of *Treponema pallidum*. Infect Immun 78:5178–5194

Cullen PA, Cameron CE (2006) Progress towards an effective syphilis vaccine: the past, present and future. Expert Rev Vaccines 5:67–80

Cumberland MC, Turner TB (1949) Rate of multiplication of *Treponema pallidum* in normal and immune rabbits. Am J Syph Gonorrhea Vener Dis 33:201–212

Ferguson TA, Green DR, Griffith TS (2002) Cell death and immune privilege. Int Rev Immunol 21:153–172

Fitzgerald TJ, Repesh LA (1985) Interactions of fibronectin with *Treponema pallidum*. Genitourin Med 61:147–155

Fitzgerald TJ, Miller JN, Sykes JA (1975) *Treponema pallidum* (Nichols strain) in tissue cultures: cellular attachment, entry, and survival. Infect Immun 11:1133–1140

Fitzgerald TJ, Johnson RC, Miller JN, Sykes JA (1977) Characterization of the attachment of *Treponema pallidum* (Nichols strain) to cultured mammalian cells and the potential relationship of attachment to pathogenicity. Infect Immun 18:467–478

Fitzgerald TJ, Repesh LA, Blanco DR, Miller JN (1984) Attachment of *Treponema pallidum* to fibronectin, laminin, collagen IV, and collagen I, and blockage of attachment by immune rabbit IgG. Br J Vener Dis 60:357–363

Fraser CM, Norris SJ, Weinstock GM, White O, Sutton GG, Dodson R, Gwinn M, Hickey EK, Clayton R, Ketchum KA, Sodergren E, Hardham JM, McLeod MP, Salzberg S, Peterson J, Khalak H, Richardson D, Howell JK, Chidambaram M, Utterback T, McDonald L, Artiach P, Bowman C, Cotton MD, Fujii C, Garland S, Hatch B, Horst K, Watthey L, Weidman J, Smith HO, Venter JC (1998) Complete genome sequence of Treponema pallidum, the syphilis spirochete. Science 281:375–388

Gerbase AC, Rowley JT, Heymann DH, Berkley SF, Piot P (1998) Global prevalence and incidence estimates of selected curable STDs. Sex Transm Infect 74(Suppl 1):S12–S16

Giacani L, Lukehart S, Centurion-Lara A (2007) Length of guanosine homopolymeric repeats modulates promoter activity of subfamily II *tpr* genes of *Treponema pallidum* ssp. *pallidum*. FEMS Immunol Med Microbiol 51:289–301

Giacani L, Godornes C, Puray-Chavez M, Guerra-Giraldez C, Tompa M, Lukehart SA, Centurion-Lara A (2009) TP0262 is a modulator of promoter activity of *tpr* subfamily II genes of *Treponema pallidum* ssp. *pallidum*. Mol Microbiol 72:1087–1099

Greene SR, Stamm LV, Hardham JM, Young NR, Frye JG (1997) Identification, sequences, and expression of *Treponema pallidum* chemotaxis genes. DNA Seq 7:267–284

Greenstein DB, Wilcox CM, Schwartz DA (1994) Gastric syphilis. Report of seven cases and review of the literature. J Clin Gastroenterol 18:4–9

Haake DA (2000) Spirochaetal lipoproteins and pathogenesis. Microbiology 146(Pt 7): 1491–1504

Haake DA, Matsunaga J (2010) *Leptospira*: a spirochaete with a hybrid outer membrane. Mol Microbiol 77:805–814

Hagman KE, Porcella SF, Popova TG, Norgard MV (1997) Evidence for a methyl-accepting chemotaxis protein gene (*mcp1*) that encodes a putative sensory transducer in virulent *Treponema pallidum*. Infect Immun 65:1701–1709

Handsfield HH, Lukehart SA, Sell S, Norris SJ, Holmes KK (1983) Demonstration of *Treponema pallidum* in a cutaneous gumma by indirect immunofluorescence. Arch Dermatol 119:677–680

Hardy PH Jr, Levin J (1983) Lack of endotoxin in *Borrelia hispanica* and *Treponema pallidum*. Proc Soc Exp Biol Med 174:47–52

Hay PE, Clarke JR, Strugnell RA, Taylor-Robinson D, Goldmeier D (1990) Use of the polymerase chain reaction to detect DNA sequences specific to pathogenic treponemes in cerebrospinal fluid. FEMS Microbiol Lett 56:233–238

Hayes NS, Muse KE, Collier AM, Baseman JB (1977) Parasitism by virulent *Treponema pallidum* of host cell surfaces. Infect Immun 17:174–186

Hira SK, Patel JS, Bhat SG, Chilikima K, Mooney N (1987) Clinical manifestations of secondary syphilis. Int J Dermatol 26:103–107

Houston S, Hof R, Francescutti T, Hawkes A, Boulanger MJ, Cameron CE (2010) Bifunctional role of the *Treponema pallidum* extracellular matrix binding adhesin Tp0751. Infect Immun 79:1386–1398

Izard J, Renken C, Hsieh CE, Desrosiers DC, Dunham-Ems S, La VC, Gebhardt LL, Limberger RJ, Cox DL, Marko M, Radolf JD (2009) Cryo-electron tomography elucidates the molecular architecture of *Treponema pallidum*, the syphilis spirochete. J Bacteriol 191:7566–7580

Jin F, Prestage GP, Kippax SC, Pell CM, Donovan BJ, Kaldor JM, Grulich AE (2005) Epidemic syphilis among homosexually active men in Sydney. Med J Aust 183:179–183

Kent ME, Romanelli F (2008) Reexamining syphilis: an update on epidemiology, clinical manifestations, and management. Ann Pharmacother 42:226–236

LaFond RE, Lukehart SA (2006) Biological basis for syphilis. Clin Microbiol Rev 19:29–49

LaFond RE, Centurion-Lara A, Godornes C, Rompalo AM, Van Voorhis WC, Lukehart SA (2003) Sequence diversity of *Treponema pallidum* subsp. *pallidum tprK* in human syphilis lesions and rabbit-propagated isolates. J Bacteriol 185:6262–6268

LaFond RE, Centurion-Lara A, Godornes C, Van Voorhis WC, Lukehart SA (2006) *TprK* sequence diversity accumulates during infection of rabbits with *Treponema pallidum* subsp. *pallidum* Nichols strain. Infect Immun 74:1896–1906

Leader BT, Hevner K, Molini BJ, Barrett LK, Van Voorhis WC, Lukehart SA (2003) Antibody responses elicited against the *Treponema pallidum* repeat proteins differ during infection with different isolates of *Treponema pallidum* subsp. *pallidum*. Infect Immun 71:6054–6057

Lee RV, Thornton GF, Conn HO (1971) Liver disease associated with secondary syphilis. N Engl J Med 284:1423–1425

Levi M, van der Poll T, Buller HR (2004) Bidirectional relation between inflammation and coagulation. Circulation 109:2698–2704

Liu CZ, Huang TF, Tsai PJ, Tsai PJ, Chang LY, Chang MC (2007) A segment of *Staphylococcus aureus* clumping factor A with fibrinogen-binding activity (ClfA221-550) inhibits platelet-plug formation in mice. Thromb Res 121:183–191

Liu J, Lin T, Botkin DJ, McCrum E, Winkler H, Norris SJ (2009) Intact flagellar motor of *Borrelia burgdorferi* revealed by cryo-electron tomography: evidence for stator ring curvature and rotor/C-ring assembly flexion. J Bacteriol 191:5026–5036

Liu J, Howell JK, Bradley SD, Zheng Y, Zhou ZH, Norris SJ (2010) Cellular architecture of *Treponema pallidum*: novel flagellum, periplasmic cone, and cell envelope as revealed by cryo electron tomography. J Mol Biol 403:546–561

Lukehart SA, Miller JN (1978) Demonstration of the *in vitro* phagocytosis of *Treponema pallidum* by rabbit peritoneal macrophages. J Immunol 121:2014–2024

Lukehart SA, Hook EW, Baker-Zander SA, Collier AC, Critchlow CW, Handsfield HH (1988) Invasion of the central nervous system by *Treponema pallidum*: implications for diagnosis and treatment. Ann Intern Med 109:855–862

Mahoney JF, Bryant KK (1934) Time element in penetration of genital mucosa by *Treponema pallidum*. J Vener Dis Inf 15:1–5

Marra C, Baker-Zander SA, Hook EWd, Lukehart SA (1991) An experimental model of early central nervous system syphilis. J Infect Dis 163:825–829

Marra CM, Gary DW, Kuypers J, Jacobson MA (1996) Diagnosis of neurosyphilis in patients infected with human immunodeficiency virus type 1. J Infect Dis 174:219–221

Marra CM, Maxwell CL, Smith SL, Lukehart SA, Rompalo AM, Eaton M, Stoner BP, Augenbraun M, Barker DE, Corbett JJ, Zajackowski M, Raines C, Nerad J, Kee R, Barnett SH (2004) Cerebrospinal fluid abnormalities in patients with syphilis: association with clinical and laboratory features. J Infect Dis 189:369–376

Matejkova P, Strouhal M, Smajs D, Norris SJ, Palzkill T, Petrosino JF, Sodergren E, Norton JE, Singh J, Richmond TA, Molla MN, Albert TJ, Weinstock GM (2008) Complete genome sequence of *Treponema pallidum* ssp. *pallidum* strain SS14 determined with oligonucleotide arrays. BMC Microbiol 8:76

McGill MA, Edmondson DG, Carroll JA, Cook RG, Orkiszewski RS, Norris SJ (2010) Characterization and serologic analysis of the *Treponema pallidum* proteome. Infect Immun 78:2631–2643

Mindel A, Tovey SJ, Timmins DJ, Williams P (1989) Primary and secondary syphilis, 20 years' experience. 2. Clinical features. Genitourin Med 65:1–3

Morgan CA, Molini BJ, Lukehart SA, Van Voorhis WC (2002) Segregation of B and T cell epitopes of *Treponema pallidum* repeat protein K to variable and conserved regions during experimental syphilis infection. J Immunol 169:952–957

Muller M, Ewert I, Hansmann F, Tiemann C, Hagedorn HJ, Solbach W, Roider J, Nolle B, Laqua H, Hoerauf H (2007) Detection of *Treponema pallidum* in the vitreous by PCR. Br J Ophthalmol 91:592–595

Noordhoek GT, Wolters EC, de Jonge ME, van Embden JD (1991) Detection by polymerase chain reaction of *Treponema pallidum* DNA in cerebrospinal fluid from neurosyphilis patients before and after antibiotic treatment. J Clin Microbiol 29:1976–1984

Norris SJ (1993) Polypeptides of *Treponema pallidum*: progress toward understanding their structural, functional, and immunologic roles. Microbiol Rev 57:750–779

Norris SJ, Charon NW, Cook RG, Fuentes MD, Limberger RJ (1988) Antigenic relatedness and N-terminal sequence homology define two classes of periplasmic flagellar proteins of *Treponema pallidum* subsp. *pallidum* and *Treponema phagedenis*. J Bacteriol 170:4072–4082

Nusbaum MR, Wallace RR, Slatt LM, Kondrad EC (2004) Sexually transmitted infections and increased risk of co-infection with human immunodeficiency virus. J Am Osteopath Assoc 104:527–535

O'Regan AW, Castro C, Lukehart SA, Kasznica JM, Rice PA, Joyce-Brady MF (2002) Barking up the wrong tree? Use of polymerase chain reaction to diagnose syphilitic aortitis. Thorax 57:917–918

Ottemann KM, Miller JF (1997) Roles for motility in bacterial-host interactions. Mol Microbiol 24:1109–1117

Radolf JD (1995) *Treponema pallidum* and the quest for outer membrane proteins. Mol Microbiol 16:1067–1073

Radolf JD, Blanco DR, Miller JN, Lovett MA (1986) Antigenic interrelationship between endoflagella of *Treponema phagedenis* biotype Reiter and *Treponema pallidum* (Nichols): molecular characterization of endoflagellar proteins. Infect Immun 54:626–634

Radolf JD, Norgard MV, Schulz WW (1989) Outer membrane ultrastructure explains the limited antigenicity of virulent *Treponema pallidum*. Proc Natl Acad Sci USA86:2051–2055

Radolf JD, Arndt LL, Akins DR, Curetty LL, Levi ME, Shen Y, Davis LS, Norgard MV (1995) *Treponema pallidum* and *Borrelia burgdorferi* lipoproteins and synthetic lipopeptides activate monocytes/macrophages. J Immunol 154:2866–2877

Raiziss GW, Severac M (1937) Rapidity with which *Spirochaeta pallida* invades the bloodstream. Arch Dermatol Syph 35:1101–1109

Rajan MS, Pantelidis P, Tong CY, French GL, Graham EM, Stanford MR (2006) Diagnosis of *Treponema pallidum* in vitreous samples using real time polymerase chain reaction. Br J Ophthalmol 90:647–648

Righarts AA, Simms I, Wallace L, Solomou M, Fenton KA (2004) Syphilis surveillance and epidemiology in the United Kingdom. Euro Surveill 9:21–25

Rivera J, Vannakambadi G, Hook M, Speziale P (2007) Fibrinogen-binding proteins of Gram-positive bacteria. Thromb Haemost 98:503–511

Riviere GR, Thomas DD, Cobb CM (1989) *In vitro* model of *Treponema pallidum* invasiveness. Infect Immun 57:2267–2271

Rolfs RT, Joesoef MR, Hendershot EF, Rompalo AM, Augenbraun MH, Chiu M, Bolan G, Johnson SC, French P, Steen E, Radolf JD, Larsen S (1997) A randomized trial of enhanced therapy for early syphilis in patients with and without human immunodeficiency virus infection. N Engl J Med 337:307–314

Rosahn PD, Gueft B, Rowe CL (1948) Experimental mouse syphilis; organ distribution of the infectious agent. Am J Syph Gonorrhea Vener Dis 32:327–336

Salazar JC, Hazlett KR, Radolf JD (2002) The immune response to infection with *Treponema pallidum*, the stealth pathogen. Microbes Infect 4:1133–1140

Sell S, Gamboa D, Baker-Zander SA, Lukehart SA, Miller JN (1980) Host response to *Treponema pallidum* in intradermally-infected rabbits: evidence for persistence of infection at local and distant sites. J Invest Dermatol 75:470–475

Sell S, Baker-Zander S, Powell HC (1982) Experimental syphilitic orchitis in rabbits: ultrastructural appearance of *Treponema pallidum* during phagocytosis and dissolution by macrophages *in vivo*. Lab Invest 46:355–364

Setubal JC, Reis M, Matsunaga J, Haake DA (2006) Lipoprotein computational prediction in spirochaetal genomes. Microbiology 152:113–121

Singh AE, Romanowski B (1999) Syphilis: review with emphasis on clinical, epidemiologic, and some biologic features. Clin Microbiol Rev 12:187–209

Sun ES, Molini BJ, Barrett LK, Centurion-Lara A, Lukehart SA, Van Voorhis WC (2004) Subfamily I *Treponema pallidum* repeat protein family: sequence variation and immunity. Microbes Infect 6:725–737

Thomas DD, Baseman JB, Alderete JF (1985) Fibronectin mediates *Treponema pallidum* cytadherence through recognition of fibronectin cell-binding domain. J Exp Med 161: 514–525

Thomas DD, Navab M, Haake DA, Fogelman AM, Miller JN, Lovett MA (1988) *Treponema pallidum* invades intercellular junctions of endothelial cell monolayers. Proc Natl Acad Sci USA85:3608–3612

Thomas DD, Fogelman AM, Miller JN, Lovett MA (1989) Interactions of *Treponema pallidum* with endothelial cell monolayers. Eur J Epidemiol 5:15–21

Tourville DR, Byrd LH, Kim DU, Zajd D, Lee I, Reichman LB, Baskin S (1976) Treponemal antigen in immunopathogenesis of syphilitic glomerulonephritis. Am J Pathol 82:479–492

Ulevitch RJ, Tobias PS (1999) Recognition of Gram-negative bacteria and endotoxin by the innate immune system. Curr Opin Immunol 11:19–22

Walker GJ, Walker DG (2007) Congenital syphilis: a continuing but neglected problem. Semin Fetal Neonatal Med 12:198–206

Walker EM, Zampighi GA, Blanco DR, Miller JN, Lovett MA (1989) Demonstration of rare protein in the outer membrane of *Treponema pallidum* subsp. *pallidum* by freeze-fracture analysis. J Bacteriol 171:5005–5011

Zoechling N, Schluepen EM, Soyer HP, Kerl H, Volkenandt M (1997) Molecular detection of *Treponema pallidum* in secondary and tertiary syphilis. Br J Dermatol 136:683–686

Chapter 2
Immunoseclusion and Chronic Infection by *Borrelia burgdorferi*

Robert D. Gilmore Jr.

2.1 Introduction

The subjects of immunoseclusion and chronic infection as pertaining to the Lyme disease pathogen, *Borrelia burgdorferi*, present challenges not only in developing suitable models for study, but also in the accurate interpretation of experimental findings in the context of human disease. The term immunoseclusion has been described as the property of a pathogen to evade host immunity by hiding itself in sites not typically afforded vasculature, also known as immunologically privileged organs or tissues. The ability of the Lyme disease bacterium to disseminate to and invade host cells and tissues may be a contributing factor for circumventing host defenses. However, any relationship between putative immunoseclusion properties of the Lyme disease agent and chronic infection has not been firmly established.

Lyme disease is the result of a tick-transmitted bacterial infection that affects thousands of people each year predominantly in North America, Europe, and Asia making it a serious public health concern. Also termed Lyme borreliosis, the infection typically begins to present clinically with the appearance of a distinctive rash at the site of the tick bite called erythema migrans (EM) and is commonly accompanied by fever, headaches, fatigue, myalgia, arthralgia, and regional lymphadenopathy (Steere 2001). If left untreated by antimicrobials, the disease can progress to a later stage with more serious manifestations including arthritis, carditis, and neurological impairment (Steere 2001). Fortunately, Lyme disease is curable in most cases with a few week (2–4) course of antibiotic treatment (Wormser et al. 2006). The disease presentation is basically similar worldwide, although there are regional

R.D. Gilmore Jr., Ph.D. (✉)
Division of Vector Borne Diseases, Centers for Disease Control and Prevention, Fort Collins, Colorado 80521, USA
e-mail: rbg9@cdc.gov

M.E. Embers (ed.), *The Pathogenic Spirochetes: Strategies for Evasion of Host Immunity and Persistence*, DOI 10.1007/978-1-4614-5404-5_2,

differences due mainly to the presence of two additional major genospecies of the causative agent in Eurasia. However, because the Lyme disease agent can only be transmitted to humans by ticks of the *Ixodes ricinis* complex which exists in distinct ecological regions (Lane et al. 1991), Lyme disease is endemic, i.e., occurs in localized foci. This fact, along with the observance of the EM and support of serological tests (when warranted), greatly aids in an accurate clinical diagnosis of Lyme disease.

The bacterial culprit that causes Lyme borreliosis is *Borrelia burgdorferi*, a spirochete as defined by its characteristic corkscrew-shaped morphology (Burgdorfer et al. 1982). In addition to *B. burgdorferi*, two other genospecies cause Lyme disease in Europe and Asia, *Borrelia garinii* and *Borrelia afzelii* (Steere 2006). Collectively, the three genospecies are known as *B. burgdorferi* sensu lato. In North America, *B. burgdorferi* sensu stricto (in the strict sense) is the sole genospecies responsible for Lyme disease and is found mainly in the northeast- and northcentral-regions of the United States where the vector is *Ixodes scapularis* and the reservoirs are small vertebrates, primarily rodents (Lane et al. 1991). In nature, *B. burgdorferi* exists exclusively inside either a vertebrate host or ticks, as the organism is incapable of living freely in an external environment. Nevertheless, the pathogen thrives in an enzootic cycle passing from one reservoir host to another by way of a tick intermediate that transfers and acquires *B. burgdorferi* during feeding on the hosts. The spirochete has developed an intricate system of genetically based survival mechanisms that allow it to adapt to each of its environments, whether mammalian or tick, in order to maintain its life cycle. *B. burgdorferi* lives within the confines of each host without causing harm, however, when transmitted to humans the scenario changes and the organism becomes pathogenic when infecting this host. Because the tick that transmits *B. burgdorferi* is very small, humans quite often are unaware that they have been bitten and subsequently become infected, even though the tick feeds for 4–5 days and borreliae do not venture out of the tick until around 48 h post-attachment (Piesman 1993; Ohnishi et al. 2001). In the absence of antibiotic treatment, *B. burgdorferi* can maintain infection in either rodent or human hosts despite the elicitation of strong immune responses, both innate and adaptive, and the processes driving this evasion ability is the subject of this book. The topics covered in this chapter are:

1. Challenges in performing studies investigating dissemination, invasion, and colonization of host organs and tissues.
2. Mechanisms of immunoseclusion including the physical structures and properties of *B. burgdorferi*, extracellular localization, and intracellular invasion.
3. Host persistence and chronic infection.

This chapter provides a review and summary of the current literature regarding the properties of *B. burgdorferi* relating to immunoseclusion and chronic infection from studies performed by many research groups utilizing animal models and cell culture. Although these models are invaluable in aiding our understanding of the complex mechanisms involved in human pathogenicity, care must be utilized when extrapolating experimental results into firm conclusions regarding the pathobiology of human Lyme borreliosis. Studies exploring borrelial mechanisms of invasion and seclusion into host tissues and cells should be regarded and interpreted in these contexts: (1) evasion of innate and/or adaptive immune responses, i.e., short-term or long-term survival or both; (2) antibiotic treated vs. untreated subjects; (3) infection in mice vs. infection in humans.

2.2 Challenges in Studying *B. burgdorferi* Host Infection, Dissemination, and Tissue Colonization

When an infected tick feeds on a mammalian host, chemical signals in the blood meal awaken the quiescent borreliae that have resided dormantly as the tick overwintered. A dynamic series of gene regulation induces borrelial gene expression essential for bacterial replication and prepares *B. burgdorferi* for migration from the tick to the host during the 4–5 day feeding period (Schwan et al. 1995; Caimano et al. 2007; Boardman et al. 2008; Tokarz et al. 2004). *B. burgdorferi* are initially deposited into the host skin at the tick bite site. It is here that the organisms encounter the first stage of the host's immune attack as the innate system assaults the borreliae by complement-mediated lysis, engulfment by phagocytic cells, and induction of pro-inflammatory cytokines (Steere et al. 2004; Samuels and Radolf 2010). The inflammation that occurs in the skin results in the distinguishing EM rash. A subpopulation of *B. burgdorferi* weathers this attack and begins dissemination to other organs and tissues by hematogenous spread (Wormser 2006). After days to weeks of incubation, dissemination of the organism can result in diverse clinical manifestations such as arthritis, carditis, and neurological conditions (Steere 2001; Marques 2010), despite the production of antibodies against several *B. burgdorferi* proteins.

It is difficult to study the pathobiology of the disease in humans because of limitations in detection assays, e.g., culture reisolation and PCR, from affected tissues. *B. burgdorferi* do not replicate to high numbers in the bloodstream, therefore clinical isolation is not routine although recovery techniques from the blood of untreated patients that are EM-positive early in infection onset have improved (Wormser et al. 2005, 1998). *B. burgdorferi* recovery from skin in an EM by culturing can be accomplished and is the "gold standard" for a diagnosis of infection. Isolation of the organism from internal tissues by culture is not practical due to the extreme insensitivity of the assay and invasive biopsy procedures that would often be required to obtain tissue samples. Synovial fluid and cerebrospinal fluid (CSF) from patients have been used for study, with borrelial DNA being detected by PCR, but culture isolation of the organism from these tissues has been rare (Marques 2010; Aguero-Rosenfeld 2008). PCR as a pathogen detection assay is useful, but limited in that it proves that bacterial DNA is present, but not whether the organisms are alive or dead, and therefore not an indicator of ongoing infection. Moreover, outstanding questions remain regarding whether the DNA represents organisms that are dead or dying, or whether the organisms are viable, but uncultivable. Finally, it is not feasible or ethical to study the biological processes of the spirochete during the course of known infection in humans without administering antibiotic treatment.

Therefore, studies on *B. burgdorferi* pathogenesis and interactions with the host tissues and cells have been relegated to animal models (in vivo) and tissue culture (in vitro).

Dissemination from the site of the tick bite and subsequent colonization of tissues occurs in both humans and the murine model of infection, but there are some differences. In mice, *B. burgdorferi* migrate to and infect heart, spleen, bladder, joint, and ear tissues, but not the central nervous system (CNS) as in humans. Murine species also do not display EM or other signs of pathology commonly associated with human Lyme disease (an exception is some inbred strains of mice will develop arthritis and carditis) despite the rodents remaining chronically infected throughout their lifespan. Therefore, mouse models are useful to study chronic infection mechanisms, but may reflect the biology of borrelial persistence in reservoir hosts as seen in nature rather than as a disease model.

2.3 Properties of *B. burgdorferi* Involved in Tissue Colonization

After exposure to *B. burgdorferi* by tick bite or needle inoculation, mice remain chronically infected throughout their life which can be demonstrated by reisolation of the spirochetes from various tissues. Such persistence is consistent with *B. burgdorferi* survival within rodent reservoir hosts that is essential for the natural perpetuation of the enzootic cycle via tick vectors. An interesting biological feature is that mice remain infected despite elicitation of a strong innate and adaptive immune response against the borreliae. The antibody response in humans following *B. burgdorferi* infection parallels that seen in mice, making mice a good experimental animal model for the study of host immunity against *B. burgdorferi* infection. Throughout the infection, in both rodents and humans, antibodies against numerous *B. burgdorferi* antigens can be detected (Dressler et al. 1993; Nowalk et al. 2006). Some have been shown to be protective and/or borreliacidal (e.g., outer surface protein C [OspC] and decorin binding protein A [DbpA]) (Preac-Mursic et al. 1992; Gilmore et al. 1996; Rousselle et al. 1998; Hanson et al. 1998; Hagman et al. 1998), yet the organism is not cleared, presumably by mechanisms employed by the spirochete for immune evasion (i.e., immune suppression, antigenic variation, and variable expression of antigen genes) (Embers et al. 2004). An additional strategy proposed for evasion is the physical seclusion of the organism away from the harmful effects of the innate and adaptive humoral response. For example, it has been postulated that *B. burgdorferi* sequesters itself from antibodies by localizing in immunocompromised areas of tissues such as the extracellular matrix (ECM), collagen, and/or intracellular niches.

The ability of *B. burgdorferi* to disseminate to and interact with host ECM components and tissues is not unexpected due to the physical attributes of the organism. First, *B. burgdorferi* is a highly motile organism having an intricate network of periplasmic flagella that aid in its spread to target sites and subsequent penetration of tissue barriers (Charon et al. 1992a, b).

Second, though *B. burgdorferi* lacks the ability to produce proteases, it can bind host proteases to assist in dissemination and penetration into host tissues. Borrelial binding and activation of host plasmin, a serine protease capable of degrading ECM, has been demonstrated. This aids the organism in penetrating cell monolayers in vitro and advancing spirochetemia in the mouse model (Hu et al. 1995; Coleman et al. 1995, 1997; Klempner et al. 1995; Fuchs et al. 1994, 1996; Grab et al. 2005). *B. burgdorferi* proteins implicated in plasminogen binding include OspA, BPBP, OspC, CRASP-1, and ErpP-A and -C (Fuchs et al. 1994; Hu et al. 1997; Brissette et al. 2009a; Lagal et al. 2006; Hallstrom et al. 2010). Additionally, several studies have shown *B. burgdorferi* can induce production in vitro of host matrix metalloproteinases (MMP), enzymes produced by numerous cells that are involved in the normal restructuring of ECM (Grab et al. 2005; Behera et al. 2004; Gebbia et al. 2001; Singh et al. 2004; Zhao et al. 2007; Perides et al. 1999). However, the effect of borrelial induction of MMPs in vivo for invasion and seclusion into host tissues is not clear; instead the evidence suggests that MMP activation by *B. burgdorferi* may lead to the pathology associated with Lyme arthritis (Hu et al. 2001; Lin et al. 2001; Heilpern et al. 2009).

Third, *B. burgdorferi* synthesizes a number of proteins that function as adhesins allowing the spirochetes to adhere to host tissue and cell receptors which may then induce signals integral for cellular internalization (Coburn et al. 2005). Borrelial proteins that bind host molecules such as fibronectin, laminin, decorin, integrins, and glycosaminoglycans have been identified and are listed in Table 2.1. Studies showing that inactivation of specific adhesin protein genes does not completely abrogate either the infectious and/or adherence properties of the organism (specifically at high infection doses), indicates that *B. burgdorferi* can synthesize numerous surface localized proteins capable of infection-mediating adherence (Li et al. 2006; Seshu et al. 2006; Weening et al. 2008; Blevins et al. 2008; Shi et al. 2008; Parveen et al. 2006). Therefore, it is suspected that *B. burgdorferi* genetically regulates a complex network of protein synthesis during mammalian infection with potential functional redundancy enabling the organism to persevere in its specific tissue and/or escape the host's production of antibodies (Caimano et al. 2007; Zhang et al. 1997; Embers et al. 2007; Coutte et al. 2009; Liang et al. 2002, 2004a; Gilmore et al. 2007, 2008; Miller and Stevenson 2006). Although not completely clear whether the organism is capable of tropism for specific tissues or selectively adheres after general dissemination, it is evident that *B. burgdorferi* has the machinery to produce a variety of proteins for tissue and cellular colonization. However, questions remain relating to the fate of the borreliae following attachment to host components that enable long-term maintenance.

2.3.1 Extracellular Locales

The ECM is a complex structure of connective tissue surrounding mammalian tissues and cells, and is composed of a fibrous network of structural proteins and proteoglycans (e.g., collagen, fibronectin, decorin, and laminin). The ability of

Table 2.1 *B. burgdorferi* outer surface adhesin proteins

Adhesin	Host receptor	References
BBK32 (fibronectin binding protein)	Fibronectin Dermatan sulfate Heparin	(Probert and Johnson 1998; Fischer et al. 2006)
DbpA (decorin binding protein A)	Decorin Glycosaminoglycans	(Fischer et al. 2003; Guo et al. 1998)
DbpB (decorin binding protein B)	Decorin	
P66	Integrin $\alpha_{IIb}\beta_3$ Integrin $\alpha_v\beta_3$	(Coburn et al. 1993; Coburn and Cugini 2003)
OspA	TrospA (tick midgut receptor) Evidence for binding to endothelial and neural cells	(Pal et al. 2004) (Rupprecht et al. 2006; Comstock et al. 1993)
BmpA (Borrelia membrane protein A)	Laminin	(Verma et al. 2009)
Bgp (Borrelia GAG-binding protein)	Glycosaminoglycans	(Cluss et al. 2004; Parveen and Leong 2000)
RevA	Fibronectin	(Brissette et al. 2009b)
BBB07	Integrin $\alpha_3\beta_1$	(Behera et al. 2008)
ErpX	Laminin	(Brissette et al. 2009c)
CRASP-1	Bone morphogenic protein 2 Collagen Fibronectin Laminin	(Hallstrom et al. 2010)

bacterial pathogens to colonize host tissues and cells requires the organism to penetrate barriers such as the ECM. *B. burgdorferi* has the ability to traverse the ECM by such physical properties as motility provided by periplasmic flagella, and the ability to utilize host proteases such as plasmin(ogen) and possibly MMPs to proteolytically bore into the tissue. This capability may serve the dual purpose of allowing *B. burgdorferi* to localize in the dense structure of collagen-rich tissue or ECM to avoid immune detection.

Studies observing *B. burgdorferi* microscopically in histopathologic sections of animal ECM or collagen-rich tissues (such as tendons, joints, or cartilage) have been limited. The Barthold laboratory has shown *B. burgdorferi* in immunohistochemically stained sections of tibiotarsal synovium and tendon from experimentally infected immunocompetent and immunodeficient mice (Barthold et al. 1993, 2006; Hodzic et al. 2008). *B. burgdorferi* attachment and colonization of collagen in vitro was demonstrated by Zambrano et al. whereby they showed that following specific binding to collagen lattices, borreliae were able to replicate and form microcolonies (Zambrano et al. 2004). Liang et al. proposed that decorin-rich tissues such as skin and joints provide a "protective niche" for *B. burgdorferi* to avoid adaptive humoral immunity by immunoseclusion due to the organism's ability to bind decorin by the surface antigens DbpA and DbpB (Liang et al. 2004b). Weening et al. extended these conclusions in a study whereby a DbpBA-deficient mutant exhibited an early colonization defect in mice, suggesting these proteins are required for resistance to the host's innate immune response (Weening et al. 2008). An independent study, however, provided evidence that although anti-DbpA antiserum can induce carditis and arthritis remission in infected mice, there was little to no effect on eliminating or reducing the numbers of borreliae in tissues (Barthold et al. 2006). These data suggested that although DbpA plays an important role in borrelial binding and colonization of ECM and host cells, there were undoubtedly other factors involved in *B. burgdorferi* tissue localization following establishment of infection. A recent review by Cabello et al. summarized the above studies, and the authors reflected on the interactive nature of *B. burgdorferi* and its adhesins with proteoglycan components of the ECM as a mechanism for the organism to establish itself within a protective habitat (Cabello et al. 2007). These authors provided a list of thought-provoking questions regarding the putative effect of *B. burgdorferi*/ECM interactions on persistent infections. Among the more interesting are: (1) What is the role, if any, for identified borrelial adhesins in persistent infections? (2) How metabolically active are *B. burgdorferi* in the ECM? (3) Does the ECM environment modulate *B. burgdorferi* gene expression to synthesize new adhesins and/or immunoprotective antigens (Cabello et al. 2007)? Other questions come to mind as well: (1) Do *B. burgdorferi* dwell and localize in connective tissues or are these locales only transient destinations as borreliae seek out host cells for colonization? (2) Does *B. burgdorferi* replicate within ECM? (3) What is the effect of *B. burgdorferi* within ECM on Lyme disease pathology? (4) Are these connective tissues resistant to permeabilization by antibiotics or antibodies thereby providing a protective site?

An interesting observation by Grab et al. indicated that *B. burgdorferi* interact with fibrocytes, fibroblast-like leukocytes that circulate in the peripheral blood. The investigators demonstrated borrelial binding to human and monkey fibrocytes in vitro and observed the organisms being taken into deep recesses of the cell membrane (Grab et al. 1999). Based on these results, they postulated a mechanism whereby *B. burgdorferi* bind to fibrocytes and are then targeted to the connective tissue. Additionally, they speculated that borreliae within the cell invaginations provide a site for immune seclusion.

Duray et al. described a human tonsillar tissue culture method in which they inoculated *B. burgdorferi* to study the invasive properties of the organism ex vivo (Duray et al. 2005). These investigators observed extracellular borreliae in the tissue and determined that the organisms were capable of replication. In addition, they described "cystic" structures formed by the borreliae and suggested they were dormant organisms that could eventually return to their spirochetal origin. More discussion on the cyst-like borreliae is provided in Sect. 2.5.

The non-human primate has been the best model to study Lyme borreliosis as it pertains to human infection. Unlike murine infection, rhesus monkeys can develop EM and a localization of *B. burgdorferi* in the CNS and peripheral nerves which mimics neuroborreliosis in humans. The CNS has been suggested as an immune privileged organ and is therefore a potential site for *B. burgdorferi* seclusion (Embers et al. 2004). *B. burgdorferi* have been observed microscopically in the CNS of monkeys where the bacteria were found localized to leptomeninges, nerve roots, and dorsal root ganglia, as well as to the connective tissue of peripheral nerves (Cadavid et al. 2000; Roberts et al. 1995). Collectively, these studies demonstrate the propensity of *B. burgdorferi* to disseminate and localize to connective tissues of various organs within a variety of susceptible hosts.

2.3.2 Internalization by Host Cells

Host cell adherence is an initial step for host colonization and proliferation of bacterial pathogens, including *B. burgdorferi*, during infection. Numerous studies have described the interactions of *B. burgdorferi* in vitro with cells of various origins, and *B. burgdorferi* produces several outer membrane proteins that facilitate binding to host cell receptors and ligands (Table 2.1). *B. burgdorferi* have been shown to adhere to human and murine fibroblasts, endothelial cells, epithelial cells, macrophages, neuronal and glial cells, fibrocytes, and lymphocytes; also tick cells, and Vero cells (Grab et al. 1999; Sambri et al. 1993; Cinco et al. 2001; Thomas et al. 1994; Leong et al. 1998; Fischer et al. 2003; Peters and Benach 1997; Garcia-Monco et al. 1989, 1991; Szczepanski et al. 1990; Thomas and Comstock 1989; Kurtti et al. 1993; Hechemy et al. 1992; Rupprecht et al. 2006; Coburn et al. 1998; Dorward et al. 1997; Livengood et al. 2008; Chmielewski and Tylewska-Wierzbanowska 2010; Wu et al. 2011; Montgomery et al. 1993). Although *B. burgdorferi* is commonly referred to as an extracellular pathogen, several investigations performed in vitro with cell culture have revealed the invasive properties of borreliae into non-professional phagocytic cells (e.g., see Figs. 2.1 and 2.2) (Hechemy et al. 1992; Chmielewski and Tylewska-Wierzbanowska 2010; Wu et al. 2011; Livengood and Gilmore 2006; Ma et al. 1991; Klempner et al. 1993; Girschick et al. 1996). Related studies have demonstrated that *B. burgdorferi* can penetrate cell junctions as a putative process for dissemination (Szczepanski et al. 1990; Comstock and Thomas 1989; Grab et al. 2009). Additionally, *B. burgdorferi* are internalized by macrophages and monocytes through phagocytosis, a critical step in the innate immune response against infection (Moore et al. 2007; Cruz et al. 2008; Shin et al. 2008; Kuhlow et al. 2005). The organisms are eventually degraded within the phagolysosome of these cells, although a study by Montgomery et al. reported the occasional survivor in mouse macrophages (Montgomery et al. 1993).

The paradigm of *B. burgdorferi* as a mainly, if not exclusive, extracellular organism has primarily been due to the inability to detect intracellular spirochetes in infected tissues in vivo, either in animal models or in humans. Nevertheless, evidence from in vitro studies demonstrating the ability of *B. burgdorferi* to invade and inhabit cells internally suggests the probability that subpopulations of borreliae localize within cells during infections. Such internalization may result in host cell damage causing pathology consistent with Lyme borreliosis, or may be a mechanism for a

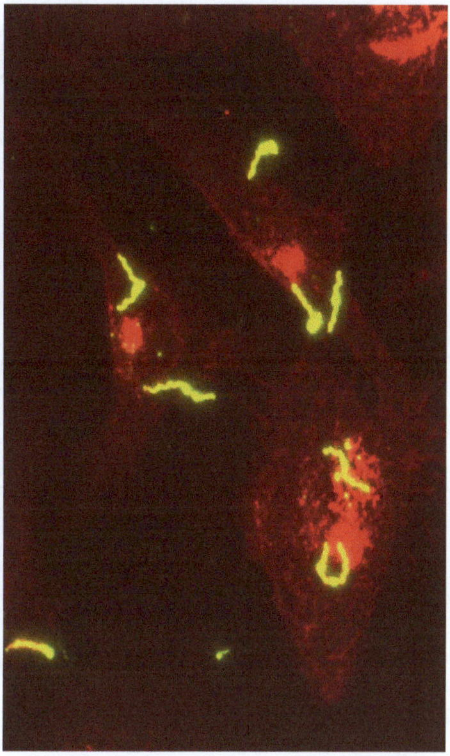

Fig. 2.1 Association of *Borrelia burgdorferi* with human cells in vitro. Confocal microscopy image of H4 neuroglial cells incubated with *B. burgdorferi*. Human cell plasma membranes were immunostained (*red*), and *B. burgdorferi* were labeled *green* with the images being merged whereby *yellow* indicates co-localization of *B. burgdorferi* with the plasma membrane. Confocal images by Jill A. Livengood

subset of the pathogen population to escape destruction from innate or adaptive host responses and/or antibiotic treatment.

Mechanisms involved in the internalization of *B. burgdorferi* into host cells post-attachment have been understudied. The first studies on borrelial adherence and invasion were performed in the late 1980s and early 1990s. A noticeable gap in this area of research ensued until recently (2006–2010) with studies by Livengood and Gilmore, Chmielewski et al., and Wu et al. reporting on borrelial invasion of eukaryotic cells (Chmielewski and Tylewska-Wierzbanowska 2010; Wu et al. 2011; Livengood and Gilmore 2006). Also, studies from the Chaconas laboratory have used intravital microscopy to visualize borrelial interactions with vasculature cells in vivo (Moriarty et al. 2008; Norman et al. 2008). The Skare laboratory's research has progressed from simple intracellular observation of *B. burgdorferi* in cells to delineating mechanisms for internalized uptake. Wu et al. investigated requirements for invasion and found that β_1-containing integrins and Src-family kinases were necessary components suggesting that *B. burgdorferi* uptake needed actin filament reorganization to gain entry. They also found that an outer surface adhesin, fibronectin binding protein BBK32, was not a requirement for the internalization process. Finally these investigators observed that *B. burgdorferi* remained viable after cellular uptake for out to 4 weeks of cocultivation (Wu et al. 2011).

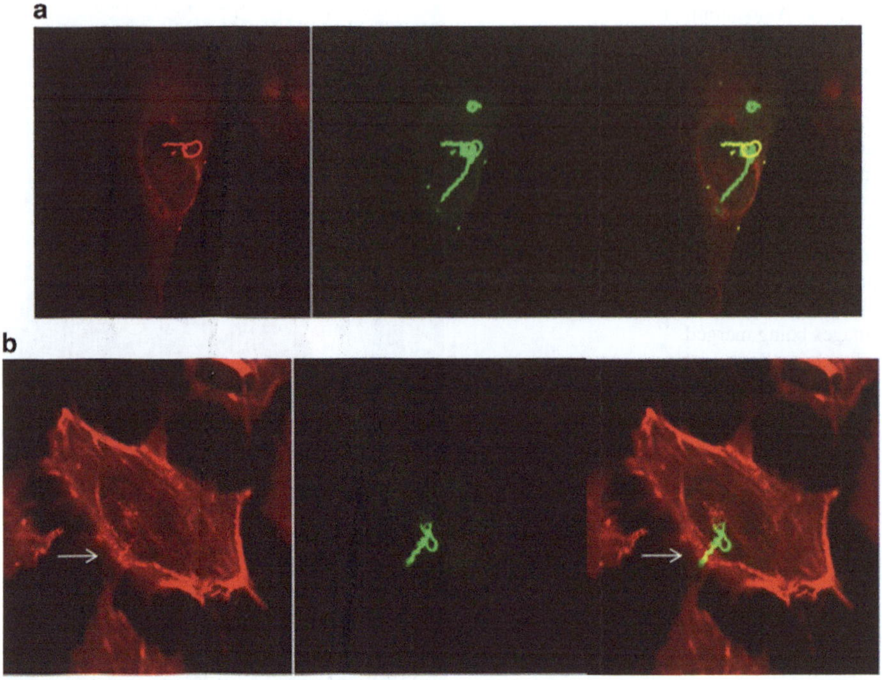

Fig. 2.2 Extra- and intracellular association of *Borrelia burgdorferi* with human cells in vitro by double immunostaining. (**a**) Human umbilical vein endothelial cells (HUVECs). (**b**) H4 neuroglial cells. *Left panels* represent extracellular spirochetes stained with a rhodamine-conjugated anti-*B. burgdorferi* antibody. *Middle panels* show the same microscopic field of spirochetes stained with a FITC-conjugated anti-*B. burgdorferi* antibody following permeabilization of fixed cells. *Right panels* show a merged image of the two stains. *Yellow*-appearing *B. burgdorferi* indicate extracellular organisms that have been stained with both rhodamine and FITC. *Green B. burgdorferi* indicate intracellular organisms. The *arrow* in (**b**) points to a portion of a spirochete that stains *yellow* indicating it is outside the cell; while the remainder of the organism is *green* inside the cell. Confocal images by Jill A. Livengood

Many mysteries relating to borrelial invasion and intracellular existence remain with the foremost question being does *B. burgdorferi* utilize intracellular environments during any phase of in vivo infection? Other questions that warrant further investigation include: (1) Does *B. burgdorferi* replicate within host cells? (2) Does *B. burgdorferi* invasion lead to host cell damage and pathological effects to the host? (3) Is *B. burgdorferi* internalization a temporary event and do borreliae eventually exit the host cells they invade? (4) Does *B. burgdorferi* undergo differential gene expression in this environment to produce new antigens or surface components for survival? The recent studies indicated above, along with the continued work with adhesins by several investigators, signal a renewed interest in understanding more about *B. burgdorferi* adherence and invasion into eukaryotic cells and the relationship to immunoseclusion and immune evasion.

2.4 *B. burgdorferi* Persistence and Chronic Infection

Webster's dictionary defines persistent as "remaining infective for a relatively long time in a vector after an initial period of incubation"; persist is defined as "to continue to exist esp. past a usual, expected, or normal time". Webster's defines chronic as "marked by a long duration or frequent recurrence". Therefore, by these definitions, the question can be raised—does *B. burgdorferi* persist (or become persistent) within humans resulting in chronic disease? The term "chronic Lyme disease" has been used in a variety of different connotations over time and has evolved to inaccurately describe a larger array of conditions that has resulted in patient diagnoses attributing chronic subjective symptoms to persistent borrelial infections that are unsubstantiated by scientifically validated and controlled laboratory tests. Various interpretations of "chronic Lyme disease" have produced diverse viewpoints on how the Lyme disease pathogen operates to cause disease and consequently how Lyme patients should be diagnosed and treated. A review of "chronic Lyme disease" detailing the applications of the term and ramifications among the interested parties is referred (Feder et al. 2007).

It is a fact that *B. burgdorferi* produces a chronic infection in rodents in nature and in experimental animals, i.e., the spirochete will persist in the animal throughout its lifetime as evidenced by being cultured from various necropsied tissues. Despite such systemic involvement, rodents do not develop disease, consistent with the natural existence of a reservoir host necessary for borreliae to perpetuate its enzootic cycle between mammalian host and tick vector. Therefore, using the mouse as a model for infection, one must be cognizant of the key difference, i.e., mice do not get sick but people do (an exception is certain laboratory strains of inbred mice that develop arthritis during infection). Borrelial infection in mice results in both innate and adaptive immune responses including a strong antibody response. As noted previously, despite elicitation of a humoral response generating antibodies that are borreliacidal, *B. burgdorferi* is not cleared and persists within the murine host. Humans share this characteristic as antibodies against several borrelial antigens can be detected by serological assays following establishment of infection. Like mice, the antibody response in humans does not result in clearance of the organisms and resolution of the disease, therefore, without antibiotic treatment, *B. burgdorferi* can presumably persist in humans for months to years (Steere et al. 2004). In humans, a 2–4 week course of antibiotics is an effective treatment for the resolution of disease manifestations in the vast majority of patients who have been accurately diagnosed with Lyme disease, as summarized by the guidelines of the Infectious Diseases Society of America (IDSA) (Wormser et al. 2006). However, following treatment a minority of patients report subjective symptoms such as arthalgia, fatigue, and impaired cognitive function. Short-term duration of the symptoms is referred to as "post-Lyme disease symptoms", while continuation of these ailments for >6 months is referred to as "post-Lyme disease syndrome" (Feder et al. 2007). The fundamental question eliciting disagreement and controversy between Lyme disease advocates and their practitioners, and infectious disease researchers, is whether *B. burgdorferi* persists in infected humans following antibiotic treatment, presumably by utilizing the putative seclusion strategies, to cause a chronic infection with debilitating effects.

The scientific evidence to date reports few cases whereby *B. burgdorferi* can be cultured from tissues obtained from clinically diagnosed Lyme disease patients following antibiotic treatment. Two European groups have reported culture isolation of *B. burgdorferi* sensu lato in humans following antibiotic treatment (Strle et al. 1996; Hunfeld et al. 2005, 2006; Preac-Mursic et al. 1989). In their study, Hunfeld et al. were able to obtain clinical reisolates from the EM site of Lyme borreliosis patients after antibiotic treatment. (The patients had EM but did not present with or develop more serious manifestations.) The investigators noted that culture reisolation is rare following the initiation of antibiotic therapy, and attributed positive cultures to treatment failure (Hunfeld et al. 2005). Their study concluded that borrelial persistence at the site of the EM was not due to acquired resistance to the antibiotic. Furthermore, these investigators pointed out that "the currently available diagnostic techniques do not reliably discriminate among possible reinfection, true endogenous relapse, and coinfection with other tick-borne pathogens. These drawbacks together with the phenomenon of resistance to therapy in individual patients undoubtedly contribute to the inconsistencies surrounding the optimal treatment regimens for Lyme borreliosis and are often misinterpreted and misused to support prolonged antibiotic treatment regimens" (Hunfeld et al. 2005, 2006).

Two separate comprehensive studies found culture-negative skin biopsy samples from EM sites following antibiotic treatment, although the patients were all culture-positive pre-treatment (Nadelman et al. 1993; Berger et al. 1992). Studies by Klempner et al. investigated the effects of antibiotic treatment on patients with a history of Lyme disease and were unable to culture *B. burgdorferi* from blood samples (12 patients) or CSF samples post-treatment (128 patients) (Wormser et al. 2006; Klempner et al. 2001). In addition, these studies evaluated blood and CSF from a total of 843 specimens and none were culture- or PCR-positive (Klempner et al. 2001; Klempner 2002). A study often cited as proof of *B. burgdorferi* surviving in patients after antibiotic treatment involved the report of positive blood cultures using a novel culture medium with Detroit tap water as an ingredient (Phillips et al. 1998). However, that study identified spirochetal forms as *B. burgdorferi* based on microscopic observation without confirmatory tests such as PCR, thereby leaving open the possibility of mistaken conclusions. Moreover, a follow-up study by an independent group sought to replicate the results by culturing in the same culture medium and found that *B. burgdorferi* were unable to grow (Marques et al. 2000). Other studies have reported similar results of *B. burgdorferi* detection, but in the end have failed to adhere to the rigorous scientific method to validate the conclusions (Wormser et al. 2006). Therefore to date, scientifically recognized proof of viable *B. burgdorferi* isolation from human patients following antimicrobial treatment is extremely rare.

A systematic approach by several researchers to investigate the existence of *B. burgdorferi* in vivo following antibiotic therapy has been applied to experimental animals. Details and comments are provided in a review article on this subject by Wormser and Schwartz (Wormser and Schwartz 2009). Three studies to determine whether *B. burgdorferi* persisted in mice after antibiotic therapy are deserving of comment. Investigations by Bockenstedt et al. and Hodzic et al. employed DNA detection (by PCR), serology, histology and/or immunohistochemistry, transmission by skin allograft, and xenodiagnosis (assessment of *B. burgdorferi* acquisition by ticks after feeding on an infected animal) in addition to culture as measurement parameters for borrelial detection (Hodzic et al. 2008; Bockenstedt et al. 2002). A third study by Yrjanainen et al. evaluated borrelial persistence in antibiotic-treated mice and detected *B. burgdorferi* by culture and PCR (Yrjanainen et al. 2007). Bockenstedt et al. concluded that spirochetes could be detected in treated mice by xenodiagnosis but for a limited time following antibiotic administration. However, the spirochetes did not survive in ticks, were not able to infect new mice, were not cultivable and did not cause disease in the treated mice. The investigators surmised that the subpopulation of *B. burgdorferi* that survived were rendered avirulent and would eventually be cleared by the host's immune defenses (Bockenstedt et al. 2002). A follow-up study by Hodzic et al. investigated the possibility of borrelial persistence after an antibiotic regimen, as they had hypothesized from their previous work that spirochetes invaded collagen during long-term infection thereby minimizing the effectiveness of antimicrobials and the immune response (Hodzic et al. 2008; Barthold et al. 2006). These investigators concluded that mice remained infected with borreliae after antibiotic treatment, but the bacteria were not cultivable nor able to replicate. In their evaluation of both studies, Wormser and Schwartz pointed out that neither study considered the pharmacodynamics of the antibiotics utilized (ceftriaxone and doxycycline) in the experimental animals as compared to the administration for humans (Wormser and Schwartz 2009). How this parameter affected the outcome and conclusions of the Bockenstedt and Hodzic research is unclear. Nevertheless, both studies and the Yrjanainen et al. study provided provocative results that raise interesting questions for further study, e.g., are the subpopulation of *B. burgdorferi* that persists after treatment merely weakened organisms that will eventually be eliminated by the host or are they metabolically quiescent organisms capable of restoration to a pathogenic state? Undoubtedly, additional research in this fascinating area is awaited.

2.5 Cystic Forms

Morphological transition of *B. burgdorferi* spirochetes in response to growth in culture medium lacking serum or other suboptimal conditions has been demonstrated (Brorson and Brorson 1997; Miklossy et al. 2008; Alban et al. 2000; Murgia et al. 2002; Murgia and Cinco 2004). The changed morphological non-motile forms that result from these conditions have been termed as cysts, cystic forms, and/or cyst-like structures. The transformation of *B. burgdorferi* from motile spirochetes to these morphologically different forms is reversible when rabbit serum is added back to the growth media (Alban et al. 2000; Murgia and Cinco 2004; Brorson and Brorson 1998). These observations have led to the proposition that *B. burgdorferi* utilizes this ability to withstand and survive temporary environments of nutrient deprivation that may occur in host tissues. It has also been speculated that these changes may allow borreliae to evade detection by the immune system. Although there is some evidence for this event in human tissue (Miklossy et al. 2008), like intracellular localization of borreliae in human cells, the existence of "cystic forms" in vivo has yet to be proven, and therefore any clinical significance has not been established. Conclusive demonstration of these structures in vivo awaits further investigations.

2.6 Summary

The ability of *B. burgdorferi* to maintain infection and cause disease within its host in the face of immune pressure is what makes the spirochete a formidable pathogen. Research has greatly improved our knowledge of the structure of the Lyme disease agent and provided new insights into the mechanisms of immune evasion. The subject of immunoseclusion, or the physical sequestering of the organism from host immunity, remains one of the more understudied areas relating to this insidious microbe. There are many unknowns in our understanding of the biology of *B. burgdorferi* as it invades and colonizes extracellularly, and likely intracellularly, within the tissues of its host. It is important to recognize that *B. burgdorferi* may utilize mechanisms to hide away from the first line of immune defense as well as the adaptive response. The spirochetes first enter the host by deposition into the skin by tick bite. The innate response can be potent in removing substantial numbers of infecting organisms and preventing further dissemination and host damage. At this early stage of infection, safe harbor into connective tissue beneath the basement membrane of dermal endothelium as well as intracellular trafficking into fibroblasts, epithelial and endothelial cells may be part of the pathogen's portfolio for survival.

Certainly, *B. burgdorferi* is capable of persisting within mammalian hosts in the absence of treatment with antimicrobials as observed in reservoir rodents in nature and experimental animal models in the laboratory. In humans, which are not natural hosts required for *B. burgdorferi's* enzootic cycle, we cannot conclude with certainty that the organism would persist indefinitely without treatment. As a "dead end host", perhaps the immune system of humans is capable of eventually clearing the pathogen, albeit with resultant sequelae from the infection. More in question is the ability of the organism to persist and continue to cause disease following antibiotic treatment. To date, the evidence is overwhelming that antibiotics are effective for Lyme disease cessation without recurrence. Whether bacteriostatic or bacteriocidal, antimicrobials may not clear every individual bacterial cell, but rather renders them susceptible to the body's immune system to "clean up" the remaining subpopulation either by reducing the invasive numbers and/or leaving them in a weakened state. Because there are no absolutes in biology, it could be argued that there may be the extremely rare instances of immunocompromised or genetically disposed individuals in which an infection could persist and cause pathology despite antimicrobial treatment. Antimicrobial treatment failure could also be a consideration for a case of persistence.

However, such hypothetical isolated cases would need to be accurately validated and separated from misinformed diagnoses in order to study and understand a scientific basis of true chronicity. As part of the immune evasion strategies for establishment and maintenance of infection, for acute and late disease, *B. burgdorferi* seemingly has the tools for physical seclusion. However, more research is needed to discover the ultimate fate of the Lyme disease pathogen after arrival to its immunosecluded destinations and how it relates to the organism's pathogenicity.

References

Aguero-Rosenfeld ME (2008) Lyme disease: laboratory issues. Infect Dis Clin North Am 22(2):301–313, vii

Alban PS, Johnson PW, Nelson DR (2000) Serum-starvation-induced changes in protein synthesis and morphology of Borrelia burgdorferi. Microbiology 146(Pt 1):119–127

Barthold SW, de Souza MS, Janotka JL, Smith AL, Persing DH (1993) Chronic Lyme borreliosis in the laboratory mouse. Am J Pathol 143(3):959–971

Barthold SW, Hodzic E, Tunev S, Feng S (2006) Antibody-mediated disease remission in the mouse model of lyme borreliosis. Infect Immun 74(8):4817–4825

Behera AK, Thorpe CM, Kidder JM, Smith W, Hildebrand E, Hu LT (2004) Borrelia burgdorferi-induced expression of matrix metalloproteinases from human chondrocytes requires mitogen-activated protein kinase and Janus kinase/signal transducer and activator of transcription signaling pathways. Infect Immun 72(5):2864–2871

Behera AK, Durand E, Cugini C, Antonara S, Bourassa L, Hildebrand E, Hu LT, Coburn J (2008) Borrelia burgdorferi BBB07 interaction with integrin alpha3beta1 stimulates production of pro-inflammatory mediators in primary human chondrocytes. Cell Microbiol 10(2):320–331

Berger BW, Johnson RC, Kodner C, Coleman L (1992) Failure of Borrelia burgdorferi to survive in the skin of patients with antibiotic-treated Lyme disease. J Am Acad Dermatol 27(1):34–37

Blevins JS, Hagman KE, Norgard MV (2008) Assessment of decorin-binding protein A to the infectivity of Borrelia burgdorferi in the murine models of needle and tick infection. BMC Microbiol 8:82

Boardman BK, He M, Ouyang Z, Xu H, Pang X, Yang XF (2008) Essential role of the response regulator Rrp2 in the infectious cycle of Borrelia burgdorferi. Infect Immun 76(9):3844–3853

Bockenstedt LK, Mao J, Hodzic E, Barthold SW, Fish D (2002) Detection of attenuated, noninfectious spirochetes in Borrelia burgdorferi-infected mice after antibiotic treatment. J Infect Dis 186(10):1430–1437

Brissette CA, Haupt K, Barthel D, Cooley AE, Bowman A, Skerka C, Wallich R, Zipfel PF, Kraiczy P, Stevenson B (2009a) Borrelia burgdorferi infection-associated surface proteins ErpP, ErpA, and ErpC bind human plasminogen. Infect Immun 77(1):300–306

Brissette CA, Bykowski T, Cooley AE, Bowman A, Stevenson B (2009b) Borrelia burgdorferi RevA antigen binds host fibronectin. Infect Immun 77(7):2802–2812

Brissette CA, Verma A, Bowman A, Cooley AE, Stevenson B (2009c) The Borrelia burgdorferi outer-surface protein ErpX binds mammalian laminin. Microbiology 155(Pt 3):863–872

Brorson O, Brorson SH (1997) Transformation of cystic forms of Borrelia burgdorferi to normal, mobile spirochetes. Infection 25(4):240–246

Brorson O, Brorson SH (1998) In vitro conversion of Borrelia burgdorferi to cystic forms in spinal fluid, and transformation to mobile spirochetes by incubation in BSK-H medium. Infection 26(3):144–150

Burgdorfer W, Barbour AG, Hayes SF, Benach JL, Grunwaldt E, Davis JP (1982) Lyme disease-a tick-borne spirochetosis? Science 216(4552):1317–1319

Cabello FC, Godfrey HP, Newman SA (2007) Hidden in plain sight: Borrelia burgdorferi and the extracellular matrix. Trends Microbiol 15(8):350–354
Cadavid D, O'Neill T, Schaefer H, Pachner AR (2000) Localization of Borrelia burgdorferi in the nervous system and other organs in a nonhuman primate model of lyme disease. Lab Invest 80(7):1043–1054
Caimano MJ, Iyer R, Eggers CH, Gonzalez C, Morton EA, Gilbert MA, Schwartz I, Radolf JD (2007) Analysis of the RpoS regulon in Borrelia burgdorferi in response to mammalian host signals provides insight into RpoS function during the enzootic cycle. Mol Microbiol 65(5):1193–1217
Charon NW, Greenberg EP, Koopman MB, Limberger RJ (1992a) Spirochete chemotaxis, motility, and the structure of the spirochetal periplasmic flagella. Res Microbiol 143(6):597–603
Charon NW, Goldstein SF, Block SM, Curci K, Ruby JD, Kreiling JA, Limberger RJ (1992b) Morphology and dynamics of protruding spirochete periplasmic flagella. J Bacteriol 174(3):832–840
Chmielewski T, Tylewska-Wierzbanowska S (2010) Interactions between Borrelia burgdorferi and mouse fibroblasts. Pol J Microbiol 59(3):157–160
Cinco M, Cini B, Murgia R, Presani G, Prodan M, Perticarari S (2001) Evidence of involvement of the mannose receptor in adhesion of Borrelia burgdorferi to monocyte/macrophages. Infect Immun 69(4):2743–2747
Cluss RG, Silverman DA, Stafford TR (2004) Extracellular secretion of the Borrelia burgdorferi Oms28 porin and Bgp, a glycosaminoglycan binding protein. Infect Immun 72(11):6279–6286
Coburn J, Cugini C (2003) Targeted mutation of the outer membrane protein P66 disrupts attachment of the Lyme disease agent, Borrelia burgdorferi, to integrin alphavbeta3. Proc Natl Acad Sci USA 100(12):7301–7306
Coburn J, Leong JM, Erban JK (1993) Integrin alpha IIb beta 3 mediates binding of the Lyme disease agent Borrelia burgdorferi to human platelets. Proc Natl Acad Sci USA 90(15):7059–7063
Coburn J, Magoun L, Bodary SC, Leong JM (1998) Integrins alpha(v)beta3 and alpha5beta1 mediate attachment of lyme disease spirochetes to human cells. Infect Immun 66(5):1946–1952
Coburn J, Fischer JR, Leong JM (2005) Solving a sticky problem: new genetic approaches to host cell adhesion by the Lyme disease spirochete. Mol Microbiol 57(5):1182–1195
Coleman JL, Sellati TJ, Testa JE, Kew RR, Furie MB, Benach JL (1995) Borrelia burgdorferi binds plasminogen, resulting in enhanced penetration of endothelial monolayers. Infect Immun 63(7):2478–2484
Coleman JL, Gebbia JA, Piesman J, Degen JL, Bugge TH, Benach JL (1997) Plasminogen is required for efficient dissemination of B. burgdorferi in ticks and for enhancement of spirochetemia in mice. Cell 89(7):1111–1119
Comstock LE, Thomas DD (1989) Penetration of endothelial cell monolayers by Borrelia burgdorferi. Infect Immun 57(5):1626–1628
Comstock LE, Fikrig E, Shoberg RJ, Flavell RA, Thomas DD (1993) A monoclonal antibody to OspA inhibits association of Borrelia burgdorferi with human endothelial cells. Infect Immun 61(2):423–431
Coutte L, Botkin DJ, Gao L, Norris SJ (2009) Detailed analysis of sequence changes occurring during vlsE antigenic variation in the mouse model of Borrelia burgdorferi infection. PLoS Pathog 5(2):e1000293
Cruz AR, Moore MW, La Vake CJ, Eggers CH, Salazar JC, Radolf JD (2008) Phagocytosis of Borrelia burgdorferi, the Lyme disease spirochete, potentiates innate immune activation and induces apoptosis in human monocytes. Infect Immun 76(1):56–70
Dorward DW, Fischer ER, Brooks DM (1997) Invasion and cytopathic killing of human lymphocytes by spirochetes causing Lyme disease. Clin Infect Dis 25(Suppl 1):S2–S8
Dressler F, Whalen JA, Reinhardt BN, Steere AC (1993) Western blotting in the serodiagnosis of Lyme disease. J Infect Dis 167(2):392–400

Duray PH, Yin SR, Ito Y, Bezrukov L, Cox C, Cho MS, Fitzgerald W, Dorward D, Zimmerberg J, Margolis L (2005) Invasion of human tissue ex vivo by Borrelia burgdorferi. J Infect Dis 191(10):1747–1754

Embers ME, Ramamoorthy R, Philipp MT (2004) Survival strategies of Borrelia burgdorferi, the etiologic agent of Lyme disease. Microbes Infect 6(3):312–318

Embers ME, Liang FT, Howell JK, Jacobs MB, Purcell JE, Norris SJ, Johnson BJ, Philipp MT (2007) Antigenicity and recombination of VlsE, the antigenic variation protein of Borrelia burgdorferi, in rabbits, a host putatively resistant to long-term infection with this spirochete. FEMS Immunol Med Microbiol 50(3):421–429

Feder HM Jr, Johnson BJ, O'Connell S, Shapiro ED, Steere AC, Wormser GP, Agger WA, Artsob H, Auwaerter P, Dumler JS et al (2007) A critical appraisal of "chronic Lyme disease". N Engl J Med 357(14):1422–1430

Fischer JR, Parveen N, Magoun L, Leong JM (2003) Decorin-binding proteins A and B confer distinct mammalian cell type-specific attachment by Borrelia burgdorferi, the Lyme disease spirochete. Proc Natl Acad Sci USA 100(12):7307–7312

Fischer JR, LeBlanc KT, Leong JM (2006) Fibronectin binding protein BBK32 of the Lyme disease spirochete promotes bacterial attachment to glycosaminoglycans. Infect Immun 74(1):435–441

Fuchs H, Wallich R, Simon MM, Kramer MD (1994) The outer surface protein A of the spirochete Borrelia burgdorferi is a plasmin(ogen) receptor. Proc Natl Acad Sci USA 91(26):12594–12598

Fuchs H, Simon MM, Wallich R, Bechtel M, Kramer MD (1996) Borrelia burgdorferi induces secretion of pro-urokinase-type plasminogen activator by human monocytes. Infect Immun 64(10):4307–4312

Garcia-Monco JC, Fernandez-Villar B, Benach JL (1989) Adherence of the Lyme disease spirochete to glial cells and cells of glial origin. J Infect Dis 160(3):497–506

Garcia-Monco JC, Fernandez Villar B, Szczepanski A, Benach JL (1991) Cytotoxicity of Borrelia burgdorferi for cultured rat glial cells. J Infect Dis 163(6):1362–1366

Gebbia JA, Coleman JL, Benach JL (2001) Borrelia spirochetes upregulate release and activation of matrix metalloproteinase gelatinase B (MMP-9) and collagenase 1 (MMP-1) in human cells. Infect Immun 69(1):456–462

Gilmore RD Jr, Kappel KJ, Dolan MC, Burkot TR, Johnson BJ (1996) Outer surface protein C (OspC), but not P39, is a protective immunogen against a tick-transmitted Borrelia burgdorferi challenge: evidence for a conformational protective epitope in OspC. Infect Immun 64(6):2234–2239

Gilmore RD Jr, Howison RR, Schmit VL, Nowalk AJ, Clifton DR, Nolder C, Hughes JL, Carroll JA (2007) Temporal expression analysis of the Borrelia burgdorferi paralogous gene family 54 genes BBA64, BBA65, and BBA66 during persistent infection in mice. Infect Immun 75(6):2753–2764

Gilmore RD Jr, Howison RR, Schmit VL, Carroll JA (2008) Borrelia burgdorferi expression of the bba64, bba65, bba66, and bba73 genes in tissues during persistent infection in mice. Microb Pathog 45(5–6):355–360

Girschick HJ, Huppertz HI, Russmann H, Krenn V, Karch H (1996) Intracellular persistence of Borrelia burgdorferi in human synovial cells. Rheumatol Int 16(3):125–132

Grab DJ, Lanners H, Martin LN, Chesney J, Cai C, Adkisson HD, Bucala R (1999) Interaction of Borrelia burgdorferi with peripheral blood fibrocytes, antigen-presenting cells with the potential for connective tissue targeting. Mol Med 5(1):46–54

Grab DJ, Perides G, Dumler JS, Kim KJ, Park J, Kim YV, Nikolskaia O, Choi KS, Stins MF, Kim KS (2005) Borrelia burgdorferi, host-derived proteases, and the blood–brain barrier. Infect Immun 73(2):1014–1022

Grab DJ, Nyarko E, Nikolskaia OV, Kim YV, Dumler JS (2009) Human brain microvascular endothelial cell traversal by Borrelia burgdorferi requires calcium signaling. Clin Microbiol Infect 15(5):422–426

Guo BP, Brown EL, Dorward DW, Rosenberg LC, Hook M (1998) Decorin-binding adhesins from Borrelia burgdorferi. Mol Microbiol 30(4):711–723
Hagman KE, Lahdenne P, Popova TG, Porcella SF, Akins DR, Radolf JD, Norgard MV (1998) Decorin-binding protein of Borrelia burgdorferi is encoded within a two-gene operon and is protective in the murine model of Lyme borreliosis. Infect Immun 66(6):2674–2683
Hallstrom T, Haupt K, Kraiczy P, Hortschansky P, Wallich R, Skerka C, Zipfel PF (2010) Complement regulator-acquiring surface protein 1 of Borrelia burgdorferi binds to human bone morphogenic protein 2, several extracellular matrix proteins, and plasminogen. J Infect Dis 202(3):490–498
Hanson MS, Cassatt DR, Guo BP, Patel NK, McCarthy MP, Dorward DW, Hook M (1998) Active and passive immunity against Borrelia burgdorferi decorin binding protein A (DbpA) protects against infection. Infect Immun 66(5):2143–2153
Hechemy KE, Samsonoff WA, Harris HL, McKee M (1992) Adherence and entry of Borrelia burgdorferi in Vero cells. J Med Microbiol 36(4):229–238
Heilpern AJ, Wertheim W, He J, Perides G, Bronson RT, Hu LT (2009) Matrix metalloproteinase 9 plays a key role in lyme arthritis but not in dissemination of Borrelia burgdorferi. Infect Immun 77(7):2643–2649
Hodzic E, Feng S, Holden K, Freet KJ, Barthold SW (2008) Persistence of Borrelia burgdorferi following antibiotic treatment in mice. Antimicrob Agents Chemother 52(5):1728–1736
Hu LT, Perides G, Noring R, Klempner MS (1995) Binding of human plasminogen to Borrelia burgdorferi. Infect Immun 63(9):3491–3496
Hu LT, Pratt SD, Perides G, Katz L, Rogers RA, Klempner MS (1997) Isolation, cloning, and expression of a 70-kilodalton plasminogen binding protein of Borrelia burgdorferi. Infect Immun 65(12):4989–4995
Hu LT, Eskildsen MA, Masgala C, Steere AC, Arner EC, Pratta MA, Grodzinsky AJ, Loening A, Perides G (2001) Host metalloproteinases in Lyme arthritis. Arthritis Rheum 44(6):1401–1410
Hunfeld KP, Ruzic-Sabljic E, Norris DE, Kraiczy P, Strle F (2005) In vitro susceptibility testing of Borrelia burgdorferi sensu lato isolates cultured from patients with erythema migrans before and after antimicrobial chemotherapy. Antimicrob Agents Chemother 49(4):1294–1301
Hunfeld KP, Ruzic-Sabljic E, Norris DE, Kraiczy P, Strle F (2006) Risk of culture-confirmed borrelial persistence in patients treated for erythema migrans and possible mechanisms of resistance. Int J Med Microbiol 296(Suppl 40):233–241
Klempner MS (2002) Controlled trials of antibiotic treatment in patients with post-treatment chronic Lyme disease. Vector Borne Zoonotic Dis 2(4):255–263
Klempner MS, Noring R, Rogers RA (1993) Invasion of human skin fibroblasts by the Lyme disease spirochete, Borrelia burgdorferi. J Infect Dis 167(5):1074–1081
Klempner MS, Noring R, Epstein MP, McCloud B, Hu R, Limentani SA, Rogers RA (1995) Binding of human plasminogen and urokinase-type plasminogen activator to the Lyme disease spirochete, Borrelia burgdorferi. J Infect Dis 171(5):1258–1265
Klempner MS, Hu LT, Evans J, Schmid CH, Johnson GM, Trevino RP, Norton D, Levy L, Wall D, McCall J et al (2001) Two controlled trials of antibiotic treatment in patients with persistent symptoms and a history of Lyme disease. N Engl J Med 345(2):85–92
Kuhlow CJ, Garcia-Monco JC, Coleman JL, Benach JL (2005) Murine microglia are effective phagocytes for Borrelia burgdorferi. J Neuroimmunol 168(1–2):183–187
Kurtti TJ, Munderloh UG, Krueger DE, Johnson RC, Schwan TG (1993) Adhesion to and invasion of cultured tick (Acarina: Ixodidae) cells by Borrelia burgdorferi (Spirochaetales: Spirochaetaceae) and maintenance of infectivity. J Med Entomol 30(3):586–596
Lagal V, Portnoi D, Faure G, Postic D, Baranton G (2006) Borrelia burgdorferi sensu stricto invasiveness is correlated with OspC-plasminogen affinity. Microbes Infect 8(3):645–652
Lane RS, Piesman J, Burgdorfer W (1991) Lyme borreliosis: relation of its causative agent to its vectors and hosts in North America and Europe. Annu Rev Entomol 36:587–609
Leong JM, Wang H, Magoun L, Field JA, Morrissey PE, Robbins D, Tatro JB, Coburn J, Parveen N (1998) Different classes of proteoglycans contribute to the attachment of Borrelia burgdorferi to cultured endothelial and brain cells. Infect Immun 66(3):994–999

Li X, Liu X, Beck DS, Kantor FS, Fikrig E (2006) Borrelia burgdorferi lacking BBK32, a fibronectin-binding protein, retains full pathogenicity. Infect Immun 74(6):3305–3313

Liang FT, Jacobs MB, Bowers LC, Philipp MT (2002) An immune evasion mechanism for spirochetal persistence in Lyme borreliosis. J Exp Med 195(4):415–422

Liang FT, Yan J, Mbow ML, Sviat SL, Gilmore RD, Mamula M, Fikrig E (2004a) Borrelia burgdorferi changes its surface antigenic expression in response to host immune responses. Infect Immun 72(10):5759–5767

Liang FT, Brown EL, Wang T, Iozzo RV, Fikrig E (2004b) Protective niche for Borrelia burgdorferi to evade humoral immunity. Am J Pathol 165(3):977–985

Lin B, Kidder JM, Noring R, Steere AC, Klempner MS, Hu LT (2001) Differences in synovial fluid levels of matrix metalloproteinases suggest separate mechanisms of pathogenesis in Lyme arthritis before and after antibiotic treatment. J Infect Dis 184(2):174–180

Livengood JA, Gilmore RD Jr (2006) Invasion of human neuronal and glial cells by an infectious strain of Borrelia burgdorferi. Microbes Infect 8(14–15):2832–2840

Livengood JA, Schmit VL, Gilmore RD Jr (2008) Global transcriptome analysis of Borrelia burgdorferi during association with human neuroglial cells. Infect Immun 76(1):298–307

Ma Y, Sturrock A, Weis JJ (1991) Intracellular localization of Borrelia burgdorferi within human endothelial cells. Infect Immun 59(2):671–678

Marques AR (2010) Lyme disease: a review. Curr Allergy Asthma Rep 10(1):13–20

Marques AR, Stock F, Gill V (2000) Evaluation of a new culture medium for Borrelia burgdorferi. J Clin Microbiol 38(11):4239–4241

Miklossy J, Kasas S, Zurn AD, McCall S, Yu S, McGeer PL (2008) Persisting atypical and cystic forms of Borrelia burgdorferi and local inflammation in Lyme neuroborreliosis. J Neuroinflammation 5:40

Miller JC, Stevenson B (2006) Borrelia burgdorferi erp genes are expressed at different levels within tissues of chronically infected mammalian hosts. Int J Med Microbiol 296(Suppl 40):185–194

Montgomery RR, Nathanson MH, Malawista SE (1993) The fate of Borrelia burgdorferi, the agent for Lyme disease, in mouse macrophages. Destruction, survival, recovery. J Immunol 150(3):909–915

Moore MW, Cruz AR, LaVake CJ, Marzo AL, Eggers CH, Salazar JC, Radolf JD (2007) Phagocytosis of Borrelia burgdorferi and Treponema pallidum potentiates innate immune activation and induces gamma interferon production. Infect Immun 75(4):2046–2062

Moriarty TJ, Norman MU, Colarusso P, Bankhead T, Kubes P, Chaconas G (2008) Real-time high resolution 3D imaging of the lyme disease spirochete adhering to and escaping from the vasculature of a living host. PLoS Pathog 4(6):e1000090

Murgia R, Cinco M (2004) Induction of cystic forms by different stress conditions in Borrelia burgdorferi. APMIS 112(1):57–62

Murgia R, Piazzetta C, Cinco M (2002) Cystic forms of Borrelia burgdorferi sensu lato: induction, development, and the role of RpoS. Wien Klin Wochenschr 114(13–14):574–579

Nadelman RB, Nowakowski J, Forseter G, Bittker S, Cooper D, Goldberg N, McKenna D, Wormser GP (1993) Failure to isolate Borrelia burgdorferi after antimicrobial therapy in culture-documented Lyme borreliosis associated with erythema migrans: report of a prospective study. Am J Med 94(6):583–588

Norman MU, Moriarty TJ, Dresser AR, Millen B, Kubes P, Chaconas G (2008) Molecular mechanisms involved in vascular interactions of the Lyme disease pathogen in a living host. PLoS Pathog 4(10):e1000169

Nowalk AJ, Gilmore RD Jr, Carroll JA (2006) Serologic proteome analysis of Borrelia burgdorferi membrane-associated proteins. Infect Immun 74(7):3864–3873

Ohnishi J, Piesman J, de Silva AM (2001) Antigenic and genetic heterogeneity of Borrelia burgdorferi populations transmitted by ticks. Proc Natl Acad Sci USA 98(2):670–675

Pal U, Li X, Wang T, Montgomery RR, Ramamoorthi N, Desilva AM, Bao F, Yang X, Pypaert M, Pradhan D et al (2004) TROSPA, an Ixodes scapularis receptor for Borrelia burgdorferi. Cell 119(4):457–468

Parveen N, Leong JM (2000) Identification of a candidate glycosaminoglycan-binding adhesin of the Lyme disease spirochete Borrelia burgdorferi. Mol Microbiol 35(5):1220–1234
Parveen N, Cornell KA, Bono JL, Chamberland C, Rosa P, Leong JM (2006) Bgp, a secreted glycosaminoglycan-binding protein of Borrelia burgdorferi strain N40, displays nucleosidase activity and is not essential for infection of immunodeficient mice. Infect Immun 74(5):3016–3020
Perides G, Tanner-Brown LM, Eskildsen MA, Klempner MS (1999) Borrelia burgdorferi induces matrix metalloproteinases by neural cultures. J Neurosci Res 58(6):779–790
Peters DJ, Benach JL (1997) Borrelia burgdorferi adherence and injury to undifferentiated and differentiated neural cells in vitro. J Infect Dis 176(2):470–477
Phillips SE, Mattman LH, Hulinska D, Moayad H (1998) A proposal for the reliable culture of Borrelia burgdorferi from patients with chronic Lyme disease, even from those previously aggressively treated. Infection 26(6):364–367
Piesman J (1993) Dynamics of Borrelia burgdorferi transmission by nymphal Ixodes dammini ticks. J Infect Dis 167(5):1082–1085
Preac-Mursic V, Weber K, Pfister HW, Wilske B, Gross B, Baumann A, Prokop J (1989) Survival of Borrelia burgdorferi in antibiotically treated patients with Lyme borreliosis. Infection 17(6):355–359
Preac-Mursic V, Wilske B, Patsouris E, Jauris S, Will G, Soutschek E, Rainhardt S, Lehnert G, Klockmann U, Mehraein P (1992) Active immunization with pC protein of Borrelia burgdorferi protects gerbils against B. burgdorferi infection. Infection 20(6):342–349
Probert WS, Johnson BJ (1998) Identification of a 47 kDa fibronectin-binding protein expressed by Borrelia burgdorferi isolate B31. Mol Microbiol 30(5):1003–1015
Roberts ED, Bohm RP Jr, Cogswell FB, Lanners HN, Lowrie RC Jr, Povinelli L, Piesman J, Philipp MT (1995) Chronic lyme disease in the rhesus monkey. Lab Invest 72(2):146–160
Rousselle JC, Callister SM, Schell RF, Lovrich SD, Jobe DA, Marks JA, Wieneke CA (1998) Borreliacidal antibody production against outer surface protein C of Borrelia burgdorferi. J Infect Dis 178(3):733–741
Rupprecht TA, Koedel U, Heimerl C, Fingerle V, Paul R, Wilske B, Pfister HW (2006) Adhesion of Borrelia garinii to neuronal cells is mediated by the interaction of OspA with proteoglycans. J Neuroimmunol 175(1–2):5–11
Sambri V, Basso F, Massaria F, Ardizzoni M, Cevenini R (1993) Adherence of Borrelia burgdorferi and Borrelia hermsii to mammalian cells in vitro. New Microbiol 16(1):43–49
Samuels DS, Radolf JD (eds.) (2010) Borrelia molecular biology, host interaction and pathogenesis. Caister Academic Press, Norwich, UK
Schwan TG, Piesman J, Golde WT, Dolan MC, Rosa PA (1995) Induction of an outer surface protein on Borrelia burgdorferi during tick feeding. Proc Natl Acad Sci USA 92(7):2909–2913
Seshu J, Esteve-Gassent MD, Labandeira-Rey M, Kim JH, Trzeciakowski JP, Hook M, Skare JT (2006) Inactivation of the fibronectin-binding adhesin gene bbk32 significantly attenuates the infectivity potential of Borrelia burgdorferi. Mol Microbiol 59(5):1591–1601
Shi Y, Xu Q, McShan K, Liang FT (2008) Both decorin-binding proteins A and B are critical for the overall virulence of Borrelia burgdorferi. Infect Immun 76(3):1239–1246
Shin OS, Isberg RR, Akira S, Uematsu S, Behera AK, Hu LT (2008) Distinct roles for MyD88 and Toll-like receptors 2, 5, and 9 in phagocytosis of Borrelia burgdorferi and cytokine induction. Infect Immun 76(6):2341–2351
Singh SK, Morbach H, Nanki T, Faber C, Baar V, Girschick HJ (2004) Differential expression of matrix metalloproteinases and cyclooxygenases in synovial cells exposed to Borrelia burgdorferi. Inflamm Res 53(12):689–696
Steere AC (2001) Lyme disease. N Engl J Med 345(2):115–125
Steere AC (2006) Lyme borreliosis in 2005, 30 years after initial observations in Lyme Connecticut. Wien Klin Wochenschr 118(21–22):625–633
Steere AC, Coburn J, Glickstein L (2004) The emergence of Lyme disease. J Clin Invest 113(8):1093–1101

Strle F, Maraspin V, Lotric-Furlan S, Ruzic-Sabljic E, Cimperman J (1996) Azithromycin and doxycycline for treatment of Borrelia culture-positive erythema migrans. Infection 24(1):64–68

Szczepanski A, Furie MB, Benach JL, Lane BP, Fleit HB (1990) Interaction between Borrelia burgdorferi and endothelium in vitro. J Clin Invest 85(5):1637–1647

Thomas DD, Comstock LE (1989) Interaction of Lyme disease spirochetes with cultured eucaryotic cells. Infect Immun 57(4):1324–1326

Thomas DD, Cadavid D, Barbour AG (1994) Differential association of Borrelia species with cultured neural cells. J Infect Dis 169(2):445–448

Tokarz R, Anderton JM, Katona LI, Benach JL (2004) Combined effects of blood and temperature shift on Borrelia burgdorferi gene expression as determined by whole genome DNA array. Infect Immun 72(9):5419–5432

Verma A, Brissette CA, Bowman A, Stevenson B (2009) Borrelia burgdorferi BmpA is a laminin-binding protein. Infect Immun 77(11):4940–4946

Weening EH, Parveen N, Trzeciakowski JP, Leong JM, Hook M, Skare JT (2008) Borrelia burgdorferi lacking DbpBA exhibits an early survival defect during experimental infection. Infect Immun 76(12):5694–5705

Wormser GP (2006) Hematogenous dissemination in early Lyme disease. Wien Klin Wochenschr 118(21–22):634–637

Wormser GP, Schwartz I (2009) Antibiotic treatment of animals infected with Borrelia burgdorferi. Clin Microbiol Rev 22(3):387–395

Wormser GP, Nowakowski J, Nadelman RB, Bittker S, Cooper D, Pavia C (1998) Improving the yield of blood cultures for patients with early Lyme disease. J Clin Microbiol 36(1):296–298

Wormser GP, McKenna D, Carlin J, Nadelman RB, Cavaliere LF, Holmgren D, Byrne DW, Nowakowski J (2005) Brief communication: hematogenous dissemination in early Lyme disease. Ann Intern Med 142(9):751–755

Wormser GP, Dattwyler RJ, Shapiro ED, Halperin JJ, Steere AC, Klempner MS, Krause PJ, Bakken JS, Strle F, Stanek G et al (2006) The clinical assessment, treatment, and prevention of lyme disease, human granulocytic anaplasmosis, and babesiosis: clinical practice guidelines by the Infectious Diseases Society of America. Clin Infect Dis 43(9):1089–1134

Wu J, Weening EH, Faske JB, Hook M, Skare JT (2011) Invasion of eukaryotic cells by Borrelia burgdorferi requires {beta}1 integrins and Src kinase activity. Infect Immun 79(3):1338–1348

Yrjanainen H, Hytonen J, Song XY, Oksi J, Hartiala K, Viljanen MK (2007) Anti-tumor necrosis factor-alpha treatment activates Borrelia burgdorferi spirochetes 4 weeks after ceftriaxone treatment in C3H/He mice. J Infect Dis 195(10):1489–1496

Zambrano MC, Beklemisheva AA, Bryksin AV, Newman SA, Cabello FC (2004) Borrelia burgdorferi binds to, invades, and colonizes native type I collagen lattices. Infect Immun 72(6):3138–3146

Zhang JR, Hardham JM, Barbour AG, Norris SJ (1997) Antigenic variation in Lyme disease borreliae by promiscuous recombination of VMP-like sequence cassettes. Cell 89(2):275–285

Zhao Z, Fleming R, McCloud B, Klempner MS (2007) CD14 mediates cross talk between mononuclear cells and fibroblasts for upregulation of matrix metalloproteinase 9 by Borrelia burgdorferi. Infect Immun 75(6):3062–3069

Part II
Immune Evasion

Chapter 3
Treponema denticola: FhbB, Dentilisin, Complement Evasion and the Paradox of Factor H Cleavage

John V. McDowell, Daniel P. Miller, Katherine L. Mallory, and Richard T. Marconi

3.1 Periodontal Disease

Periodontitis, one of the most common afflictions of man, is a broad clinical term used to describe a range of acquired or inherited disorders with a common clinical outcome: inflammation and destruction of the periodontium (the connective tissue that supports the teeth), alveolar bone resorption, and tooth loss (Fig. 3.1). While early stage disease (gingivitis) can be reversed through treatment and improved oral hygiene, advanced periodontal disease cannot. In healthy individuals the pocket depth between the tooth and gums is typically 1–3 mm. In periodontal pockets the pocket depth increases significantly resulting in pronounced changes in the physiochemical properties of the subgingival crevice. The Centers for Disease Control (CDC) and Prevention and the American Academy of Periodontology define moderate periodontal disease as "having at least two teeth with interproximal attachment loss of 4 mm or more OR at least two teeth with 5 mm or more of pocket depth at interproximal sites". Severe disease is defined as "having at least two teeth with interproximal attachment loss of 6 mm or more AND at least one tooth with 5 mm or more of pocket depth at interproximal sites" (http://www.nidcr.nih.gov). While estimates from different studies vary, data compiled through the National Health and Nutrition Examination Survey (NHANES) estimate that in the US, 10% of the adult population (>35 years of age) have periodontal disease and 11% of those over the age of 50 have severe periodontal disease. In the US, approximately 25% of individuals 65 years of age and older are edentulous (lacking natural teeth) with periodontal disease being the primary cause. In underdeveloped parts of the world the incidence of periodontal disease in adults may be as high as 60%.

J.V. McDowell • D.P. Miller, Ph.D • K.L. Mallory • R.T. Marconi, Ph.D. (✉)
Virginia Commonwealth University, Department of Microbiology and Immunology, Richmond, VA 980678
e-mail: rmarconi@vcu.edu

M.E. Embers (ed.), *The Pathogenic Spirochetes: Strategies for Evasion of Host Immunity and Persistence*, DOI 10.1007/978-1-4614-5404-5_3,

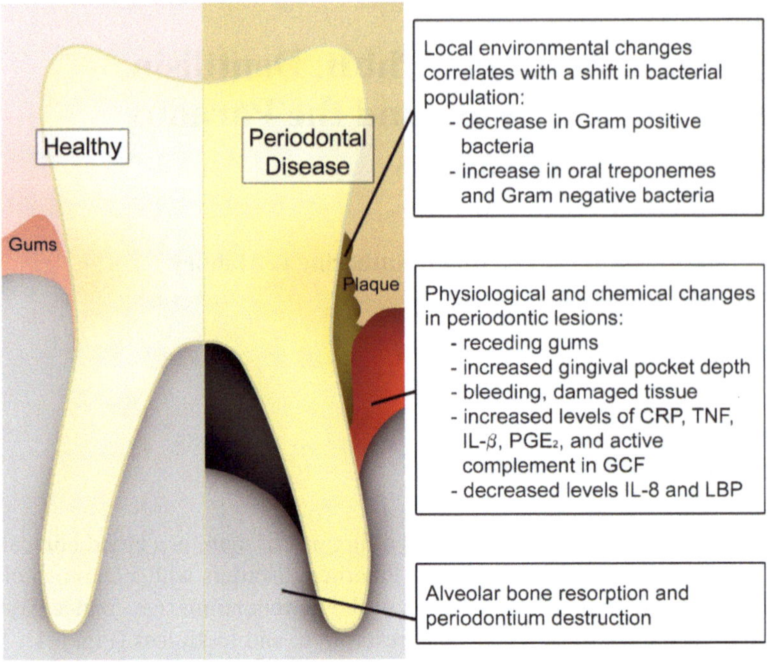

Fig. 3.1 Clinical transitions associated with the development of periodontal disease. A view of healthy tissue/tooth is presented on the *left*. Depicted on the *right* are changes that occur with periodontal disease. The *boxes* provide a brief summary of the bacteriological, physiological, chemical and structural changes that develop

Low socioeconomic status, age, hormonal state, poor oral hygiene, smoking, and alcohol consumption are major risk factors for periodontal disease (Gautam et al. 2011; Kumar and Faizuddin 2011; Shepherd 2011). Smokers are three times more likely to develop gum disease and smoking during treatment correlates with failed or compromised treatment outcomes. Periodontitis is more common in men than women and in individuals with diabetes mellitus and impaired fasting glucose. The economic costs associated with treating and maintaining dental/oral health and its associated conditions are staggering. According to the National Institutes for Dental and Craniofacial Research (Dental, Oral and Craniofacial Data Resource Center; Oral Health Report 2002) dental visits and oral health problems result in approximately 164 million lost work hours each year at a cost of 100 billion dollars.

The health consequences of periodontal disease are not confined to the oral cavity. Periodontal disease has been postulated to be a risk factor for thromboembolic events, myocardial infarction, atherosclerosis, cardiovascular disease, adverse pregnancy outcomes, and certain oral and gastrointestinal cancers (reviewed in (Pihlstrom et al. 2005)). While there are conflicting reports regarding a direct correlation of periopathogens and systemic disease, several studies have demonstrated the presence of periopathogens, including the spirochete, *Treponema denticola,* in atherosclerotic plaque (Toyofuku et al. 2011; Raffaelli et al. 2010; Romano et al. 2007).

3.1.1 The Microbial Etiology of Periodontal Disease

Periodontal disease has been described as a "polymicrobial disruption of host homeostasis" (Darveau 2010). Under healthy conditions, the oral cavity is home to hundreds of bacterial species that exist in a symbiotic relationship with their host. As plaque develops, the bacterial population in the gingival crevice increases in both number and diversity with a transition from predominantly Gram-positive bacteria to Gram-negatives and spirochetes (exclusively *Treponemes*) (Paster et al. 2006). The microbial etiology of periodontal disease is complex with over 500 species of bacteria represented in plaque, 76 of which belong to the genus *Treponema* (Paster et al. 1991, 2001; Dewhirst et al. 2000). Of the oral spirochetes, only *T. denticola* and nine other species have been successfully cultivated. In a healthy gingival environment, *T. denticola* constitutes a relatively small percentage of the total bacteria. However, as periodontitis develops, *T. denticola* becomes a dominant species (Ellen and Galimanas 2005). Because of its abundance and localization at the plaque–tissue interface, its production of potent proteolytic enzymes, and its ability to circumvent complement mediated killing and dysregulate complement, *T. denticola* is considered to be a central contributor to periodontal disease (Ellen 2006; Dashper et al. 2011).

3.2 *T. denticola*, Serum Resistance and Complement

T. denticola survives in an anatomical niche that is bathed in gingival crevicular fluid (GCF). GCF, which pools around the teeth, is derived from serum and local tissue exudate. Proteome analyses of GCF collected from healthy individuals revealed that 57% of the proteins identified in GCF are also found in serum while 43% are unique to GCF (Carneiro et al. 2012). GCF recovered from periodontal disease patients is rich in immune effectors including activated complement, C-reactive protein (CRP), proinflammatory cytokines, and others (reviewed in (Darveau 2010)). The ability of *T. denticola* to survive exposure to GCF indicates that it must be capable of evading complement mediated destruction. *T. denticola* has been demonstrated to have an "intermediate" serum resistance phenotype. It thrives in 25% human serum but is killed by 40% serum (McDowell et al. 2011). The intermediate serum resistance level of *T. denticola* is consistent with, and sufficient for, survival in the levels of complement encountered in the GCF. In contrast to *T. denticola*, *Treponema vincentii* is highly sensitive to serum mediated killing (Marconi RT unpublished data). To date, the serum sensitivities of other oral treponeme species have not been assessed.

3.2.1 The Complement System

The complement system consists of more than 30 proteins that function to eliminate non-self entities such as bacterial pathogens. Specific complement proteins bind to the surface of bacteria triggering an enzymatic cascade that activates and amplifies the complement system (Boackle 1990). Activation of the complement cascade results in formation of the membrane attack complex (MAC), deposition of opsonins (C3b and iC3b) on the surface of pathogens, production of anaphylatoxins (C3a and C5a), and stimulation of antibody production. Complement activity must be regulated or damage to host cells and tissues can occur. Factor H (FH), a highly abundant 155 kDa serum glycoprotein (400–800 mg ml^{-1} serum), regulates the alternative complement cascade. It controls C3b by: (1) serving as a cofactor for factor I (a serine protease that cleaves C3b), (2) accelerating decay of preformed C3 convertase (C3bBb), and (3) blocking the formation of new C3 convertase by competing with factor B for binding to C3b (Ruddy and Austen 1969, 1971; Whaley and Ruddy 1976). FH regulation of complement amplification is not strictly confined to the alternative complement system. FH also binds to and inhibits classical pathway activation by the monomeric CRP (mCRP)-C1q complex (Jarva et al. 1999; Mihlan et al. 2009; Ji et al. 2006a, b). It is important to note that CRP levels are highly elevated in serum and GCF of patients with severe periodontal disease increasing from nearly undetectable levels to as high as 500 mg L^{-1} (Pepys and Hirschfield 2003; Haba et al. 2011; Megson et al. 2010; Tuter et al. 2007), hence CRP-mediated complement regulation (or lack thereof) could be an important factor in defining the immunological environment in periodontal pockets (Kodovazenitis et al. 2011; Maekawa et al. 2011; Gomes-Filho et al. 2011). Later in this chapter, we explore the possibility that *T. denticola* plays a central role in periodontal disease progression by dysregulating complement control mechanisms in the subgingival crevice by depleting local FH.

To understand the molecular basis of *T. denticola* serum resistance and the role that FH cleavage plays in disease progression, it is important to consider the structure and known functional activities of FH. FH is composed of 20 imperfect repeat domains (50–60 amino acids) referred to as complement control protein (CCP) domains. Different CCP domains mediate different functional activities. The binding sites for host-derived regulatory molecules relevant to this discussion are indicated in Fig. 3.2. In addition, the interaction sites for spirochetal FH binding proteins are also indicated (Zipfel 2009; Zipfel and Skerka 1999; Zipfel et al. 2002; Okemefuna et al. 2010).

3.2.2 Mechanisms of Complement Evasion Employed by *T. denticola*

T. denticola employs both passive and active mechanisms to evade complement mediated killing. Passive protection mechanisms include cell aggregation and biofilm formation. The structural nature of cell aggregates, biofilms and subgingival plaque

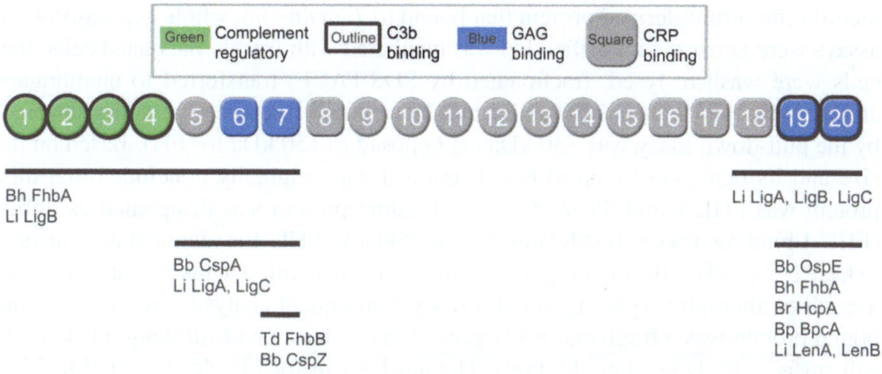

Fig. 3.2 Structure and functional organization of FH. The schematic depicts the general organization of FH which consists of 20 complement control protein (CCP) domains. The CCP domains are labeled 1 through 20 (N-terminal to C-terminal). Additional information pertaining to the functional activities of individual CCP domains is indicated in the *inset*. The binding sites for spirochetal FH binding proteins (McDowell et al. 2005a; Hovis et al. 2006a; Hartmann et al. 2006; Hellwage et al. 2001; Kraiczy et al. 2004; Stevenson et al. 2007; Schott et al. 2010; Grosskinsky et al. 2009; Castiblanco-Valencia et al. 2012) are indicated *below* the FH schematic. Bacterial genus/species abbreviations are as follows: Bh *Borrelia hermsii*, Li *Leptospira interrogans*, Bb *Borrelia burgdorferi*, Br *Borrelia recurrentis*, Bp *Borrelia parkeri*, and Td *Treponema denticola*

provide a physical barrier against immune effector cells, complement and antibody. Active mechanisms of complement evasion have also been identified and are the focus of this chapter. *T. denticola* binds FH to its surface and exploits its negative regulatory activity to downregulate and inhibit complement at the cell surface (McDowell et al. 2005a, 2009, 2011). *T. denticola* also produces dentilisin, a complement degrading protease (Yamazaki et al. 2006). Dentilisin, a multi-subunit chymotrypsin-like protease (CTLP) complex, has been demonstrated to cleave C3, C3b, and FH (McDowell et al. 2011; Yamazaki et al. 2006; Fenno et al. 1998a, b). The cleavage of key complement proteins could profoundly affect the immunological environment of the subgingival crevice. Several excellent reviews are available that detail the multiple functions that dentilisin plays in *T. denticola* pathogenesis and periodontal disease progression (Dashper et al. 2011; Fenno 2012; Ishihara 2010).

3.2.3 Identification of the T. denticola FH Binding Protein

Using indirect immunofluorescence assays, McDowell and colleagues demonstrated the deposition of a human serum-derived protein on the *T. denticola* cell surface that was detected with anti-FH antibody (McDowell et al. 2005a). Subsequent affinity ligand binding immunoblot (ALBI) assays demonstrated that the interaction occurred through a ~12 kDa *T. denticola* protein. ALBI assays were performed by separating the proteins of *T. denticola* cell lysates through SDS-PAGE. The proteins were transferred to membranes and the membranes were incubated with purified human FH. Bound FH was then detected using anti-FH antiserum. In an effort to identify the serum-derived protein that bound to *T. denticola*, whole cell absorption assays were performed. Purified FH was incubated with freshly harvested cells; the cells were washed, lysed, fractionated by SDS-PAGE, transferred to membranes and then screened with anti-FH antiserum. Interestingly, the size of protein recovered by the pull-down assay was ~50 kDa (as opposed to 150 kDa for FH). Based on its size and its detection by anti-FH antiserum it was originally concluded that this protein was FHL-1 and the *T. denticola* binding protein was designated as FhbB (FHL-1 binding protein B (McDowell et al. 2005a)). FHL-1 is identical in sequence to CCP1–7 of FH differing only in its four C-terminal amino acids that are added as a result of alternative splicing of FH mRNA. Subsequent analyses revealed that the bound protein was a fragment of FH generated by cleavage of full length FH at the cell surface (McDowell et al. 2009). The mechanism for FH cleavage and its biological implications are discussed in detail below.

The gene that encodes FhbB was identified in part using a bioinformatics based approach. The *T. denticola* genome sequence was scanned for all genes predicted to encode lipoproteins in the 10–14 kDa range (McDowell et al. 2007). The potential list of candidates was narrowed by focusing on those that possess high probability coiled-coil heptad repeat sequences (identified with the COILS program) (Lupas et al. 1991). Earlier studies demonstrated that this motif is present in most *Borrelial* FH binding proteins (McDowell et al. 2004, 2005b, 2007; Hovis et al. 2006a). Coiled-coil heptad repeats are defined by the periodicity of hydrophobic, non-polar residues occurring at the a and d positions of the a-g_n heptad and not by a conserved primary sequence (Lupas 1996). By screening for this motif the potential list of candidate genes was reduced to nine. Recombinant proteins were generated for each and tested for FH binding using ALBI assays (McDowell et al. 2007). ORF TDE0108 was found to encode an 11.4 kDa lipidated FH binding protein and was designated as *fhbB* (McDowell et al. 2005a). It is the sole FH binding protein produced by *T. denticola* and it has no discernible sequence homology with other bacterial FH binding proteins. FhbB is universal among *T. denticola* isolates but not found in other oral treponemes (McDowell et al. 2007). Phylogenetic sequence analyses of *fhbB* from nearly 30 *T. denticola* isolates revealed the existence of three distinct major FhbB phyletic types. Recombinant proteins representing each phyletic type have been generated and FH binding verified (Marconi RT unpublished data). The FH binding and cleavage phenotype has not been demonstrated in any other oral spirochete. Table 3.1 provides a summary of the spirochetal FH binding proteins identified to date and their general properties.

3.2.4 FhbB, Dentilisin and Serum Resistance

The contributions of FhbB and dentilisin in *T. denticola* serum resistance have been determined using *fhbB* (Td35405ΔfhbB) and dentilisin (Td35405-CCE) isogenic mutants derived from the *T. denticola* 35405 parental wild-type strain (Td35405wt). Each strain was incubated in 25% human serum (complement certified) and bacterial killing monitored at 2 h intervals using dark-field microscopy. The serum sensitivity of Td35405ΔfhbB was markedly increased relative to the Td35405wt. Cell survival rates for the Td35405wt and Td35405ΔfhbB cells were 85% and 15%, respectively. No significant killing was observed with heat inactivated serum demonstrating that killing occurs through a complement dependent mechanism. In contrast to Td35405ΔfhbB, Td35405-CCE retained wild-type serum resistance levels indicating that dentilisin is not required for complement evasion. From these analyses, it was concluded that FH binding by FhbB is the primary mechanism of serum resistance.

3.2.5 *Cleavage of FH by T. denticola*

The interaction between *T. denticola* and FH is unique and distinct from other pathogens. *T. denticola* is the only pathogen that has been demonstrated to cleave surface bound FH (McDowell et al. 2009). The direct involvement of dentilisin in the cleavage process was demonstrated using *fhbB* and dentilisin mutants (Td35405ΔfhbB and Td35405-CCE, respectively). Actively growing cultures of strains Td35405wt, Td35405ΔfhbB, and Td35405-CCE were incubated with purified human FH. Aliquots were removed at 10-min intervals, assessed by SDS-PAGE, transferred to membranes, and screened with anti-FH antibody. With the Td35405wt strain, complete degradation of the input FH occurred by 30 min. In contrast, no cleavage was observed with Td35405-CCE (which lacks dentilisin). Td35405ΔfhbB cleaved FH but with significantly reduced kinetics. The dentilisin cleavage sites within FH have not been mapped but based on its reported activity and specificity in cleaving prolyl-phenylalanine and prolyl-leucine peptide bonds, cleavage products of 1.8, 2.3, 6.6, 12.5, 18.8, 29, and 66 kDa are predicted. This assumes that all sites are accessible and complete cleavage occurs. Indeed, a subset of these fragments is observed after incubation of *T. denticola* with purified FH (Marconi, R.T.; unpublished data). However, additional fragments were also seen. These could result from partial cleavage at one of more sites. In addition, there may be other proteases that cleave FH after its initial cleavage by dentilisin. Identification of the cleavage sites within FH will be an important step in assessing the possible functional activities of the cleavage products. A schematic diagram depicting the sequence of events in FH cleavage is presented in Fig. 3.3.

3.2.6 *Molecular Interactions Between FH and FhbB*

The interaction of FH with pathogenic spirochetes that cause Lyme disease and relapsing fever is dependent on long range intra-molecular interactions in the binding proteins. Truncation analyses revealed that intact N- and C-terminal domains of the *Borrelia burgdorferi* OspE, CspZ, and CspA proteins and the *Borrelia hermsii* FhbA protein are required for maximal FH binding (McDowell et al. 2005b; Hovis et al. 2006a; Metts

Table 3.1 Properties of Spirochetal FH Binding Proteins

Species	Protein	Ligand(s)	Comments	References
Treponema denticola	FhbB	FH, ~50 kDa FH cleavage product	Smallest bacterially produced FH binding protein (11.4 kDa); required for serum resistance; its negatively charged face binds to CCP7 of human FH; atomic structure determined at 1.7 Å resolution; thermodynamically stable	(McDowell et al. 2005a, 2007, 2011)
Borrelia burgdorferi s.l.	OspE	FH, FHR1	Binds CCP19–20 of FH (binding site on FHR1 is assumed to be CCP4–5 due to its strong homology with CCP19–20 of FH); belongs to a plasmid encoded, polymorphic gene family; individual strains may produce 3 or more distinct paralogs; paralogs show some mammalian species specific FH binding; binds plasminogen; over-expression in a serum-sensitive CspA mutant Borrelia increases serum resistance	(Hellwage et al. 2001; Seling et al. 2010; Hovis et al. 2006b; Kenedy et al. 2009)
B. burgdorferi s.l.	CspA	FH, FHL-1	Atomic structure has been determined; not expressed during mammalian infection; may function in complement resistance in fed ticks	(McDowell et al. 2005b, 2006; Kraiczy et al. 2004; Kenedy et al. 2009; Cordes et al. 2005)

(continued)

Table 3.1 (continued)

Species	Protein	Ligand(s)	Comments	References
B. burgdorferi s.l.	CspZ	FH, FHL-1	*B. burgdorferi* CspZ binds FH and FHL-1 while CspZ derived from other *Borrelia* species does not; all CspZ variants bind to an unidentified 60 kDa serum protein that is produced by diverse mammals; encoded by a variable linear plasmid of 28 kb	(Hartmann et al. 2006; Rogers et al. 2009; Rogers and Marconi 2007)
Borrelia hermsii	FhbA	FH, FHR1	Binds CCP6–8, 19–20; only FH binding protein produced by *B. hermsii;* encoded by a gene carried on a 200 kb linear plasmid; antibody responses to FhbA are diagnostic for relapsing fever; binds unidentified serum proteins	(Hovis et al. 2004, 2006a; McDowell et al. 2003; Rossmann et al. 2007)
Borrelia recurrentis	HcpA	FH, FHR-1	Binds to CCP19–20 of FH, binding site on FHR-1 is unknown; also binds plasmin/plasminogen; expression of HcpA in a heterologous strain protects against complement-mediated killing; the closely related relapsing fever spirochete, *Borrelia duttoni,* also binds FH but the binding protein has not been identified	(Grosskinsky et al. 2009; Meri et al. 2006)
Borrelia parkeri	BpcA	FH, FHR1	Binds CCP19–20 of FH (binding site on FHR1 has not been identified, assumed to be CCP4–5); also binds plasminogen	(Schott et al. 2010)

(continued)

Table 3.1 (continued)

Species	Protein	Ligand(s)	Comments	References
Leptospira interrogans	LenA, LenB	FH, FHR1	Only two members of a 6 member paralogous family to bind FH; binds CCP19–20 (binding site on FHR1 has not been identified, although assumed to be CCPs 4–5 which share strong homology to CCP19–20 of FH); binds laminin	(Stevenson et al. 2007; Verma et al. 2006)
L. interrogans	LigA, LigB, LigC	FH, FHL-1, FHR-1	All three Lig proteins bind CCP5–7 and CCP18–20; Lig B also binds CCP1–4; FH binding is localized to the N terminus of LigB and the C-termini of LigA and LigB; bind complement regulator C4bp also	(Castiblanco-Valencia et al. 2012; Silva et al. 2007; Faisal et al. 2008, 2009; Yan et al. 2009)

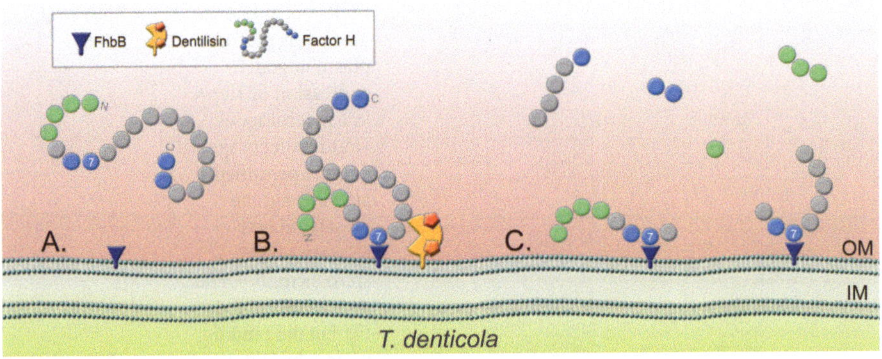

Fig. 3.3 The process of FH cleavage by *T. denticola*. The central players involved in FH binding and cleavage are indicated. The individual CCP domains of FH are indicated by *circles*. CCP domains 1 through 4 which bind C3b and play a central role in mediating complement regulation are indicated in *green*. CCP domains that have been demonstrated to bind to glycosaminoglycans (CCP6, 7, 19 and 20) are indicated in *blue* (the remaining CCP domains are indicated in *gray*). The inner and outer membranes (IM and OM, respectively) of *T. denticola* are presented *across the bottom* of the figure. (**a**) Depicts FhbB presented on the outer surface of *T. denticola* with unbound FH in the gingival crevicular fluid (GCF). (**b**) Depicts the binding of FH (via CCP7) to FhbB in proximity to dentilisin. (**c**) Depicts two possible interactions; one in which the 50 kDa FH cleavage product retains CCP1-4 and one in which is does not

et al. 2003; Hartmann et al. 2006; Rossmann et al. 2006). In addition, it was demonstrated through binding analyses with internal deletion mutants and single amino acid substitution mutants that the residues and domains that form the binding interface are discontinuous (McDowell et al. 2004). The molecular basis of the FH-FhbB interaction has also been investigated through truncation analysis. Recombinant FhbB proteins harboring N- or C-terminal truncations of differing length were purified and tested for their ability to bind FH (McDowell et al. 2009). Binding was assessed using ELISA and ALBI assay approaches. FH binding to FhbB proved to be highly sensitive to even short C-terminal truncations. Removal of as few as the C-terminal seven amino acids completely eliminated FH binding. Truncation of the N-terminus (22 amino acids) from the mature protein also eliminated binding. Shorter N-terminal truncations were not assessed so the minimal truncation that results in the loss of FH binding is not yet known. Consistent with the apparent conformational nature of the FH binding sites of spirochetal FH binding proteins, comparative sequence analyses of OspE, CspZ, CspA, FhbA and FhbB have not identified conserved sequence motifs that might be directly involved in binding.

3.2.7 Determination of the Molecular Structure of FhbB

The recent determination of the atomic structure of FhbB has provided new insight into the molecular basis of the FhbB-FH interaction and the minimal determinants required for FH binding (Miller et al. 2011, 2012). FhbB crystals were generated through hanging-drop vapor diffusion and the structure determined at a resolution of 1.7 Å. Two molecules were present in the asymmetric unit indicating dimer formation. The structure of the FhbB monomer is unique consisting of an $\alpha1\alpha2\beta1\alpha3\beta2$ fold. Dali analyses revealed that a similar structure has not been previously reported for any other protein. FhbB is an elongated and thin protein with a hydrophobic core and well defined negatively ($\alpha1$-$\alpha2$) and positively charged ($\beta1$-$\alpha3$-$\beta2$) surfaces. Circular dichroism thermal denaturation-renaturation analyses revealed that FhbB is thermodynamically stable. Only a small decrease in alpha helical content of FhbB was observed at 90°C and it was fully restored upon cooling. Similarly, the FH binding-OspE protein of *B. burgdorferi* is also highly thermostable (Marconi, R.T., unpublished data). The high thermostability of FhbB and OspE may explain why these proteins retain their ability to bind FH even after boiling in SDS-PAGE and electrophoresis under denaturing conditions (McDowell et al. 2005a; Metts et al. 2003). The inherently high stability and compact nature of FhbB may protect it from degradation in the hostile environment of periodontal pockets.

As mentioned above, FhbB crystallized as a dimer. However, several lines of evidence suggest that the dimer is not biologically relevant (Miller et al. 2012). The FhbB dimer lacks inter-monomer hydrogen bonds and the interface has an extensive water network. Analytical ultracentrifugation analyses demonstrated that dimerization only occurs at non-physiological protein concentrations and consistent with this, the K_D for the dimer is relatively weak (217 ± 40.5 μM). These properties, coupled with the fact that monomeric FhbB efficiently binds FH, suggest that the biologically active form of FhbB is a monomer.

3.2.8 Identification of Specific Residues that Influence FH-FhbB Binding: A Potential Model for the Molecular Interaction

Earlier random and site-directed mutagenesis analyses identified residues at widely separated positions in the FhbB primary sequence that influence FH binding (McDowell et al. 2009). With the determination of the atomic structure of FhbB, informed site-directed mutagenesis studies designed to identify FH-FhbB contact points became possible. Towards this goal, solvent accessible, charged amino acids were substituted with alanine to yield recombinant FhbB proteins with single or double amino acid substitutions. The impact of these mutations on FH binding was assessed by ELISA and ALBI assay approaches (Miller et al. 2012). Alanine substitution of residues E45, K47, H50, D58, E62, and E66 which are presented on the $\alpha 1\alpha 2$-negatively charged surface of FhbB abolished or attenuated FH binding. In contrast, alanine substitution of K77, K78, R93, and K94 on the $\beta 1\alpha 3\beta 2$-positively charged surface had no effect on FH binding. Residues identified to influence FH binding through alanine substitution were also replaced with residues of the same charge but different side chain length. These studies were performed to determine if subtle alterations in the ion cloud influence the FhbB-FH interaction (Miller et al. 2012). In contrast to the alanine substitution mutants, the E45D, D58E, E62D, and E66D mutants bound FH. However, binding was significantly reduced relative to wild-type. Collectively, the mutational analyses presented here and in earlier studies indicate that the primary interaction site for FH is on the $\alpha 1\alpha 2$ face and that the interaction requires a strict electrostatically defined binding interface.

3.2.9 Identification of Domains and Residues of FH that Are Involved in Binding FhbB

The majority of bacterial FH binding proteins interact with CCP6–8 or CCP19–20 (reviewed in (Ferreira et al. 2010), see Fig. 3.2). To identify the interaction site for FhbB, FH constructs spanning CCP1–7, CCP8–15, and CCP16–20 were generated in *Pichia pastoris* (Giannakis et al. 2003) and tested for binding (McDowell et al. 2005a). The CCP1–7 construct, but not the CCP8–15 or CCP19–20 constructs, bound to *T. denticola*. Evidence that the interaction occurs specifically with CCP7 came from binding analyses with a CCP1–7 construct that harbored alanine substitutions at positions Arg387, Lys388, Arg404, and Lys405 (designated as the H7AB construct (McDowell et al. 2005a; Giannakis et al. 2003)). Arg387 and Lys388 reside within the linker between CCP6 and CCP7, and Arg404 and Lys405 are contained within the CCP7 hyper-variable loop. The H7AB construct did not bind to FhbB. Potential intermolecular interactions between FH and FhbB were assessed through computer assisted docking analyses using the Z-Dock algorithm. $D58_{FhbB}$-$R387_{FH}$ and $E62_{FhbB}$-$K405_{FH}$ salt bridges were predicted (Fig. 3.4). These predictions are consistent with the inability of the H7AB construct to bind to FhbB and are in excellent agreement with the FhbB site-directed mutagenesis analyses that demonstrated that residues D58 and E62 are required for FH binding.

Further evidence for the interaction of FhbB with CCP7 has come from competitive binding analyses. CCP7 serves as an interaction site for several host-derived molecules including CRP and glycosaminoglycans including heparin (Giannakis et al. 2003). FH binding to glycosaminoglycans presented on host cell surfaces and within the extracellular matrix plays an important role in preventing targeting of host tissues by complement. It has been demonstrated that heparin inhibits the binding of FH to several different FH binding proteins including FhbB (McDowell et al. 2005a). More recent studies with sucrose octasulfate (SOS), a homogenous glycosaminoglycan analog (Prosser et al. 2007), also support the FhbB-CCP7 interaction (Miller et al. 2012). Co-crystalization of SOS and CCP6–8 revealed that SOS binds within a large, positively charged groove present in CCP7. $K405_{FH}$ was found to hydrogen bond with SOS. As detailed above, residue $K405_{FH}$ was found in the binding analyses with the H7AB construct to be a residues important for the FH-FhbB interaction. Miller et al. demonstrated that SOS inhibits FH binding to FhbB in a dose-dependent manner (Miller et al. 2012) suggesting that the binding site for FhbB also lies with the positively charged groove of CCP6–8 and that the binding sites of SOS and FhbB partially overlap.

3.2.10 The Kinetics of FH Cleavage

The kinetics of the FH-FhbB interaction have been assessed using surface plasmon resonance (Miller et al. 2012). The K_D for the interaction was determined to be 1.51 ± 0.02 μM, a value similar to that reported for the interaction of FH with the complement proteins, C3b and C3d (Morgan et al. 2011; Wu et al. 2009). The association (k_a) and dissociation (k_d) constants of the FH:FhbB interaction were determined to be $7.56 \pm 0.26 \times 10^4$ M^{-1} s^{-1} and 0.11 ± 0.01 s^{-1}, respectively, indicating a rapid off rate. Rapid release of cleaved FH by *T. denticola* would allow for the binding and positioning of a new molecule of FH and its subsequent cleavage by dentilisin ultimately leading to FH depletion. Organisms that bind FH solely for the purpose of complement evasion would benefit by tighter binding and slower release. Consistent with this, K_D values determined for the interaction of FH with other bacterial FH binding proteins are typically in the nanomolar range.

3.2.11 The Paradox of FH Binding and Cleavage by T. denticola

While numerous pathogens exploit surface bound FH for immune evasion, the role of the FH–*T. denticola* interaction in periodontal disease pathogenesis is more complex. Since FH binding serves to protect against complement mediated attack, the biological rationale for its cleavage by dentilisin is somewhat of a paradox. Intuitively, if FH is depleted then the expectation would be that *T. denticola* would be more susceptible to complement mediated killing. It is possible that the benefits of FH depletion outweigh the negatives. Consider the possible environmental and

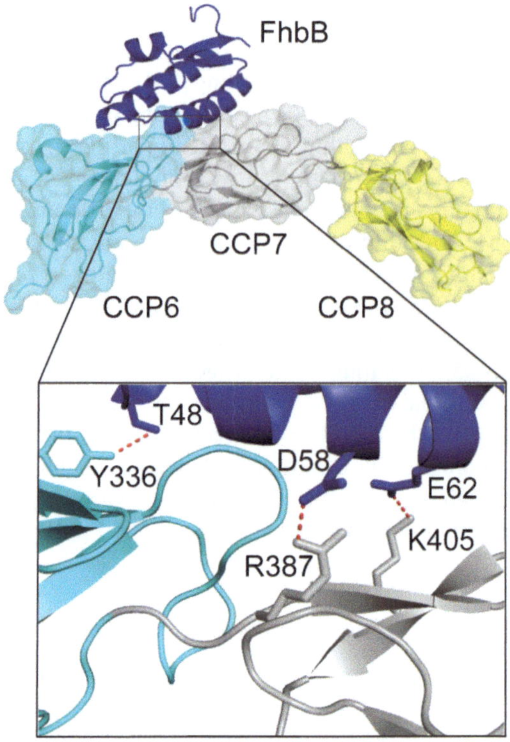

Fig. 3.4 Model for the molecular interactions between FH and *T. denticola* FhbB. The FhbB and FH interaction was predicted by computational modeling. FhbB (*dark blue*) binds to the previously identified glycosaminoglycan binding site between CCP6 (*light blue*) and CCP7 (*gray*). Specific amino acids involved in the molecular interaction are *highlighted* in the window

immunological changes that would occur in periodontal pockets as a consequence of local FH depletion. Host control of complement activation would be disrupted. Decay of the C3bBb convertase complex and factor I mediated cleavage of C3b would be inhibited resulting in unchecked C3b production. Without FH, mCRP would not be able to negatively regulate C5 convertase complex formation and terminal pathway activation. These conditions would lead to C3b deposition on host surfaces causing tissues to be targeted by the immune system. The ensuing tissue destruction would benefit periopathogens by increasing nutrient availability and by creating a larger anaerobic growth zone (periodontal pocket) where the bacteria can thrive. FH depletion would also inhibit the clearance of damaged cells from periodontal pockets. It has been demonstrated that FH bound to apoptotic particles increase phagocytosis and uptake by human macrophages (Mihlan et al. 2009). The inhibition of clearance of damaged cells would thus hinder the regeneration of healthy tissue. Pro-inflammatory cytokine production (TNF, IL-1, IL-6, and others) would also increase as a result of FH depletion further promoting favorable conditions for the growth of *T. denticola*.

The phenomenon of FH depletion would only occur once the *T. denticola* population size crosses a threshold such that the rate of FH cleavage in periodontal pockets is faster than its replenishment. Hence, FH binding could serve different purposes during the progression of disease. During early disease when high level outgrowth of the *T. denticola* population has not yet occurred, there would be sufficient levels of FH to bind to *T. denticola* and protect against complement mediated killing. It is precisely during this stage of disease when *T. denticola* would be most vulnerable to complement mediated killing since passive mechanisms of protection such as the physical barrier provided by dense plaque would be less effective. However, during late stage disease when *T. denticola* is highly abundant and the plaque biofilm is matured, the tables could turn. Now survival of the population as a whole is not dependent on complement evasion through FH binding, but rather there is more benefit to be derived from FH depletion. The overall "concern" about complement evasion could be moot if the cleavage product that remains associated with the *T. denticola* surface retains complement regulatory ability. If CCP1–4 are present in the fragment and can bind C3b, then decay acceleration and cofactor activities will be retained (Gordon et al. 1995). If CCP1–4 are not part of the fragment perhaps other active mechanisms of protection compensate. As alluded to above, *T. denticola* can cleave C3 and C3b through a FH independent, dentilisin dependent mechanism. While this activity does not protect *T. denticola* against complement mediated killing in in vitro serum sensitivity assays, it could be more active and effective in vivo (McDowell et al. 2005a; Yamazaki et al. 2006).

3.2.12 FhbB: A Possible T. denticola Targeting Periodontal Disease Vaccine

There is a pressing need for new and improved preventive strategies and non-invasive therapeutic approaches for periodontal disease. Preventing periodontal disease would represent a significant step forward in promoting good oral health and in decreasing the risk of other periodontal disease-associated systemic disorders. The *T. denticola* FhbB protein is an attractive candidate for vaccine development. By targeting *T. denticola* it may be possible to block the cascade of events triggered by FH depletion that culminate in disease. FH binding proteins in general have received considerable attention as possible vaccine candidates due to their potential to protect through multiple synergistic mechanisms including antibody mediated killing and increased complement sensitivity (Beernink et al. 2006, 2011; Seib et al. 2010). As detailed above, the inherent properties of FhbB are well suited for its use as a vaccine. Recombinant FhbB is thermodynamically stable, highly antigenic, easily produced, and small in size. FhbB is also constitutively expressed, surface exposed, and universally produced by all *T. denticola* strains. Recent data from our lab demonstrate that antibody to FhbB has potent bactericidal activity (Marconi RT unpublished data). It has been demonstrated that single site-directed mutations can be introduced into FhbB that block its FH binding ability without altering its structure or thermostability (Miller et al. 2012). As demonstrated for the *Neisseria meningitidis* fHbp protein (Beernink et al. 2011), an fHbp mutant that does not bind FH elicits antibodies with higher bactericidal activity than antibodies generated against FH binding fHbp. This is most likely due to the fact that non-binding variants would not be saturated by endogenous FH when used as an immunizing antigen.

3.3 Conclusions

Periodontal disease remains a major human health concern worldwide. Current strategies to prevent or treat disease are insufficient. While the microbial etiology of periodontal disease is complex, it is clear that species of the red microbial complex (a taxonomic cluster consisting of *Porphyromonas gingivalis*, *Tannerella forsythia*, and *T. denticola*) are central players in the disease process (Socransky et al. 1998). *P. gingivalis* and *T. denticola* are considered to be "keystone species" (Hajishengallis 2010). Whether such status is warranted for *Tannerella forysthia* remains to be determined. Keystone species are organisms that "serve(s) an essential function for the entire community, similar to a differentiated cell serving a function for an entire tissue". *P. gingivalis* subverts complement receptor 3 and C5a anaphylatoxin receptor signaling as a means to increase its fitness in the hostile inflammatory environment of periodontal pockets. As reviewed here, it is our hypothesis that *T. denticola* facilitates the outgrowth of periopathogens by dysregulating the control of complement activation; a process that alters the "immunological environment" of the subgingival crevice. Critical questions remain to be answered including the composition of the dentilisin-generated cleavage products and whether or not these fragments possess complement regulatory activity.

References

Beernink PT, Leipus A, Granoff DM (2006) Rapid genetic grouping of factor h-binding protein (genome-derived neisserial antigen 1870), a promising group B meningococcal vaccine candidate. Clin Vaccine Immunol 13:758–763

Beernink PT, Shaughnessy J, Braga EM, Liu Q, Rice PA, Ram S, Granoff DM (2011) A meningococcal factor H binding protein mutant that eliminates factor H binding enhances protective antibody responses to vaccination. J Immunol 186:3606–3614

Boackle RJ (1990) The complement system. Immunol Ser 50:143–171

Carneiro LG, Venuleo C, Oppenheim FG, Salih E (2012) Proteome data set of human gingival crevicular fluid from healthy periodontium sites by multidimensional protein separation and mass spectrometry. J Periodontal Res 47:248–262

Castiblanco-Valencia MM, Fraga TR, Silva LB, Monaris D, Abreu PA, Strobel S, Jozsi M, Isaac L, Barbosa AS (2012) Leptospiral immunoglobulin-like proteins interact with human complement regulators factor H, FHL-1, FHR-1, and C4BP. J Infect Dis 205:995–1004

Cordes FS, Roversi P, Kraiczy P, Simon MM, Brade V, Jahraus O, Wallis R, Skerka C, Zipfel PF, Wallich R, Lea SM (2005) A novel fold for the factor H-binding protein BbCRASP-1 of Borrelia burgdorferi. Nat Struct Mol Biol 12:276–277

Darveau RP (2010) Periodontitis: a polymicrobial disruption of host homeostasis. Nat Rev Microbiol 8:481–490

Dashper SG, Seers CA, Tan KH, Reynolds EC (2011) Virulence factors of the oral spirochete Treponema denticola. J Dent Res 90(6):691–703

Dewhirst FE, Tamer MA, Ericson RE, Lau CN, Levanos VA, Boches SK, Galvin JL, Paster BJ (2000) The diversity of periodontal spirochetes by 16S rRNA analysis. Oral Microbiol Immunol 15:196–202

Ellen RP (2006) Pathogenic Treponema: molecular and cellular biology. Caister Academic Press, Norfolk

Ellen RP, Galimanas VB (2005) Spirochetes at the forefront of periodontal infections. Periodontol 2000 38:13–32

Faisal SM, Yan W, Chen CS, Palaniappan RU, McDonough SP, Chang YF (2008) Evaluation of protective immunity of Leptospira immunoglobulin like protein A (LigA) DNA vaccine against challenge in hamsters. Vaccine 26:277–287

Faisal SM, Yan W, McDonough SP, Mohammed HO, Divers TJ, Chang YF (2009) Immune response and prophylactic efficacy of smegmosomes in a hamster model of leptospirosis. Vaccine 27:6129–6136

Fenno JC (2012) Treponema denticola interactions with host proteins. J Oral Microbiol 4

Fenno JC, Hannam PM, Leung WK, Tamura M, Uitto VJ, McBride BC (1998a) Cytopathic effects of the major surface protein and the chymotrypsin like protease of Treponema denticola. Infect Immun 66:1869–1877

Fenno JC, Wong GW, Hannam PM, McBride BC (1998b) Mutagenesis of outer membrane virulence determinants of the oral spirochete Treponema denticola. FEMS Microbiol Lett 163:209–215

Ferreira VP, Pangburn MK, Cortes C (2010) Complement control protein factor H: the good, the bad, and the inadequate. Mol Immunol 47:2187–2197

Gautam DK, Jindal V, Gupta SC, Tuli A, Kotwal B, Thakur R (2011) Effect of cigarette smoking on the periodontal health status: a comparative, cross sectional study. J Indian Soc Periodontol 15:383–387

Giannakis E, Jokiranta TS, Male DA, Ranganathan S, Ormsby RJ, Fischetti VA, Mold C, Gordon DL (2003) A common site within factor H SCR 7 responsible for binding heparin, C-reactive protein and streptococcal M protein. Eur J Immunol 33:962–969

Gomes-Filho IS, Freitas Coelho JM, da Cruz SS, Passos JS, Teixeira de Freitas CO, Aragao Farias NS, Amorim da Silva R, Silva Pereira MN, Lima TL, Barreto ML (2011) Chronic periodontitis and C-reactive protein levels. J Periodontol 82:969–978

Gordon DL, Kaufman RM, Blackmore TK, Kwong J, Lublin DM (1995) Identification of complement regulatory domains in human factor H. J Immunol 155:348–356

Grosskinsky S, Schott M, Brenner C, Cutler SJ, Kraiczy P, Zipfel PF, Simon MM, Wallich R (2009) *Borrelia recurrentis* employs a novel multifunctional surface protein with anti-complement, anti-opsonic and invasive potential to escape innate immunity. PLoS One 4:e4858

Haba D, Teslaru S, Ungureanu D, Hodorog D, Alecu C, Benghiac AG, Zetu L, Ancuta C, Ancuta E, Nemtoi A, Iordache C (2011) Evaluation of serum and gingival crevicular fluid C-reactive protein and IL-6 levels in patients with periodontitis and transient ischemic attacks. Rom J Morphol Embryol 52:1243–1247

Hajishengallis G (2010) Complement and periodontitis. Biochem Pharmacol 80:1992–2001

Hartmann K, Corvey C, Skerka C, Kirschfink M, Karas M, Brade V, Miller JC, Stevenson B, Wallich R, Zipfel PF, Kraiczy P (2006) Functional characterization of BbCRASP-2, a distinct outer membrane protein of *Borrelia burgdorferi* that binds host complement regulators factor H and FHL-1. Mol Microbiol 61:1220–1236

Hellwage J, Meri T, Heikkila T, Alitalo A, Panelius J, Lahdenne P, Seppala IJT, Meri S (2001) The complement regulator factor H binds to the surface protein OspE of *Borrelia burgdorferi*. J Biol Chem 276:8427–8435

Hovis KM, McDowell JV, Griffin L, Marconi RT (2004) Identification and characterization of a linear-plasmid-encoded factor H-binding protein (FhbA) of the relapsing fever spirochete *Borrelia hermsii*. J Bacteriol 186:2612–2618

Hovis KM, Jones JP, Sadlon T, Raval G, Gordon DL, Marconi RT (2006a) Molecular analyses of the interaction of *Borrelia hermsii* FhbA with the complement regulatory proteins factor H and factor H-like protein 1. Infect Immun 74:2007–2014

Hovis KM, Tran E, Sundy CM, Buckles E, McDowell JV, Marconi RT (2006b) Selective binding of Borrelia burgdorferi OspE paralogs to factor H and serum proteins from diverse animals: possible expansion of the role of OspE in Lyme disease pathogenesis. Infect Immun 74:1967–1972

Ishihara K (2010) Virulence factors of Treponema denticola. Periodontol 2000 54:117–135

Jarva H, Jokirant TS, Hellwage J, Zipfel PF, Meri S (1999) Regulation of complement activation by C-reactive protein: targeting then complement regulatory activity of factor H by an interaction with short consensus repeat domains 7 and 8–11. J Immunol 163:3957–3962
Ji SR, Wu Y, Potempa LA, Liang YH, Zhao J (2006a) Effect of modified C-reactive protein on complement activation: a possible complement regulatory role of modified or monomeric C-reactive protein in atherosclerotic lesions. Arterioscler Thromb Vasc Biol 26:935–941
Ji SR, Wu Y, Potempa LA, Qiu Q, Zhao J (2006b) Interactions of C-reactive protein with low-density lipoproteins: implications for an active role of modified C-reactive protein in atherosclerosis. Int J Biochem Cell Biol 38:648–661
Kenedy MR, Vuppala SR, Siegel C, Kraiczy P, Akins DR (2009) CspA-mediated binding of human factor H inhibits complement deposition and confers serum resistance in Borrelia burgdorferi. Infect Immun 77:2773–2782
Kodovazenitis G, Pitsavos C, Papadimitriou L, Deliargyris EN, Vrotsos I, Stefanadis C, Madianos PN (2011) Periodontal disease is associated with higher levels of C-reactive protein in non-diabetic, non-smoking acute myocardial infarction patients. J Dent 39:849–854
Kraiczy P, Hellwage J, Skerka C, Becker H, Kirschfink M, Simon MS, Brade V, Zipfel PF, Wallich R (2004) Complement resistance of *Borrelia burgdorferi* correlates with the expression of BbCRASP-1, a novel linear plasmid-encoded surface protein that interacts with human factor H and FHL-1 and is unrelated to Erp proteins. J Biol Chem 279:2421–2429
Kumar V, Faizuddin M (2011) Effect of smoking on gingival microvasculature: a histological study. J Indian Soc Periodontol 15:344–348
Lupas A (1996) Coiled coils: new structures and new functions. Trends Biochem Sci 96:375–382
Lupas A, Van Dyke M, Stock J (1991) Predicting coiled coils from protein sequences. Science 252:1162–1164
Maekawa T, Tabeta K, Kajita-Okui K, Nakajima T, Yamazaki K (2011) Increased expression of C-reactive protein gene in inflamed gingival tissues could be derived from endothelial cells stimulated with interleukin-6. Arch Oral Biol 56:1312–1318
McDowell JV, Tran E, Hamilton D, Wolfgang J, Miller K, Marconi RT (2003) Analysis of the ability of spirochete species associated with relapsing fever, avian borreliosis, and epizootic bovine abortion to bind factor H and cleave c3b. J Clin Microbiol 41:3905–3910
McDowell JV, Wolfgang J, Senty L, Sundy CM, Noto MJ, Marconi RT (2004) Demonstration of the involvement of outer surface protein E coiled coil structural domains and higher order structural elements in the binding of infection-induced antibody and the complement-regulatory protein, factor H. J Immunol 173:7471–7480
McDowell JV, Lankford J, Stamm L, Sadlon T, Gordon DL, Marconi RT (2005a) Demonstration of factor H-like protein 1 binding to Treponema denticola, a pathogen associated with periodontal disease in humans. Infect Immun 73:7126–7132
McDowell JV, Harlin ME, Rogers EA, Marconi RT (2005b) Putative coiled-coil structural elements of the BBA68 protein of Lyme disease spirochetes are required for formation of its factor H binding site. J Bacteriol 187:1317–1323
McDowell JV, Hovis KM, Zhang H, Tran E, Lankford J, Marconi RT (2006) Evidence that the BBA68 protein (BbCRASP-1) of the Lyme disease spirochetes does not contribute to factor H-mediated immune evasion in humans and other animals. Infect Immun 74:3030–3034
McDowell JV, Frederick J, Stamm L, Marconi RT (2007) Identification of the gene encoding the FhbB protein of *Treponema denticola*, a highly unique factor H-like protein 1 binding protein. Infect Immun 75:1050–1054
McDowell JV, Huang B, Fenno JC, Marconi RT (2009) Analysis of a unique interaction between the complement regulatory protein factor H and the periodontal pathogen Treponema denticola. Infect Immun 77:1417–1425
McDowell JV, Frederick J, Miller DP, Goetting-Minesky MP, Goodman H, Fenno JC, Marconi RT (2011) Identification of the primary mechanism of complement evasion by the periodontal pathogen, Treponema denticola. Mol Oral Microbiol 26:140–149

Megson E, Fitzsimmons T, Dharmapatni K, Mark Bartold P (2010) C-reactive protein in gingival crevicular fluid may be indicative of systemic inflammation. J Clin Periodontol 37:797–804

Meri T, Cutler SJ, Blom AM, Meri S, Jokiranta TS (2006) Relapsing fever spirochetes Borrelia recurrentis and B. duttonii acquire complement regulators C4b-binding protein and factor H. Infect Immun 74:4157–4163

Metts MS, McDowell JV, Theisen M, Hansen PR, Marconi RT (2003) Analysis of the OspE determinants involved in binding of factor H and OspE-targeting antibodies elicited during *Borrelia burgdorferi* infection in mice. Infect Immun 71:3587–3596

Mihlan M, Stippa S, Jozsi M, Zipfel PF (2009) Monomeric CRP contributes to complement control in fluid phase and on cellular surfaces and increases phagocytosis by recruiting factor H. Cell Death Differ 16:1630–1640

Miller DP, McDowell JV, Bell JK, Marconi RT (2011) Crystallization of the factor H-binding protein, FhbB, from the periopathogen Treponema denticola. Acta Crystallogr Sect F Struct Biol Cryst Commun 67:678–681

Miller DP, Bell JK, McDowell JV, Conrad DH, Burgner JW, Heroux A, Marconi RT (2012) Structure of factor H binding protein B (FhbB) of the periopathogen, Treponema denticola: insights into the progression of periodontal disease. J Biol Chem 287:12715–12722

Morgan HP, Jiang J, Herbert AP, Kavanagh D, Uhrin D, Barlow PN, Hannan JP (2011) Crystallographic determination of the disease-associated T1184R variant of complement regulator factor H. Acta Crystallogr D Biol Crystallogr 67:593–600

Okemefuna AI, Nan R, Miller A, Gor J, Perkins SJ (2010) Complement factor H binds at two independent sites to C-reactive protein in acute phase concentrations. J Biol Chem 285:1053–1065

Paster BJ, Dewhirst FE, Weisburg WG, Tordoff LA, Fraser GJ, Hespell RB, Stanton TB, Zablen L, Mandelco L, Woese CR (1991) Phylogenetic analysis of the spirochetes. J Bacteriol 173:6101–6109

Paster BJ, Boches SK, Galvin JL, Ericson RE, Lau CN, Levanos VA, Sahasrabudhe A, Dewhirst FE (2001) Bacterial diversity in human subgingival plaque. J Bacteriol 183:3770–3783

Paster BJ, Olsen I, Aas JA, Dewhirst FE (2006) The breadth of bacterial diversity in the human periodontal pocket and other oral sites. Periodontol 2000 42:80–87

Pepys MB, Hirschfield GM (2003) C-reactive protein: a critical update. J Clin Invest 111:1805–1812

Pihlstrom BL, Michalowicz BS, Johnson NW (2005) Periodontal diseases. Lancet 366:1809–1820

Prosser BE, Johnson S, Roversi P, Herbert AP, Blaum BS, Tyrrell J, Jowitt TA, Clark SJ, Tarelli E, Uhrin D, Barlow PN, Sim RB, Day AJ, Lea SM (2007) Structural basis for complement factor H linked age-related macular degeneration. J Exp Med 204:2277–2283

Raffaelli L, Santangelo R, Falchetti P, Galluccio F, Luciani N, Anselmi A, Nowzari H, Verdugo F, Fadda G, D'Addona A (2010) Examination of periodontal pathogens in stenotic valve specimens and in whole blood samples in patients affected by aortic valve stenosis and chronic periodontitis. Int J Immunopathol Pharmacol 23:561–566

Rogers EA, Marconi RT (2007) Delineation of species-specific binding properties of the CspZ protein (BBH06) of Lyme disease spirochetes: evidence for new contributions to the pathogenesis of *Borrelia* spp. Infect Immun 75:5272–5281

Rogers EA, Abdunnur SV, McDowell JV, Marconi RT (2009) Comparative analysis of the properties and ligand binding characteristics of CspZ, a factor H binding protein, derived from Borrelia burgdorferi isolates of human origin. Infect Immun 77:4396–4405

Romano F, Barbui A, Aimetti M (2007) Periodontal pathogens in periodontal pockets and in carotid atheromatous plaques. Minerva Stomatol 56:169–179

Rossmann E, Kitiratschky V, Hofmann H, Kraiczy P, Simon MM, Wallich R (2006) *Borrelia burgdorferi* complement regulator-acquiring surface protein 1 of the Lyme disease spirochetes is expressed in humans and induces antibody responses restricted to nondenatured structural determinants. Infect Immun 74:7024–7028

Rossmann E, Kraiczy P, Herzberger P, Skerka C, Kirschfink M, Simon MM, Zipfel PF, Wallich R (2007) Dual binding specificity of a *Borrelia hermsii*-associated complement regulator-acquiring

surface protein for factor H and plasminogen discloses a putative virulence factor of relapsing fever spirochetes. J Immunol 178:7292–7301
Ruddy S, Austen KF (1969) C3 inactivator of man. I. Hemolytic measurement by the inactivation of cell-bound C3. J Immunol 102:533–543
Ruddy S, Austen KF (1971) C3b inactivator of man. II. Fragments produced by C3b inactivator cleavage of cell-bound or fluid phase C3b. J Immunol 107:742–750
Schott M, Grosskinsky S, Brenner C, Kraiczy P, Wallich R (2010) Molecular characterization of the interaction of Borrelia parkeri and Borrelia turicatae with human complement regulators. Infect Immun 78:2199–2208
Seib KL, Oriente F, Adu-Bobie J, Montanari P, Ferlicca F, Giuliani MM, Rappuoli R, Pizza M, Delany I (2010) Influence of serogroup B meningococcal vaccine antigens on growth and survival of the meningococcus in vitro and in ex vivo and in vivo models of infection. Vaccine 28:2416–2427
Seling A, Siegel C, Fingerle V, Jutras BL, Brissette CA, Skerka C, Wallich R, Zipfel PF, Stevenson B, Kraiczy P (2010) Functional characterization of Borrelia spielmanii outer surface proteins that interact with distinct members of the human factor H protein family and with plasminogen. Infect Immun 78:39–48
Shepherd S (2011) Alcohol consumption a risk factor for periodontal disease. Evid Based Dent 12:76
Silva EF, Medeiros MA, McBride AJ, Matsunaga J, Esteves GS, Ramos JG, Santos CS, Croda J, Homma A, Dellagostin OA, Haake DA, Reis MG, Ko AI (2007) The terminal portion of leptospiral immunoglobulin-like protein LigA confers protective immunity against lethal infection in the hamster model of leptospirosis. Vaccine 25:6277–6286
Socransky S, Haffajee A, Cugini M, Smith C, Kent RJ (1998) Microbial complexes in subgingival plaque. J Clin Periodontol 25:134–144
Stevenson B, Choy HA, Pinne M, Rotondi ML, Miller MC, Demoll E, Kraiczy P, Cooley AE, Creamer TP, Suchard MA, Brissette CA, Verma A, Haake DA (2007) Leptospira interrogans endostatin-like outer membrane proteins bind host fibronectin, laminin and regulators of complement. PLoS One 2:e1188
Toyofuku T, Inoue Y, Kurihara N, Kudo T, Jibiki M, Sugano N, Umeda M, Izumi Y (2011) Differential detection rate of periodontopathic bacteria in atherosclerosis. Surg Today 41:1395–1400
Tuter G, Kurtis B, Serdar M (2007) Evaluation of gingival crevicular fluid and serum levels of high-sensitivity C-reactive protein in chronic periodontitis patients with or without coronary artery disease. J Periodontol 78:2319–2324
Verma A, Hellwage J, Artiushin S, Zipfel PF, Kraiczy P, Timoney JF, Stevenson B (2006) LfhA, a novel factor H-binding protein of Leptospira interrogans. Infect Immun 74:2659–2666
Whaley K, Ruddy S (1976) Modulation of the alternative complement pathways by beta 1 H globulin. J Exp Med 144:1147–1163
Wu J, Wu YQ, Ricklin D, Janssen BJ, Lambris JD, Gros P (2009) Structure of complement fragment C3b-factor H and implications for host protection by complement regulators. Nat Immunol 10:728–733
Yamazaki T, Miyamoto M, Yamada S, Okuda K, Ishihara K (2006) Surface protease of Treponema denticola hydrolyzes C3 and influences function of polymorphonuclear leukocytes. Microbes Infect 8:1758–1763
Yan W, Faisal SM, McDonough SP, Divers TJ, Barr SC, Chang CF, Pan MJ, Chang YF (2009) Immunogenicity and protective efficacy of recombinant Leptospira immunoglobulin-like protein B (rLigB) in a hamster challenge model. Microbes Infect 11:230–237
Zipfel PF (2009) Complement and immune defense: from innate immunity to human diseases. Immunol Lett 126:1–7
Zipfel PF, Skerka C (1999) FHL-1/reconectin: a human complement and immune regulator with cell-adhesive function. Immunol Today 20:135–141
Zipfel PF, Skerka C, Hellwage J, Jokiranta ST, Meri S, Brade V, Kraiczy P, Noris M, Remuzzi G (2002) Structure-function studies of the complement system. Biochem Soc Trans 30:971–978

Chapter 4
Borrelial Complement-Binding Proteins

Peter Kraiczy and Reinhard Wallich

4.1 Introduction

Spirochetes causing Lyme disease or relapsing fever exploit an array of sophisticated means to establish infection and to combat innate and adaptive immune responses of the human host. One central immune escape mechanism of serum-resistant spirochetes involves the inactivation of host complement attack through binding of distinct complement regulatory proteins, including factor H (CFH), factor H-like protein 1 (FHL1), factor H-related protein 1 (CFHR1), CFHR2, CFHR5, C4b-binding protein (C4Bp), and/or C1 esterase inhibitor (C1-Inh) to the cell surface (Fig. 4.1) (Alitalo et al. 2001; Bhide et al. 2009; Grosskinsky et al. 2009; Haupt et al. 2007; Hellwage et al. 2001; Kraiczy et al. 2001a, b; McDowell et al. 2003a; Pietikainen et al. 2010; Rossmann et al. 2007; Schott et al. 2010; Siegel et al. 2010; Stevenson et al. 2002; Meri et al. 2006).

Lyme disease spirochetes acquire complement regulators via bacterial surface molecules of three distinct and genetically unrelated groups collectively termed complement regulator-acquiring surface proteins or CRASPs (Kraiczy et al. 2001a, b; Wallich et al. 2005). Individual *Borrelia* species can produce up to five different CRASP molecules and depending on their genetic composition, different combinations of these molecules are expressed on the borrelial outer surface (Fig. 4.1a). The three groups of CRASPs differ not only in their binding capabilities for distinct complement regulators and other serum proteins but also in their

P. Kraiczy, Ph.D. (✉)
Institute of Medical Microbiology and Infection Control, University Hospital of Frankfurt am Main, Paul-Ehrlich-Str. 40, D-60596 Frankfurt/Main, Germany
e-mail: Kraiczy@em.uni-frankfurt.de

R. Wallich, Ph.D.
Institute of Immunology, University of Heidelberg, Im Neuenheimer Feld 305, D-69120 Heidelberg, Germany
e-mail: wallich@uni-hd.de

M.E. Embers (ed.), *The Pathogenic Spirochetes: Strategies for Evasion of Host Immunity and Persistence*, DOI 10.1007/978-1-4614-5404-5_4,

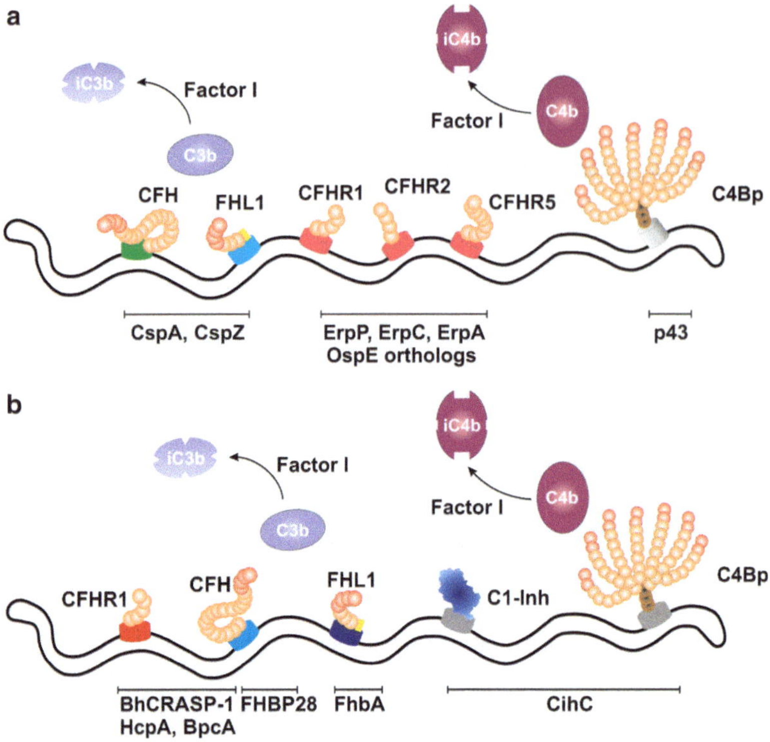

Fig. 4.1 Interaction of outer surface proteins of Lyme disease (**a**) or relapsing fever (**b**) spirochetes with distinct human complement regulators of the alternative and classical pathways. *Borreliae* produce various complement regulator-binding proteins to escape the destructive attack of human complement at multiple time points during the course of mammalian infection (initial, established, and persistent infection). Complement-binding proteins display different binding properties to CFH, FHL1, CFHR1, CFHR2, CFHR5, C1-Inh, and C4Bp-binding protein (C4Bp) (see text and tables)

protein sequences and structures, gene locations, gene regulatory mechanisms, and expression patterns during the tick-mammal infection cycle (Kraiczy et al. 2001a, 2002, 2003, 2004a; McDowell et al. 2003a; Wallich et al. 2005; Alitalo et al. 2004, 2005; Brissette et al. 2009; Bykowski et al. 2007; Coleman et al. 2008; Cordes et al. 2005; Hallström et al. 2010; Hartmann et al. 2006; Rogers et al. 2009; von Lackum et al. 2005).

Relapsing fever is a recurrent febrile infection caused by various *Borrelia* spirochetes transmitted either by lice (epidemic relapsing fever) or by ticks (endemic relapsing fever). Tick- and louse-borne pathogens, i.e., *Borrelia hermsii*, *Borrelia parkeri*, *Borrelia duttonii,* and *Borrelia recurrentis*, bind distinct complement regulators of the alternative, classical, and lectin pathways via various outer surface proteins (Fig. 4.1b), thereby contributing to complement resistance of these emerging human pathogens (Grosskinsky et al. 2009, 2010; McDowell et al. 2003a; Rossmann et al. 2007; Schott et al. 2010; Meri et al. 2006; Hovis et al. 2004, 2006a).

4.1.1 Complement Regulatory Proteins

The complement system forms a first line of defense of the human innate immune system that is immediately and directly activated upon entry of microorganisms (Walport 2001). In general, direct activation of complement via the alternative, the mannose-binding lectin pathway (activated by the surface composition of the pathogen) and the classical pathway (activated by specific antibodies or other surface moieties) results in the formation and deposition of C3b, a key component of the complement cascade. C3b deposition on microbial cell surfaces leads to opsonization and flags the cell for phagocytosis. If activation progresses, the C5 convertase is generated and initiates the terminal pathway and the assembly of the terminal complement complex (TCC) causing complement-mediated bacteriolysis. Since C3b is generated by all three pathways, it would be beneficial for invading microorganisms to inhibit complement activation at the central point of the cascade. Under normal conditions, the activation of complement effector molecules is tightly controlled on the surface of intact host cells and biomembranes by a number of fluid-phase regulators (e.g., CFH, FHL1, CFHR1, CFHR2, CFHR5, C4Bp, and C1-Inh) and integral membrane, surface-attached glycoproteins (e.g., CR1 (complement receptor 1)/CD35, CR2 (complement receptor 2)/CD21, MCP (membrane cofactor protein)/CD46, DAF (decay-accelerating factor)/CD55, and protectin/CD59). The concerted action of regulators and inhibitors ensures cell and tissue integrity (Harris et al. 1999; Zipfel 2009; Zipfel and Skerka 2009).

The alternative pathway regulators, CFH and FHL1, are encoded by the human *cfh* gene and generated by alternative splicing (Zipfel 2009; Zipfel and Skerka 2009). Both proteins are structurally related and fold into repetitive protein domains termed short consensus repeats (SCRs). CFH is a 150 kDa glycoprotein that is composed of 20 SCR domains. FHL1 is a 42 kDa glycoprotein, comprised of the seven amino-terminal SCRs of CFH plus four unique amino acids (SFTL) at the C-terminus. CFH and FHL1 act as cofactors for factor I-mediated degradation of C3b and support the dissociation (decay-accelerating activity) of the C3 convertase, C3bBb (Zipfel 2009). Furthermore, the human CFH family includes six "factor H-related" proteins (CFHR1, CFHR2, CFHR3, CFHR4A, CFHR4B, and CFHR5) which are encoded by distinct genes (McRae et al. 2001; Skerka and Zipfel 2008). The encoded proteins show extensive sequence similarities to CFH (Józsi and Zipfel 2008). CFHR1 consists of five SCRs and two glycosylated forms occur, the 37 kDa CFHR1α and the 43 kDa CFHR1β protein (Estaller et al. 1991; Skerka et al. 1991). CFHR1 is a complement regulator that blocks C5 convertase activity as well as assembly and membrane insertion of the terminal components (Heinen et al. 2009). CFHR2 is composed of four SCRs and is found in plasma as a non-glycosylated 24 kDa form (CFHR2) and a glycosylated 29 kDa form (CFHR2α) (Skerka et al. 1992). The function of CFHR2 is poorly understood. The 65 kDa CFHR5 protein is comprised of nine SCRs and displays cofactor activity for factor I-mediated inactivation of C3b (McRae et al. 2001, 2005). CFHR-5 can inhibit the activity of the fluid-phase C3 convertase.

A key fluid-phase complement regulator of the classical and lectin pathways is the 570 kDa C4b-binding protein (C4Bp). This molecule is comprised of seven identical α-chains and a single short β-chain arranged in a spider-like fashion (Blom 2002; Nagasawa et al. 1982; Scharfstein et al. 1978). C4Bp acts as cofactor for factor I-mediated inactivation/degradation of C4b to C4d, and interferes with the assembly and decay of the C3 convertase (C4bC2a) of the classical and lectin pathways (Fujita et al. 1978; Fujita and Nussenzweig 1979).

C1-Inh is a 105 kDa protein that regulates the initial steps of the classical pathway by inhibiting serine proteases including C1r and C1s of the classical complement pathway and MASP-1 and MASP-2 of the lectin pathway (Davis et al. 2008).

4.2 Summary of Current Findings

4.2.1 Complement-Binding Proteins of Lyme Disease Spirochetes

4.2.1.1 Complement Regulator-Acquiring Surface Protein 1 (CspA)

CRASP-1 or CspA of *Borrelia burgdorferi* is a surface-exposed lipoprotein encoded by the *cspA* gene on a linear plasmid of 54 kbp, lp54 (Table 4.1) (Wallich et al. 2005; Kraiczy et al. 2004a; Schutzer et al. 2011; Fraser et al. 1997). Sequence analysis of a number of borrelial genomes revealed that *cspA* belongs to a large paralogous gene family, PFam54. These genes show great sequence divergence and functional linear replicons apparently containing tandem arrays between 6 (*B. burgdorferi* strain JD1) and 12 (*Borrelia afzelii* strain PKo) intact PFam54 genes (Wallich et al. 2005; Schutzer et al. 2011; Fraser et al. 1997; Casjens et al. 2000; Wywial et al. 2009).

CspA orthologs of other Lyme disease spirochetes (e.g., *B. burgdorferi*, *B. afzelii*, *Borrelia spielmanii*, and *Borrelia bavariensis*) are capable of binding to human complement regulators CFH and/or FHL1. However, differences in their binding properties and interaction with distinct SCR domains of complement regulators could be observed (Table 4.1). Interestingly, CspA of *B. burgdorferi* exclusively binds to two distinct regions of CFH (SCR7 and SCR19–20), whereas other CspA orthologs bind to different CFH domains (Wallich et al. 2005; Kraiczy et al. 2004a; Herzberger et al. 2009; van Burgel et al. 2010). Moreover, CspA of *B. burgdorferi*, *B. afzelii*, and *B. spielmanii* exhibit strong binding capacities to CFH and FHL1, whereas BGA66 of *B. bavariensis* binds weakly to both complement regulators. Furthermore, BgCRASP-1α and BgCRASP-1β of *Borrelia garinii* as well as BGA71 of *B. bavariensis* exhibit very low binding affinities to FHL1 in ELISA. Differences in the amino acid sequences of the proposed CFH/FHL1-binding region(s) of CspA orthologs may explain the impaired ability of these molecules to interact with individual host complement regulators. All other PFam54 protein family members of *B. burgdorferi*, *B. afzelii*, *B. garinii*, and *B. bavariensis* fail to bind CFH and/or FHL1 (van Burgel et al. 2010; Kraiczy et al. 2004b, 2006; McDowell et al. 2005; P. Kraiczy unpublished data).

Table 4.1 Characteristics of CRASP proteins of Lyme disease spirochetes

Complement binding protein	Genospecies	Strain	Synonyms/ other designations	Gene name/ORF	Interaction with complement regulator(s)	Binding capacity	Binding region of host protein	Interacting CRASP domain	Imparts complement resistance	References
CspA CRASP-1	*B. burgdorferi*	B31 LW2 ZS7	BBA68 BbCRASP-1 ZS7.A68 FHBP	*cspA* *bba68*	CFH FHL1	++ ++ +	SCR7 SCR19–20 SCR7	C-terminus, three putative binding regions	Yes	(Kraiczy et al. 2001a, 2004a; Brooks et al 2005 McDowell et al. 2005)
CRASP-1	*B. afzelii*	MMS	BaCRASP-1	*cspA* *mmsa71*	CFH FHL1	++ ++	SCR5–7 SCR5–7	C-terminus	Yes	(Wallich et al. 2005)
CRASP-1	*B. spielmanii*	A14S	BsCRASP-1	NA	CFH FHL1	++ ++	SCR5–7 SCR5–7	C-terminus	Yes	(Herzberger et al. 2007, 2009)
BGA66	*B. bavariensis*	PBi	None	*bga66*	CFH FHL1	+ +	SCR6–7 SCR6–7	NA	NA	(van Burgel et al. 2010)
BGA71	*B. bavariensis*	PBi	None	*bga71*	FHL1	±	SCR6–7	NA	NA	(van Burgel et al. 2010)
BgCRASP-1α	*B. garinii* ST6	ZQ1	None	*zqa68*	FHL1	±	NA	NA	NA	(Wallich et al. 2005)
BgCRASP-1β	*B. garinii* ST6	ZQ1	None	*zqa71*	FHL1	±	NA	NA	NA	(Wallich et al. 2005)
CspZ CRASP-2	*B. burgdorferi*	B31 LW2	BBH06 BbCRASP-2	*cspZ* *bbh06*	CFH FHL1	++ ++	SCR6–7 SCR6–7	C-terminus, three putative binding regions	Yes	(Hartmann et al. 2006; Siegel et al. 2008, Rogers and Marconi 2007)
CRASP-2	*B. afzelii*	FEM1-D15	NA	NA	CFH FHL1	++ ++	SCR6–7 SCR6–7	NA	NA	(Kraiczy et al. 2001a)
CRASP-2	*B. spielmanii*	A14S	NA	NA	CFH FHL1	++ ++	NA	NA	NA	(Herzberger et al. 2007)

(continued)

Table 4.1 (continued)

Complement binding protein	Genospecies	Strain	Synonyms/ other designations	Gene name/ORF	Interaction with complement regulator(s)	Binding capacity	Binding region of host protein	Interacting CRASP domain	Imparts complement resistance	References
CRASP-3	*B. burgdorferi*	B31	ErpP BBN38 BbCRASP-3	*erpP*	CFH[a] CFHR1 CFHR2 CFHR5	++ ++ ++ ±	SCR19–20 SCR3–5 NA NA	C-terminus, different regions C-terminus NA NA	No[b]	(Haupt et al. 2007; Siegel et al. 2010; Alitalo et al. 2004; Kraiczy et al. 2003; Metts et al. 2003; Seling et al. 2010; Kenedy and Akins 2011)
CRASP-3	*B. afzelii*	FEM1-D15	NA	NA	FHL1	++	SCR6–7	NA	NA	(Kraiczy et al. 2001a)
CRASP-3	*B. spielmanii*	A14S	BsCRASP-3	NA	CFH	++	SCR20	NA	NA	(Herzberger et al. 2007)
CRASP-3	*B. spielmanii*	PC-Eq17	Erp60	*erp60*	CFH CFHR1	++	SCR20	NA	NA	(Seling et al. 2010)
CRASP-3	*B. spielmanii*	TIsar2	Erp61	*erp61*	CFH CFHR1	++	SCR20	NA	NA	(Seling et al. 2010)
CRASP-3	*B. spielmanii*	TIsar2	Erp62	*erp62*	CFH CFHR1	++	SCR20	NA	NA	(Seling et al. 2010)
CRASP-3	*B. spielmanii*	TIsar3	Erp63	*erp63*	No	–	No	NA	NA	(Seling et al. 2010)
CRASP-3	*B. lusitaniae*	MT-M8	OspE	*ospE*	CFH	±	NA	NA	NA	(Dieterich et al. 2010)
CRASP-3	*B. garinii*	40/97	OspE	*ospE*	CFH	++			Partial	(Alitalo et al. 2005)
CRASP-4	*B. burgdorferi*	B31 LW2	ErpC	*erpC*	CFH[a] CFHR1 CFHR2	++ + ++	SCR19–20 SCR3–5 NA	C-terminus	No[b]	(Haupt et al. 2007) Hammerschmidt et al. 2012)
CRASP-4	*B. afzelii*	FEM1-D15	NA	NA	CFH	++	SCR19–20	NA	NA	(Kraiczy et al. 2001a)

CRASP-5	*B. burgdorferi*	B31 297	ErpA ErpI ErpN BBP38 BBL39 BbCRASP-5 OspE p21	*erpA* *erpI* *erpN* *ospE*	CFH[a] CFHR1 CFHR2 CFHR5	++ ++ ++ ±	SCR19–20	C-terminus, different regions	No[b]	(Haupt et al. 2007; Hellwage et al. 2001; Kraiczy et al. 2001a; McDowell et al. 2003a; Siegel et al. 2010; Alitalo et al. 2004, 2005; Hovis et al. 2006a; Metts et al. 2003)
CRASP-5	*B. afzelii*	FEM1-D15	NA	NA	CFH	++	SCR19–20	NA	NA	(Kraiczy et al. 2001a)

[a]CFH binds to recombinant CRASP molecules but not to the native, surface-exposed molecule

[b]CRASP-3, -4, and -5 might imparts complement resistance under certain conditions

NA not available, *ST* serotype

++, strong binding; +, weak binding; ±, very weak binding; –, no binding

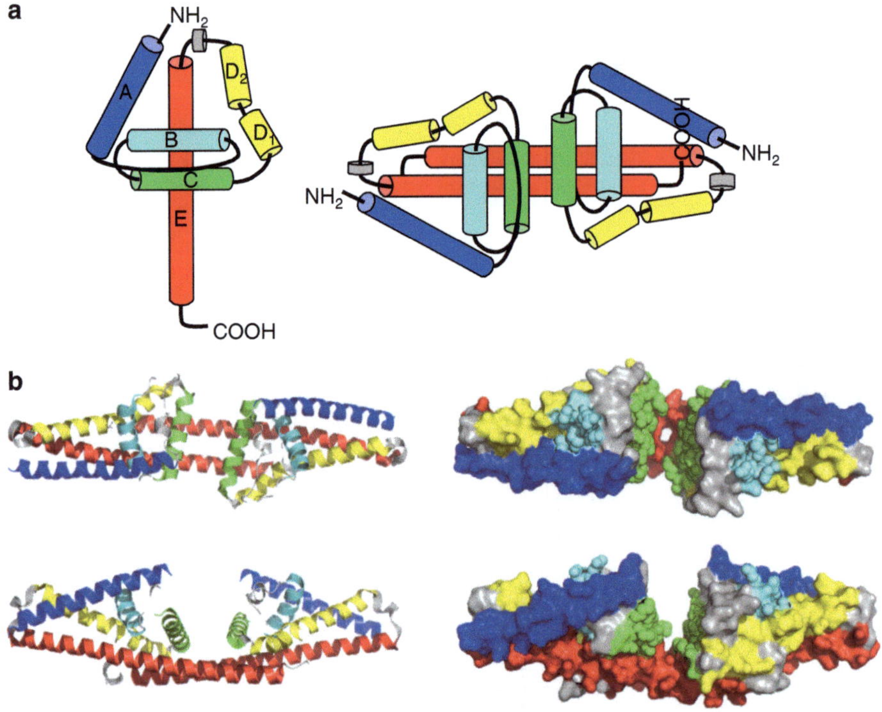

Fig. 4.2 Structure of CspA. Schematic representation of the topology of the CspA monomer and homodimer (**a**). The dimer at the *right panel* is shown in the "top" view. Helices (A to E) are labeled from the N-terminus to the C-terminus. Ribbon representation of the CspA dimer in the "top" and "side" views (**b**). Helices are *colored* as shown in **a**

Among all borrelial CRASP molecules identified so far the atomic structure was only reported for CspA (Cordes et al. 2005). The three-dimensional analysis of CspA (*B. burgdorferi* strain ZS7) reveals five crossing α-helices (A to E) connected by short 3_{10} turns and short loops (Fig. 4.2) (Cordes et al. 2005). The α-helices form a "helical-lollipop" structure with a pyramidal core and an additional C-terminal helix protruding outwards from the center. Two CspA monomers are closely assembled together by their C-terminal helices forming a large cleft at the dimer interface (Cordes et al. 2005). The extensive buried surface of the interface is typical for a biologically relevant dimer and for the interaction of complement regulators. Computer models of CspA derived from *B. afzelii* or *B. spielmanii* suggest that these molecules can also form homodimeric complexes (unpublished data). Interestingly, the architecture of the cleft has approximately the assumed dimension for accommodating a single SCR domain implying that the interdimeric cleft forms the CFH/FHL1-binding site (Cordes et al. 2006). In vitro mutagenesis studies indicated that distinct amino acid residues at the interdimeric cleft mediate binding of complement regulators, whereas amino acid substitutions in the C-terminal region of CspA lead to loss of CFH and FHL1 binding as well as disruption of the CspA dimer (Kraiczy et al. 2009). Hence, the C-terminus of CspA is responsible for structure maintenance rather than being involved directly in the interaction with complement regulators. Coiled-coil structures have also been hypothesized to be involved in binding of CFH (McDowell et al. 2005). However, CspA does not contain any coiled-coil structures (Cordes et al. 2005). It has been shown that substitutions of charged residues as well as the negatively charged heparin influence binding of CFH and/or FHL1 to CspA, suggesting that their interaction is mainly influenced by ionic attractions (Kraiczy et al. 2004a; P. Kraiczy unpublished data).

CFH and FHL1 bound to surface-expressed CspA of *B. burgdorferi*, *B. afzelii*, and *B. spielmanii* impart resistance of borrelial cells to complement-mediated bacteriolysis and enable spirochetes to grow in the presence of human serum (Brooks et al. 2005; Kenedy et al. 2009, P. Kraiczy unpublished data). In contrast, CspA-deficient mutants or borrelial strains naturally lacking the lp54 plasmid, e.g., *B. burgdorferi* B313 and *B. spielmanii* PSigII are highly susceptible to human complement indicating that CspA alone is sufficient for mediating resistance to human serum (Hartmann et al. 2006; Brooks et al. 2005; Kenedy et al. 2009; Herzberger et al. 2007). Notably, interaction of CspA with CFH seems to be directly correlated with an invasive ribosomal spacer type (RST) 3 previously shown to be a genetic marker for hematogenous dissemination and invasiveness of *B. burgdorferi* (Rogers et al. 2009; Iyer et al. 2001; Wormser et al. 2008). Thus, its role in pathogenesis is likely significant.

Concerning the natural mammal-tick infectious cycle, almost all spirochetes transmitted during tick feeding produce CspA on their surface while 2 weeks after transmission to mammals, *cspA* transcripts are no longer detectable (Bykowski et al. 2007; von Lackum et al. 2005; Lederer et al. 2005; McDowell et al. 2006; Wallich et al. 2003). Consequently, Lyme disease patients produce low levels of CspA-specific antibodies that could only be detected in assays where native CspA preparations are employed (McDowell et al. 2006; Rossmann et al. 2006). Transfer of CspA-specific immune sera protects immunodeficient SCID mice against subsequent experimental *B. burgdorferi* infection (Wallich et al. 2003). Interestingly, expression of *cspA* can also be observed when spirochetes are incubated in the presence of various animal sera (Kisova-Vargova et al. 2011).

4.2.1.2 Complement Regulator-Acquiring Surface Protein 2 (CspZ)

The outer surface 23.2 kDa lipoprotein CspZ (in the literature also referred to as BbCRASP-2 or BBH06) is expressed by various borrelial isolates, including *B. burgdorferi* strain B31, and is encoded by the *cspZ* gene on a linear plasmid lp28-3 (Hartmann et al. 2006; Siegel et al. 2008). CspZ binds human complement regulators, CFH and FHL1 (Table 4.1). *CspZ* is distinct from *cspA* and all other known genes encoding for borrelial complement-binding proteins (Hartmann et al. 2006; Fraser et al. 1997; Casjens et al. 2000; Kraiczy et al. 2008). Comprehensive sequence analysis revealed that the *cspZ* gene is conserved (identity ≥97.2%) among genospecies associated with Lyme disease, e.g., *B. burgdorferi*, *B. afzelii*, *B. garinii*, *B. spielmanii*, and *B. bavariensis* as well as *Borrelia bissettii* (Kraiczy et al. 2008; Rogers and Marconi 2007).

Alignments of the predicted CspZ protein sequences showed extensive similarities and a conserved binding region for complement regulators (Rogers et al. 2009; Kraiczy et al. 2008; Rogers and Marconi 2007). Solely CspZ derived from *B. burgdorferi* specifically interacts with complement regulators, whereas all other CspZ orthologs of *B. garinii*, *B. afzelii*, and *B. spielmanii* lack CFH and FHL1 binding capabilities (Rogers et al. 2009; Rogers and Marconi 2007, P. Kraiczy unpublished data). Computer modeling of several CspZ orthologs incapable of binding to CFH revealed that a four-amino acid insertion in a loop between two α-helices as well as point mutations at three positions could provoke destabilization of the entire molecular architecture (Rogers et al. 2009). Four putative CFH/FHL1-binding regions were mapped and a structurally sensitive domain (aa position 204–211) located in binding region 4 was identified as the contiguous binding domain (Hartmann et al. 2006; Siegel et al. 2008). In addition, amino acid substitutions located within and outside the four putative binding regions could influence binding of CFH and FHL1 to CspZ indicating that multiple contact sites are involved in the CspZ-CFH/FHL1 interaction. Most likely, electrostatic and hydrophobic forces may contribute to the interaction of CspZ with CFH and FHL1.

During the natural mammalian-tick-infection cycle, spirochetes fail to express *cspZ* in ticks (Bykowski et al. 2007). Following transmission to the mammalian host only some spirochetes produce CspZ at the entrance site. By 2-weeks post infection *cspZ* transcription significantly increases to higher levels as compared to in vitro cultured spirochetes. Lyme disease patients as well as infected mice elicit a strong antibody response to CspZ indicating substantial production during natural infection (Bykowski et al. 2007; Rogers et al. 2009; Kraiczy et al. 2008). Moreover, antibody titers are higher in patients with late-stage disease as compared to titers in sera obtained from early Lyme disease patients. This finding supports the notion that spirochetes produce CspZ throughout dissemination and persistence in humans and infected animals; this may contribute to the pathogenesis of Lyme disease (Coleman et al. 2008; Rogers et al. 2009). Mice immunized with recombinant CspZ elicit a strong antibody response but immunization fails to protect mice from infection with needle-inoculated spirochetes (Coleman et al. 2008; Rogers et al. 2009). However, the low abundance of CspZ at the surface of in vitro cultivated spirochetes and/or binding of CFH to CspZ could abrogate antigen–antibody formation in the presence of anti-CspZ antibodies.

Experimentally, ectopic expression of CspZ in spirochetes lacking additional complement-binding proteins imparts resistance to complement-mediated bacteriolysis (Hartmann et al. 2006; Siegel et al. 2008). Contradictory data have recently been published using a CspZ-deficient *B. burgdorferi* strain that produces additional CFH/FHL1-binding proteins, such as CspA, ErpA as well as ErpP (Coleman et al. 2008). Survival of this particular *B. burgdorferi* isolate in human serum led these authors to speculate that CspZ is dispensable for complement resistance of *borreliae*.

4.2.1.3 Complement Regulator-Acquiring Surface Proteins 3 (ErpP)

The surface-exposed 19.8-kDa ErpP protein (in the literature also referred to as BbCRASP-3 or BBN38) belongs to the OspE/F paralogous protein family PF162 and is encoded by the *erpP* gene that is organized in a bi-cistronic locus on the circular plasmid cp32-9 of *B. burgdorferi* B31 (Table 4.1) (Fraser et al. 1997; Casjens et al. 1997, 2000; Stevenson et al. 1996, 2001). All *B. burgdorferi* strains examined so far carry multiple cp32 plasmids, each of which contains a unique mono- or bi-cistronic *erp* locus (Stevenson and Miller 2003). Associated *erpP* loci were identified in *B. burgdorferi* strains N40 (*p21*), 297 (*p21*), Sh-2-82 (*erp47*), and BL206 (*erpPQ*) (Stevenson and Miller 2003; Brissette et al. 2008). Hence, spirochetes are able to produce a variety of distinct Erp proteins sharing between 16 and 100% amino acid sequence identity. In addition, numerous related cp32 plasmids were also found in other Lyme disease-associated spirochetes, including *B. afzelii*, *B. garinii*, *B. bavariensis*, *B. spielmanii,* and *B. lusitaniae* (Brissette et al. 2008; Stevenson et al. 2006; P. Kraiczy unpublished data).

Recombinant ErpP protein binds human CFH via the C-terminal SCR20 domain (Kraiczy et al. 2001a). In addition, ErpP binds to host complement regulators CFHR1, CFHR2, and CFHR5 through their C-terminal SCR domain(s) (which are >90% identical to the C-terminal SCR20 domain of CFH) (Haupt et al. 2007; Siegel et al. 2010). Amino (N)- and carboxyl (C)-terminally truncated ErpP molecules abrogate binding of CFH, suggesting that discontinuous or conformational domains are required for binding of CFH (Kraiczy et al. 2003; Metts et al. 2003). Unexpectedly, spirochetes producing ErpP on the outer surface bind complement regulators CFHR1, CFHR2, and CFHR5 but not CFH, suggesting that the binding properties of native ErpP differ from the purified molecule (Siegel et al. 2010). Other membrane proteins proximate to ErpP could impair possible interactions with the large CFH protein and thus permit binding of the smaller CFHR1, CFHR2, and CFHR5 molecules. Furthermore, several lines of evidence indicate that ErpP as well as other Erp proteins form homodimers on the spirochete surface as demonstrated for CspA (Siegel et al. 2010; Cordes et al. 2005; Stevenson et al. 2006). Functionally, ErpP may contribute to immune evasion, but not exclusively. Heterologous production of ErpP in a *B. garinii* strain lacking CRASPs did not protect against complement-mediating bacteriolysis suggesting that ErpP may act in concert with CspA and/or CspZ (Siegel et al. 2010).

ErpP, ErpC, and ErpA (see below) are produced during mammalian infection and are repressed during tick colonization (Akins et al. 1995; Hefty et al. 2001; McDowell et al. 2001a, b; Miller and Stevenson 2006; Miller et al. 2003; Stevenson et al. 1995, 1998; Wallich et al. 1995). Throughout mammalian infection, spirochetes continue to synthesize Erp proteins although expression levels vary significantly (Miller and Stevenson 2006; Miller et al. 2003).

Host adaptive immune systems produce robust antibody responses against these lipoproteins (Akins et al. 1995; Miller et al. 2003; Stevenson et al. 1998;

Wallich et al. 1995; Lam et al. 1994; Nguyen et al. 1994). However, Erp-specific antibodies fail to effectively protect mice from spirochetal infection (Nguyen et al. 1994; Hefty et al. 2002).

4.2.1.4 Complement Regulator-Acquiring Surface Protein 4 (ErpC)

The 18.5 kDa ErpC surface protein (in the literature also referred to as BbCRASP-4) also belongs to the OspE/F paralogous protein family PF162 (Table 4.1). The *erpC* gene is located on the circular cp32-2 plasmid of *B. burgdorferi* B31 and organized in a bi-cistronic locus together with the *erpD* gene (Casjens et al. 1997, 2000; Stevenson et al. 1996, 2001). Associated genes sharing a high degree of homology with *erpC* have also been identified in *B. burgdorferi* strains BL206 (*erpC*), N40 (*ospE*), Sh-2-82 (*erp43*), and 297 (*elpA2*) (Stevenson et al. 2001; Stevenson and Miller 2003; Brissette et al. 2008).

Only fragmentary data on ErpC expression of spirochetes throughout the enzootic cycle are currently available; however, the conserved 5′ region of *erpC* led to the conclusion that *erpC* transcription is controlled by the same regulatory factors as for *erpP* or *erpA/I/N* (see below). Sera obtained from infected animals and Lyme disease patients show that ErpC elicits a robust antibody response during disseminated infection indicating that ErpC might be expressed by spirochetes throughout the mammalian infection (Stevenson et al. 1998). Similarly, specific antibodies against the ErpC homolog of strain N40 (OspE) have been detected in serum samples of tick-infected mice and of Lyme disease patients (Das et al. 1997). Contradictory data have been reported for the ErpC homolog of strain 297 (ElpA2) indicating that tick-infected mice fail to produce specific antibodies during long-term infections (Hefty et al. 2001). However, host-adapted spirochetes cultivated within dialysis membrane chambers implanted into the peritoneal cavities of rats revealed that regulation of the *elpA2* gene is enhanced at elevated temperature suggesting that this protein might be expressed at the spirochetal outer surface during mammalian infection.

Unlike ErpP and ErpA, recombinant ErpC binds to CFHR1 with higher affinity as compared to CFH (Haupt et al. 2007). Further studies revealed that CRASP-deficient *B. garinii* cells producing ErpC on their surface did bind CFHR1 and CFHR2 but not CFH (Hammerschmidt et al. 2012). C-terminal truncations of ErpC homologs of *B. burgdorferi* strains N40 and LW2 as well as *B. afzelii* strain FEM1-D15 completely abolished CFH binding, indicating that this particular region is relevant for binding of complement regulators (Kraiczy et al. 2003). Finally, additional discontinuous regions or higher-order structural determinants might participate in binding of these host proteins to ErpC. Further investigation is needed to precisely identify the ErpC domains associated with direct interaction with human complement regulators. Conversely, CFH binds ErpC via its C-terminal domain SCR20 (Kraiczy et al. 2001a). Sequence analysis showed high homology of SCR20 of CFH to the C-terminal SCR domains of CFHR1, CFHR2, and CFHR5, indicating that ErpC binds to these different complement regulators via their C-terminal region.

Finally, heterologous production of ErpC in a serum-sensitive strain failed to confer complement resistance (Hammerschmidt et al. 2012) indicating that Erp plays, if at all, a subordinate role in facilitating escape from complement-mediated killing.

4.2.1.5 Complement Regulator-Acquiring Surface Protein 5 (ErpA)

Alternative names occasionally found in publications for the 17.7 kDa ErpA protein include BbCRASP-5, ErpI, ErpN, BBP38, and BBL39. Type strain B31 harbors three identical *erpA* genes located on cp32-1, cp32-5, and cp32-8 (Stevenson et al. 2001; Stevenson 2001). This strong redundancy among the *erp* genes has also been found in other *B. burgdorferi* isolates. In addition, ErpA orthologs could also be identified in other borrelial genospecies and thus have been grouped into the Erp/OspE paralogous protein family PF162.

Erp proteins show a high content of the charged amino acids lysine and glutamate. Lysine residues are located in five putative binding regions and are proposed to be required for the interaction of OspE with human CFH (Alitalo et al. 2004, 2005). Truncations of the N- and C-terminus of ErpA completely disrupt binding of CFH, suggesting that linear, discontinuous, and/or higher-ordered elements including coiled-coil structures constitute the CFH-binding site (Alitalo et al. 2004, 2005; Kraiczy et al. 2003; Metts et al. 2003; McDowell et al. 2004). The contribution of individual amino acid residues in binding of CFH and CFHR1 has been clearly demonstrated for the ErpA/C/P homolog Erp63 of *B. spielmanii* strain TIsar3 (Seling et al. 2010). Exchange of a single amino acid residue completely abolished the ability of this Erp to bind CFH and CFHR1. However, the ability of Erp63 to bind the host-derived plasminogen was not affected, suggesting that the molecular architecture of ErpA was not altered and that distinct amino acid residue(s) directly interact with CFH and CFHR1. Participation of numerous charged residues (lysine or histidine) in binding of different complement regulators and the fact that heparin (a polyanionic molecule) inhibits binding of ErpA as well as ErpP to CFH strongly implicate that binding is critically influenced by ionic interactions (Alitalo et al. 2004; Kraiczy et al. 2003; Seling et al. 2010). In contrast, *Borrelia* strains harboring CRASP encoding genes *erpA*, *erpC,* and *erpP* fail to bind CFH to their surface. Instead, they interact with complement regulators CFHR1, CFHR2, and CFHR5 (Siegel et al. 2010). Overall, these findings warrant continuative investigation of Erp protein architecture and protein–protein interactions to define the functions of these ubiquitous, infection-associated borrelial outer surface proteins.

The function of these members of the Erp protein family is still not well understood. Exclusive expression of ErpA on the spirochetal surface did not confer complement resistance (Siegel et al. 2010; Hartmann et al. 2006; Brooks et al. 2005). However, concurrent expression of ErpA and ErpP in a CRASP-deficient *B. burgdorferi* strain significantly increased spirochetal resistance against human complement, suggesting that in combination both proteins exhibit a synergistic effect on complement resistance (Kenedy and Akins 2011). Other members of the Erp protein family, e.g., ErpX and ErpY of type strain B31 as well as Erp proteins of

B. lusitaniae similarly bind CFH. Currently, it is not clear whether those interactions are of biological relevance and play a role in immune evasion of *borreliae* against innate immunity (Dieterich et al. 2010; Stevenson 2002). It has been shown that *B. valaisiana*, *B. andersonii*, *B. bissettii*, *B. turdi*, *B. tanukii* as well as *B. japonica* produce CFH-interacting proteins (Bhide et al. 2009; McDowell et al. 2003a). Apparently, some of these CFH-binding proteins belong to the Erp protein family as demonstrated by cross-reactivity with anti-OspE sera (McDowell et al. 2003a).

4.2.1.6 Other Complement-Binding Proteins of Lyme Disease Spirochetes

C4Bp, the main fluid-phase regulator of the classical pathway, has recently been demonstrated to bind to the surface of serum-resistant *B. burgdorferi* and *B. afzelii* strains as well as serum-sensitive *B. garinii* strains (Pietikainen et al. 2010). In this study, a putative C4Bp-binding protein of approximately 43 kDa, tentatively designated as p43 could be identified in serum-resistant *B. burgdorferi* strain B31 and serum-sensitive *B. garinii* strains g46 and g50. Apparently, binding of C4Bp did not completely protect *borreliae* from the destructive attack of complement, thus the biological impact of C4Bp binding for evading complement-mediated lysis remains to be determined.

4.2.2 Complement-Binding Proteins of Relapsing Fever Spirochetes

4.2.2.1 *Borrelia hermsii*-Associated Complement Regulator-Acquiring Surface Proteins (FhbA and BhCRASP-1)

Two distinct CFH-binding proteins *of B. hermsii* have been identified, termed FhbA and BhCRASP-1 (Rossmann et al. 2007; Hovis et al. 2004; McDowell et al. 2003b). FhbA of *B. hermsii* strain YOR specifically binds both CFH and FHL1 via their SCR domains SCR1–7 and SCR16–20 (Hovis et al. 2004) (Table 4.2). In contrast, CFH binding to BhCRASP-1 of *B. hermsii* strain HS1 is exclusively associated with the C-terminal domain SCR20 of CFH (Rossmann et al. 2007). Furthermore, BhCRASP-1 also binds to CFHR1 but not to FHL1. Thus, BhCRASP-1 and FhbA clearly display distinct biological activities in that both show similar binding potential for CFH, but different capacities to interact with FHL1 or CFHR1.

CFH associated with BhCRASP-1 maintains its regulatory activity and controls C3b deposition and C3-convertase activity (Rossmann et al. 2007). BhCRASP-1-bound CFH was shown to be more efficient in mediating C3b conversion than CFH bound to either CspA or ErpP of Lyme disease spirochetes under similar conditions. Furthermore, CFH bound to CspA is up to ten times more efficient in factor I-mediated C3b conversion when compared to ErpP (Kraiczy et al. 2003, 2004a).

Table 4.2 Characteristics of complement-binding proteins of relapsing fever spirochetes

Complement binding protein	Genospecies	Strain	Synonyms/ other designations	Gene name/ORF	Interaction with complement regulator(s)	Binding capacity	Binding region of host protein	CRASP interacting domain	Imparts complement resistance	References
BhCRASP-1	*B. hermsii*	HS1	None	*cspA*	CFH CFHR1	++ ++	SCR20	C-terminus	Yes	(Rossmann et al. 2007)
FhbA	*B. hermsii*	YOR	FHBP19	*fhbA*	CFH FHL1	++ +	NA	Multiple regions	NA	(Hovis et al. 2004, 2008; McDowell et al. 2003b)
HcpA	*B. recurrentis*	A1 A17	None	*hcpA*	CFH CFHR1	++ ++	SCR19–20 SCR3–5	Multiple regions	Yes	(Grosskinsky et al. 2009)
CihC	*B. recurrentis*	A1	None	*cihC*	C4Bp C1-Inh	++ ++	NA	Central region	Yes	(Grosskinsky et al. 2010)
FHBP28	*B. parkeri*	RML	None		CFH	++	NA	Multiple regions	NA	(McDowell et al. 2003b)
BpcA	*B. parkeri*	RML	None	*bpcA*	CFH CFHR1	++ ++	SCR19–20 SCR3–5	Multiple regions	Yes	(Schott et al. 2010)
CihC	*B. duttonii*	La	BDU_1	*cihC*	C4Bp C1-Inh	++ ++	NA	NA	NA	(Meri et al. 2006; Grosskinsky et al. 2010)

NA not available
++, strong binding; +, weak binding

Thus, acquisition of CFH molecules by surface-exposed BhCRASP-1 leads to enhanced complement-regulatory activity, a process that would allow relapsing fever spirochetes to evade recognition and clearance by the innate immune system. Production of BhCRASP-1 of relapsing fever *borreliae* in serum-sensitive Lyme disease spirochete *B. burgdorferi* B313 results in an increased resistance of the parental strain to complement-mediated killing, suggesting the involvement of BhCRASP-1 in immune evasion of *B. hermsii* in vivo (Rossmann et al. 2007). Deletions of either the N-terminal (fragment spanning residues 51–185) or the C-terminal (fragment spanning residues 21–173) portion of BhCRASP-1 completely abrogated CFH binding, indicating that the CFH-binding site of BhCRASP-1 consists of a higher ordered structure rather than a contiguous linear domain. In contrast, Hovis and colleagues proposed that determinants of the C-terminal domain of FhbA excluding the ten-amino acid C-terminal tail are important in CFH/FHL1 binding (Hovis et al. 2006b, 2008). In this context, the recent identification of two genetically distinct groups of *B. hermsii* is of particular interest (Porcella et al. 2005). Together with the proposition of two clusters of CFH-binding proteins found in the *B. hermsii* genome, the combined studies thus propose a differential association of FhbA and BhCRASP-1 within these two subgroups.

4.2.2.2 *Borrelia parkeri* Complement Regulator-Acquiring Surface Protein (BpcA)

B. parkeri and *B. turicatae* represent closely related spirochetes in North America that are suspected to cause tick-borne relapsing fever in humans (Dworkin et al. 2002). The 18-kDa outer surface protein BpcA (*B. parkeri c*omplement regulator-binding protein A) of *B. parkeri* binds to the human complement regulators CFH and CFHR1 (Table 4.2) (Schott et al. 2010). The bacteria-bound CFH maintained its ability to downregulate complement activation. Adsorption experiments with human plasma and live *B. parkeri* organisms demonstrated that binding of CFH to BpcA is associated with its C-terminal domains SCR19–20 indicating that selective binding of CFH to BpcA occurs in vivo and thus might be of relevance for bacterial virulence. Acquisition of CFH molecules by surface-exposed BpcA results in enhanced complement-regulatory activity by which relapsing fever spirochetes can evade clearance by the immune system of the mammalian host. In addition, *B. burgdorferi* strain B313, deficient in CFH-binding CRASP proteins, resists complement-mediated killing upon transformation with *bpcA* strongly, suggesting its involvement in immune avoidance (Schott et al. 2010).

BpcA with either N-terminal or C-terminal truncations led to complete loss of CFH binding, suggesting that conformational determinants rather than contiguous linear structural elements are required for the interaction with CFH. This finding is reminiscent of the CFH-binding potential as reported for *B. hermsii* strain HS1 (Rossmann et al. 2007). Interestingly, the homologous protein BtcA of the closely related spirochete *B. turicatae* failed to bind to CFH and CFHR1. Moreover, the observed difference in the coiled-coil elements that were predicted in the BpcA and BtcA sequences in combination with the loss of the CFH/CFHR1 binding site in BtcA suggests an important role of these structures for their differential interaction with complement regulators of various species. The importance of coiled-coil elements in the formation of a putative CFH-binding site has also been hypothesized for the *B. burgdorferi* complement regulator binding protein, OspE (McDowell et al. 2004).

4.2.2.3 The Multifunctional CFH-Binding HcpA Protein of *Borrelia recurrentis*

B. recurrentis is the etiologic agent of louse-borne relapsing fever and is transmitted to humans by the infected body louse *Pediculus humanus*. *B. recurrentis* spirochetes produce a multifunctional 17-kDa outer surface lipoprotein, termed HcpA (*H*uman *c*omplement regulator and *p*lasminogen-binding protein A), that specifically interacts with CFH and/or CFHR1 (Table 4.2) (Grosskinsky et al. 2009). The CFH-binding site of HcpA was determined by using mutants with either N-terminal (HcpA48–175, HcpA78–175) or C-terminal (HcpA18–145, HcpA18–115) truncations. Independent of the location and length of the truncations, all mutants of HcpA caused complete abrogation of CFH binding, indicating that structural determinants required for CFH binding are defined by conformation rather than contiguous linear elements. This finding is reminiscent of the observed CFH-binding capabilities for the CspA and ErpP protein of *B. burgdorferi* (Kraiczy et al. 2003, 2004a; Hovis et al. 2006a).

Plasma adsorption experiments and surface plasmon resonance analyses revealed that binding of CFH to HcpA is exclusively associated with its C-terminal domain, SCR20. This finding is further substantiated by the fact that HcpA bound CFHR1 whose C-terminal domains are an almost identical to SCR19–20 of CFH. Notably, HcpA is also capable of binding CFHR2, whose function remains yet to be determined (Zipfel et al. 2002). HcpA-bound CFH retains its regulatory capacity and controls both C3b deposition and C3-convertase activity, resulting in enhanced complement regulatory activity. This process is expected to increase resistance of louse-borne relapsing fever spirochetes to complement attack, an assumption supported by findings with *B. hermsii* (Rossmann et al. 2007).

B. recurrentis also binds plasminogen via HcpA and upon processing to enzymatically active plasmin by human uPA (urokinase-type plasminogen activator), the surface-bound protease was shown to degrade the physiological substrate fibrinogen (Grosskinsky et al. 2009). *Staphylococcus aureus* employs host plasmin to proteolytically cleave both surface-bound IgG as well as C3b, and thereby confers resistance to innate and adaptive immune defense processes (Rooijakkers et al. 2005). HcpA-bound plasmin leads to a decrease of C3b molecules on the surface of *B. recurrentis* and thus may represent a novel immune escape mechanism, by which spirochetes interfere with complement deposition on their cell surface and moreover could overcome phagocytosis. In line with this assumption, *hcpA*-transformed *B. burgdorferi* B313 cells exhibit a significant reduction of C3b molecules deposited on their surface when compared to the parental *B. burgdorferi* B313 strain (Grosskinsky et al. 2009). Modulation of opsonic molecules such as C3b seems an elaborate strategy for bacterial survival and may account for the extraordinary virulence of *B. recurrentis*. Although the biological significance of HcpA expression by *B. recurrentis* has still to be elucidated, the present finding that the *B. burgdorferi* B313 mutant, deficient in CFH-binding proteins, resists complement-mediated killing after transformation with an *hcpA* containing shuttle vector strongly suggests the involvement of HcpA in immune evasion of *B. recurrentis*.

4.2.2.4 Outer Membrane Lipoprotein of *Borrelia recurrentis* Acquiring Complement Regulators C4Bp and C1-Inh (CihC)

B. recurrentis produces a 40 kDa surface receptor, termed CihC (*C*omplement *inh*ibition via *C*4Bp), that specifically binds C4Bp and C1-Inh, two major serum-derived inhibitors of the lectin and classical complement pathways (Table 4.2). Genetic and molecular analyses revealed that CihC of *B. recurrentis* is a potential lipoprotein. A highly homologous receptor for C4Bp was also found in the African tick-borne relapsing fever spirochete *B. duttonii* (Grosskinsky et al. 2010; Lescot et al. 2008). Upon binding to the pathogen's surface, C4Bp retained its cofactor activity for factor I-mediated C4b inactivation. The correlation between serum resistance and cell surface binding of functionally active C4Bp has also been reported before for relapsing fever spirochetes (Meri et al. 2006). Southern blot analysis and BLASTN searches on databases revealed a putative homolog of *cihC* only in *B. duttonii* but not in other spirochetal species, suggesting that the gene which encodes this C4Bp receptor is unique to these two *Borrelia* species. The *cihC* gene was localized to a 190 kbp linear plasmid adjacent to the *hcpA* gene encoding the CFH-binding receptor (Grosskinsky et al. 2010). Similarly, the *B. hermsii* gene that encodes the CFH-binding protein FhbA maps to the large linear plasmid of 220 kbp (Hovis et al. 2004).

C4Bp and C1-Inh binding was not abrogated by N-terminal (amino acid residues 20–121) or C-terminal (amino acid residues 191–356) truncations indicating that both C4Bp and C1-Inh bind to a central region of CihC. Moreover, C4Bp binding is independent of ionic strength, suggesting a hydrophobic interaction between CihC and its ligand. Similar findings have been reported before for other human pathogenic microorganisms. For example, interaction of the *Yersinia enterocolitica* Ail receptor with C4Bp was also found to be less sensitive to salt (Kirjavainen et al. 2008). Moreover, C4Bp receptors like Por1A of *Neisseria gonorrhoeae* (Ram et al. 2001a, b), UspA1/2 of *Moraxella catarrhalis* (Nordstrom et al. 2004), OmpA of *Escherichia. coli* (Prasadarao et al. 2002), and the M-proteins of *Streptococcus pyogenes* bind C4Bp in a nonionic fashion.

In addition to the observed anti-complement activity, *B. recurrentis*-exposed C4Bp may exhibit another biological activity relevant for spirochetal serum resistance as indicated by the fact that C4Bp circulates in plasma as a complex with protein S that in turn binds to negatively charged phospholipids on membranes (Dahlback and Stenflo 1981; Schwalbe et al. 1990a, b). Thus, C4Bp potentially promotes adhesion and subsequently hematogenous dissemination by simultaneously binding to

B. recurrentis and endothelial cells. This assumption is supported by the recent observation that the related fibronectin and glycosaminoglycan binding protein, BBK32, of *B. burgdorferi* mediates endothelial interactions in vivo, thereby facilitating microvascular interactions (Norman et al. 2008). Similarly, CihC of *B. recurrentis* and *B. duttonii* were shown to bind fibronectin and thus might be involved in the dissemination process of relapsing fever spirochetes (Grosskinsky et al. 2010).

The assumption that CihC of *B. recurrentis* and probably also CihC/BDU_1 of *B. duttonii* are critically involved in their escape from complement-mediated lysis is supported by the finding that ectopic expression of CihC in the serum-sensitive *B. burgdorferi* strain B313 led to a significant increase in resistance to complement-mediated lysis. Moreover, binding of C1-Inh, the major inhibitor of several pathways of inflammation in humans, to CihC could be observed (Grosskinsky et al. 2010). However, the actual role of CihC in the pathogenesis of louse-borne relapsing fever will only be elucidated by in vivo studies in a relevant mouse model (Larsson et al. 2009). Complement resistance in a *cihC*-transformed *B. burgdorferi* strain was detected in the presence of non-immune factor B-depleted human serum, indicating that the lectin/classical pathway of complement activation might be triggered by *Borrelia* structures other than specific antibodies. Thus, C1q and the C1 complex may bind to the surface of *B. recurrentis* in the absence of specific antibodies. Moreover, recognition molecules specific for the lectin pathway (i.e., MBLs and ficolins) were shown to bind to borrelial carbohydrates and activate MASPs (Ben-Menachem et al. 2003; Ma et al. 2009; Stubs et al. 2009; Fujita et al. 2004). MASP-2 is the enzyme component that, like C1s in the classical pathway, cleaves the complement components C4 and C2 to form the C3 convertase C4bC2a, common to the activation of both the lectin and classical pathways. Together with the fact that *B. recurrentis* also produces HcpA, these findings suggest that the potential of louse-borne relapsing fever spirochetes to interfere with both the classical and alternative pathways contributes to their high resistance and pathogenicity in humans.

4.3 Future Directions

The ubiquity of complement-binding proteins among Lyme disease and relapsing fever spirochetes points toward important clues in the natural infectious cycle of these emerging pathogens. The strong affinities and binding capacities of the functionally related spirochetal molecules for diverse human complement regulatory proteins strongly support their hypothesized roles in immune evasion. Although the three-dimensional structure of a single CRASP molecule has been determined and structural determinants involved in interactions of CRASPs with complement regulators have been identified, it remains to be determined whether these molecules increase complement resistance in humans. Continued studies of borrelial CRASP molecules will further clarify their role(s) in spirochetal infections and the pathogenesis of Lyme disease as well as relapsing fever. Likewise, these investigations will be helpful to provide insight into the immune evasion mechanisms of many other infectious microorganisms.

Acknowledgments We want to thank all of our colleagues whose research has contributed to an increased understanding of the role of borrelial complement-binding proteins in immune evasion of these emerging pathogens. Many thanks to Brian Stevenson for critical review of the manuscript and fruitful discussions, and Arno Koenigs for help with the graphics. Research in the authors' labs was supported by the Deutsche Forschungsgemeinschaft (Kr3383/1-2, Wa533/7-1, and Wa533/8-1).

References

Akins DR, Porcella SF, Popova TG, Shevchenko D, Baker SI, Li M, Norgard MV, Radolf JD (1995) Evidence for in vivo but not in vitro expression of a *Borrelia burgdorferi* outer surface protein F (OspF) homologue. Mol Microbiol 18(3):507–520

Alitalo A, Meri T, Ramo L, Jokiranta TS, Heikkila T, Seppala IJT, Oksi J, Viljanen M, Meri S (2001) Complement evasion by *Borrelia burgdorferi*: serum-resistant strains promote C3b inactivation. Infect Immun 69(6):3685–3691

Alitalo A, Meri T, Chen T, Lankinen H, Cheng Z-Z, Jokiranta TS, Seppala IJT, Lahdenne P, Hefty PS, Akins DR et al (2004) Lysine-dependent multipoint binding of the *Borrelia burgdorferi* virulence factor outer surface protein E to the C terminus of factor H. J Immunol 172(10):6195–6201

Alitalo A, Meri T, Comstedt P, Jeffery L, Tornberg J, Strandin T, Lankinen H, Bergström S, Cinco M, Vuppala SR et al (2005) Expression of complement factor H binding immunoevasion proteins in *Borrelia garinii* isolated from patients with neuroborreliosis. Eur J Immunol 35(10):3043–3053

Ben-Menachem G, Kubler-Kielb J, Coxon B, Yergey A, Schneerson R (2003) A newly discovered cholesteryl galactoside from *Borrelia burgdorferi*. Proc Natl Acad Sci USA 100(13):7913–7918

Bhide MR, Escudero R, Camafeita E, Gil H, Jado I, Anda P (2009) Complement factor H binding by different Lyme disease and relapsing fever *Borrelia* in animals and human. BMC Res Notes 2:134

Blom AM (2002) Structural and functional studies of complement inhibitor C4b-binding protein. Biochem Soc Trans 30(Pt 6):978–982

Brissette CA, Cooley AE, Burns LH, Riley SP, Verma A, Woodman ME, Bykowski T, Stevenson B (2008) Lyme borreliosis spirochete Erp proteins, their known host ligands, and potential roles in mammalian infection. Int J Med Microbiol 298(Suppl 1):257–267

Brissette CA, Haupt K, Barthel D, Cooley AE, Bowman A, Skerka C, Wallich R, Zipfel PF, Kraiczy P, Stevenson B (2009) *Borrelia burgdorferi* infection-associated surface proteins ErpP, ErpA, and ErpC bind human plasminogen. Infect Immun 77(1):300–306

Brooks CS, Vuppala SR, Jett AM, Alitalo A, Meri S, Akins DR (2005) Complement regulator-acquiring surface protein 1 imparts resistance to human serum in *Borrelia burgdorferi*. J Immunol 175(5):3299–3308

Bykowski T, Woodman ME, Cooley AE, Brissette CA, Brade V, Wallich R, Kraiczy P, Stevenson B (2007) Coordinated expression of *Borrelia burgdorferi* complement regulator-acquiring surface proteins during the Lyme disease spirochete's mammal-tick infection cycle. Infect Immun 75(9):4227–4236

Casjens, S., R. van Vugt, K. Tilly, P. A. Rosa, and B. Stevenson (1997) Homology throughout the multiple 32-kilobase circular plasmids present in Lyme disease spirochetes. J Bacteriol 179(1):217-227

Casjens S, Palmer N, van Vugt R, Huang WM, Stevenson B, Rosa P, Lathigra R, Sutton G, Peterson J, Dodson RJ et al (2000) A bacterial genome in flux: the twelve linear and nine circular extrachromosomal DNAs in an infectious isolate of the Lyme disease spirochete *Borrelia burgdorferi*. Mol Microbiol 35(3):490–516

Coleman AS, Yang X, Kumar M, Zhang X, Promnares K, Shroder D, Kenedy MR, Anderson JF, Akins DR, Pal U (2008) *Borrelia burgdorferi* complement regulator-acquiring surface protein 2 does not contribute to complement resistance or host infectivity. PLoS One 3(8):3010e

Cordes FS, Roversi P, Kraiczy P, Simon MM, Brade V, Jahraus O, Wallis R, Skerka C, Zipfel PF, Wallich R et al (2005) A novel fold for the factor H-binding protein BbCRASP-1 of *Borrelia burgdorferi*. Nat Struct Mol Biol 12(3):276–277

Cordes FS, Kraiczy P, Roversi P, Simon MM, Brade V, Jahraus O, Wallis R, Goodstadt L, Ponting CP, Skerka C et al (2006) Structure-function mapping of BbCRASP-1, the key complement factor H and FHL-1 binding protein of *Borrelia burgdorferi*. Int J Med Microbiol 296(Suppl 40):177–184

Dahlback B, Stenflo J (1981) High molecular weight complex in human plasma between vitamin K-dependent protein S and complement component C4b-binding protein. Proc Natl Acad Sci USA 78(4):2512–2516

Das S, Barthold SW, Giles SS, Montgomery RR, Telford SR III, Fikrig E (1997) Temporal pattern of *Borrelia burgdorferi* p21 expression in ticks and the mammalian host. J Clin Invest 99(5):987–995

Davis AE III, Mejia P, Lu F (2008) Biological activities of C1 inhibitor. Mol Immunol 45(16): 4057–4063

Dieterich R, Hammerschmidt C, Richter D, Skerka C, Wallich R, Matuschka FR, Zipfel PF, Kraiczy P (2010) Inadequate binding of immune regulator factor H is associated with sensitivity of *Borrelia lusitaniae* to human complement. Infect Immun 78(11):4467–4476

Dworkin MS, Schwan TG, Anderson DE Jr (2002) Tick-borne relapsing fever in North America. Med Clin North Am 86(2):417–433, viii–ix

Estaller C, Koistinen V, Schwaeble W, Dierich MP, Weiss EH (1991) Cloning of the 1.4-kb mRNA species of human complement factor H reveals a novel member of the short consensus repeat family related to the carboxy terminal of the classical 150-kDa molecule. J Immunol 146(9): 3190–3196

Fraser CM, Casjens S, Huang WM, Sutton GG, Clayton R, Lathigra R, White O, Ketchum KA, Dodson R, Hickey EK et al (1997) Genomic sequence of a Lyme disease spirochaete, *Borrelia burgdorferi*. Nature 390(6660):580–586

Fujita T, Nussenzweig V (1979) The role of C4-binding protein and beta 1 H in proteolysis of C4b and C3b. J Exp Med 150(2):267–276

Fujita T, Gigli I, Nussenzweig V (1978) Human C4-binding protein. II. Role in proteolysis of C4b by C3b-inactivator. J Exp Med 148(4):1044–1051

Fujita T, Matsushita M, Endo Y (2004) The lectin-complement pathway—its role in innate immunity and evolution. Immunol Rev 198:185–202

Grosskinsky S, Schott M, Brenner C, Cutler SJ, Kraiczy P, Zipfel PF, Simon MM, Wallich R (2009) *Borrelia recurrentis* employs a novel multifunctional surface protein with anti-complement, anti-opsonic and invasive potential to escape innate immunity. PLoS One 4(3):e4858

Grosskinsky S, Schott M, Brenner C, Cutler SJ, Simon MM, Wallich R (2010) Human complement regulators C4b-binding protein and C1 esterase inhibitor interact with a novel outer surface protein of *Borrelia recurrentis*. PLoS Negl Trop Dis 4(6):e698

Hallström T, Haupt K, Kraiczy P, Hortschansky P, Wallich R, Skerka C, Zipfel PF (2010) Complement regulator-acquiring surface protein 1 of *Borrelia burgdorferi* binds to human bone morphogenic protein 2, several extracellular matrix proteins, and plasminogen. J Infect Dis 202(3):490–498

Hammerschmidt C, Hallström T, Skerka C, Wallich R, Stevenson B, Zipfel PF, Kraiczy P (2012) Contribution of the infection-associated complement regulator-acquiring surface protein 4 (ErpC) to complement resistance of *Borrelia burgdorferi*. Clin Dev Immunol 2012:349657.

Harris CL, Rushmere NK, Morgan BP (1999) Molecular and functional analysis of mouse decay accelerating factor (CD55). Biochem J 341(Pt 3):821–829

Hartmann K, Corvey C, Skerka C, Kirschfink M, Karas M, Brade V, Miller JC, Stevenson B, Wallich R, Zipfel PF et al (2006) Functional characterization of BbCRASP-2, a distinct outer membrane protein of *Borrelia burgdorferi* that binds host complement regulators factor H and FHL-1. Mol Microbiol 61(5):1220–1236

Haupt K, Kraiczy P, Wallich R, Brade V, Skerka C, Zipfel P (2007) Binding of human factor H-related protein 1 to serum-resistant *Borrelia burgdorferi* is mediated by borrelial complement regulator-acquiring surface proteins. J Infect Dis 196(1):124–133

Hefty PS, Jolliff SE, Caimano MJ, Wikel SK, Radolf JD, Akins DR (2001) Regulation of OspE-related, OspF-related, and Elp lipoproteins of *Borrelia burgdorferi* strain 297 by mammalian host-specific signals. Infect Immun 69(6):3618–3627

Hefty PS, Brooks CS, Jett AM, White GL, Wikel SK, Kennedy RC, Akins DR (2002) OspE-related, OspF-related, and Elp lipoproteins are immunogenic in baboons experimentally infected with *Borrelia burgdorferi* and in human lyme disease patients. J Clin Microbiol 40(11):4256–4265

Heinen, S., A. Hartmann, N. Lauer, U. Wiehl, H. M. Dahse, S. Schirmer, K. Gropp, T. Enghardt, R. Wallich, S. Halbich, M. Mihlan, U. Schlotzer-Schrehardt, P. F. Zipfel, and C. Skerka (2009) Factor H-related protein 1 (CFHR-1) inhibits complement C5 convertase activity and terminal complex formation. Blood 114(12):2439–2447

Hellwage J, Meri T, Heikkila T, Alitalo A, Panelius J, Lahdenne P, Seppala IJT, Meri S (2001) The complement regulator factor H binds to the surface protein OspE of *Borrelia burgdorferi*. J Biol Chem 276(11):8427–8435

Herzberger P, Siegel C, Skerka C, Fingerle V, Schulte-Spechtel U, van Dam A, Wilske B, Brade V, Zipfel PF, Wallich R et al (2007) Human pathogenic *Borrelia spielmanii* sp. nov. resists complement-mediated killing by direct binding of immune regulators factor H and factor H-like protein 1. Infect Immun 75(10):4817–4825

Herzberger P, Siegel C, Skerka C, Fingerle V, Schulte-Spechtel U, Wilske B, Brade V, Zipfel PF, Wallich R, Kraiczy P (2009) Identification and characterization of the factor H and FHL-1 binding complement regulator-acquiring surface protein 1 of the Lyme disease spirochete *Borrelia spielmanii* sp. nov. Int J Med Microbiol 299(2):141–154

Hovis KM, McDowell JV, Griffin L, Marconi RT (2004) Identification and characterization of a linear-plasmid-encoded factor H-binding protein (FhbA) of the relapsing fever spirochete *Borrelia hermsii*. J Bacteriol 186(9):2612–2618

Hovis KM, Tran E, Sundy CM, Buckles E, McDowell JV, Marconi RT (2006a) Selective binding of *Borrelia burgdorferi* OspE paralogs to factor H and serum proteins from diverse animals: possible expansion of the role of OspE in Lyme disease pathogenesis. Infect Immun 74(3):1967–1972

Hovis KM, Jones JP, Sadlon T, Raval G, Gordon DL, Marconi RT (2006b) Molecular analyses of the interaction of *Borrelia hermsii* FhbA with the complement regulatory proteins factor H and factor H-like protein 1. Infect Immun 74(4):2007–2014

Hovis KM, Freedman JC, Zhang H, Forbes JL, Marconi RT (2008) Identification of an antiparallel coiled-coil/loop domain required for ligand binding by the *Borrelia hermsii* FhbA protein: additional evidence for the role of FhbA in the host-pathogen interaction. Infect Immun 76(5):2113–2122

Iyer R, Liveris D, Adams A, Nowakowski J, McKenna D, Bittker S, Cooper D, Wormser GP, Schwartz I (2001) Characterization of *Borrelia burgdorferi* isolated from erythema migrans lesions: interrelationship of three molecular typing methods. J Clin Microbiol 39(8):2954–2957

Józsi M, Zipfel PF (2008) Factor H family proteins and human diseases. Trends Immunol 29(8):380–387

Kenedy MR, Akins DR (2011) The OspE-related proteins inhibit complement deposition and enhance serum resistance of *Borrelia burgdorferi*, the Lyme disease spirochete. Infect Immun 79(4):1451–1457

Kenedy MR, Vuppala SR, Siegel C, Kraiczy P, Akins DR (2009) CspA-mediated binding of human factor H inhibits complement deposition and confers serum resistance in *Borrelia burgdorferi*. Infect Immun 77(7):2773–2782

Kirjavainen V, Jarva H, Biedzka-Sarek M, Blom AM, Skurnik M, Meri S (2008) *Yersinia enterocolitica* serum resistance proteins YadA and Ail bind the complement regulator C4b-binding protein. PLoS Pathog 4(8):e1000140

Kisova-Vargova L, Mucha R, Cernanka D, Bhide M (2011) Host-dependent differential expression of factor H binding proteins, their affinity to factor H and complement evasion by Lyme and relapsing fever borreliae. Vet Microbiol 148(2–4):341–347

Kraiczy P, Skerka C, Brade V, Zipfel PF (2001a) Further characterization of complement regulator-acquiring surface proteins of *Borrelia burgdorferi*. Infect Immun 69(12):7800–7809

Kraiczy P, Skerka C, Kirschfink M, Brade V, Zipfel Peter F (2001b) Immune evasion of *Borrelia burgdorferi* by acquisition of human complement regulators FHL-1/reconectin and Factor H. Eur J Immunol 31(6):1674–1684

Kraiczy P, Skerka C, Zipfel PF, Brade V (2002) Complement regulator-acquiring surface proteins of *Borrelia burgdorferi*: a new protein family involved in complement resistance. Wien Klin Wochenschr 114(13–14):568–573

Kraiczy P, Hellwage J, Skerka C, Kirschfink M, Brade V, Zipfel PF, Wallich R (2003) Immune evasion of *Borrelia burgdorferi*: mapping of a complement-inhibitor factor H-binding site of BbCRASP-3, a novel member of the Erp protein family. Eur J Immunol 33(3):697–707

Kraiczy P, Hellwage J, Skerka C, Becker H, Kirschfink M, Simon MM, Brade V, Zipfel PF, Wallich R (2004a) Complement resistance of *Borrelia burgdorferi* correlates with the expression of BbCRASP-1, a novel linear plasmid-encoded surface protein that interacts with human factor H and FHL-1 and is unrelated to Erp proteins. J Biol Chem 279(4):2421–2429

Kraiczy P, Hartmann K, Hellwage J, Skerka C, Kirschfink M, Brade V, Zipfel PF, Wallich R, Stevenson B (2004b) Immunological characterization of the complement regulator factor H-binding CRASP and Erp proteins of *Borrelia burgdorferi*. Int J Med Microbiol 293(37):152–157

Kraiczy P, Rossmann E, Brade V, Simon MM, Skerka C, Zipfel PF, Wallich R (2006) Binding of human complement regulators FHL-1 and factor H to CRASP-1 orthologs of *Borrelia burgdorferi*. Wien Klin Wochenschr 118(21–22):669–676

Kraiczy P, Seling A, Brissette CA, Rossmann E, Hunfeld KP, Bykowski T, Burns LH, Troese MJ, Cooley AE, Miller JC et al (2008) *Borrelia burgdorferi* complement regulator-acquiring surface protein 2 (CspZ) as a serological marker of human Lyme disease. Clin Vaccine Immunol 15(3):484–491

Kraiczy P, Hanssen-Hubner C, Kitiratschky V, Brenner C, Besier S, Brade V, Simon MM, Skerka C, Roversi P, Lea SM et al (2009) Mutational analyses of the BbCRASP-1 protein of *Borrelia burgdorferi* identify residues relevant for the architecture and binding of host complement regulators FHL-1 and factor H. Int J Med Microbiol 299(4):255–268

Lam TT, Nguyen TP, Montgomery RR, Kantor FS, Fikrig E, Flavell RA (1994) Outer surface proteins E and F of *Borrelia burgdorferi*, the agent of Lyme disease. Infect Immun 62(1):290–298

Larsson C, Lundqvist J, van Rooijen N, Bergstrom S (2009) A novel animal model of *Borrelia recurrentis* louse-borne relapsing fever borreliosis using immunodeficient mice. PLoS Negl Trop Dis 3(9):e522

Lederer S, Brenner C, Stehle T, Gern L, Wallich R, Simon MM (2005) Quantitative analysis of *Borrelia burgdorferi* gene expression in naturally (tick) infected mouse strains. Med Microbiol Immunol 194(1–2):81–90

Lescot M, Audic S, Robert C, Nguyen TT, Blanc G, Cutler SJ, Wincker P, Couloux A, Claverie JM, Raoult D et al (2008) The genome of *Borrelia recurrentis*, the agent of deadly louse-borne relapsing fever, is a degraded subset of tick-borne *Borrelia duttonii*. PLoS Genet 4(9):e1000185

Ma YJ, Doni A, Hummelshoj T, Honore C, Bastone A, Mantovani A, Thielens NM, Garred P (2009) Synergy between ficolin-2 and pentraxin 3 boosts innate immune recognition and complement deposition. J Biol Chem 284(41):28263–28275

McDowell JV, Sung SY, Labandeira-Rey M, Skare JT, Marconi RT (2001a) Analysis of mechanisms associated with loss of infectivity of clonal populations of *Borrelia burgdorferi* B31MI. Infect Immun 69(6):3670–3677

McDowell JV, Sung SY, Price G, Marconi RT (2001b) Demonstration of the genetic stability and temporal expression of select members of the lyme disease spirochete OspF protein family during infection in mice. Infect Immun 69(8):4831–4838

McDowell JV, Wolfgang J, Tran E, Metts MS, Hamilton D, Marconi RT (2003a) Comprehensive analysis of the factor H binding capabilities of *Borrelia* species associated with Lyme disease: delineation of two distinct classes of factor H binding proteins. Infect Immun 71(6): 3597–3602

McDowell JV, Tran E, Hamilton D, Wolfgang J, Miller K, Marconi RT (2003b) Analysis of the ability of spirochete species associated with relapsing fever, avian borreliosis, and epizootic bovine abortion to bind factor H and cleave C3b. J Clin Microbiol 41(8):3905–3910
McDowell JV, Wolfgang J, Senty L, Sundy CM, Noto MJ, Marconi RT (2004) Demonstration of the involvement of outer surface protein E coiled-coil structural domains and higher order structural elements in the binding of infection-induced antibody and the complement-regulatory protein, factor H. J Immunol 173(12):7471–7480
McDowell JV, Harlin ME, Rogers EA, Marconi RT (2005) Putative coiled-coil structural elements of the BBA68 protein of Lyme disease spirochetes are required for formation of its factor H binding site. J Bacteriol 187(4):1317–1323
McDowell JV, Hovis KM, Zhang H, Tran E, Lankford J, Marconi RT (2006) Evidence that the BBA68 protein (BbCRASP-1) of the Lyme disease spirochetes does not contribute to factor H-mediated immune evasion in humans and other animals. Infect Immun 74(5):3030–3034
McRae JL, Cowan PJ, Power DA, Mitchelhill KI, Kemp BE, Morgan BP, Murphy BF (2001) Human factor H-related protein 5 (FHR-5). A new complement-associated protein. J Biol Chem 276(9):6747–6754
McRae JL, Duthy TG, Griggs KM, Ormsby RJ, Cowan PJ, Cromer BA, McKinstry WJ, Parker MW, Murphy BF, Gordon DL (2005) Human factor H-related protein 5 has cofactor activity, inhibits C3 convertase activity, binds heparin and C-reactive protein, and associates with lipoprotein. J Immunol 174(10):6250–6256
Meri T, Cutler SJ, Blom AM, Meri S, Jokiranta TS (2006) Relapsing fever spirochetes *Borrelia recurrentis* and *B. duttonii* acquire complement regulators C4b-binding protein and factor H. Infect Immun 74(7):4157–4163
Metts MS, McDowell JV, Theisen M, Hansen PR, Marconi RT (2003) Analysis of the OspE determinants involved in binding of factor H and OspE-targeting antibodies elicited during *Borrelia burgdorferi* infection in mice. Infect Immun 71(6):3587–3596
Miller JC, Stevenson B (2006) *Borrelia burgdorferi erp* genes are expressed at different levels within tissues of chronically infected mammalian hosts. Int J Med Microbiol 296(Suppl 40):185–194
Miller JC, von Lackum K, Babb K, McAlister JD, Stevenson B (2003) Temporal analysis of *Borrelia burgdorferi* Erp protein expression throughout the mammal-tick infectious cycle. Infect Immun 71(12):6943–6952
Nagasawa S, Mizuguchi K, Ichihara C, Koyama J (1982) Limited chymotryptic cleavage of human C4-binding protein: isolation of a carbohydrate-containing core domain and an active fragment. J Biochem 92(4):1329–1332
Nguyen TP, Lam TT, Barthold SW, Telford SR III, Flavell RA, Fikrig E (1994) Partial destruction of *Borrelia burgdorferi* within ticks that engorged on OspE- or OspF-immunized mice. Infect Immun 62(5):2079–2084
Nordstrom T, Blom AM, Forsgren A, Riesbeck K (2004) The emerging pathogen *Moraxella catarrhalis* interacts with complement inhibitor C4b binding protein through ubiquitous surface proteins A1 and A2. J Immunol 173(7):4598–4606
Norman MU, Moriarty TJ, Dresser AR, Millen B, Kubes P, Chaconas G (2008) Molecular mechanisms involved in vascular interactions of the Lyme disease pathogen in a living host. PLoS Pathog 4(10):e1000169
Pietikainen J, Meri T, Blom AM, Meri S (2010) Binding of the complement inhibitor C4b-binding protein to Lyme disease *Borreliae*. Mol Immunol 47(6):1299–1305
Porcella SF, Raffel SJ, Anderson DE Jr, Gilk SD, Bono JL, Schrumpf ME, Schwan TG (2005) Variable tick protein in two genomic groups of the relapsing fever spirochete *Borrelia hermsii* in western North America. Infect Immun 73(10):6647–6658
Prasadarao NV, Blom AM, Villoutreix BO, Linsangan LC (2002) A novel interaction of outer membrane protein A with C4b binding protein mediates serum resistance of *Escherichia coli* K1. J Immunol 169(11):6352–6360
Ram S, Cullinane M, Blom AM, Gulati S, McQuillen DP, Boden R, Monks BG, O'Connell C, Elkins C, Pangburn MK et al (2001a) C4bp binding to porin mediates stable serum resistance of *Neisseria gonorrhoeae*. Int Immunopharmacol 1(3):423–432

Ram S, Cullinane M, Blom AM, Gulati S, McQuillen DP, Monks BG, O'Connell C, Boden R, Elkins C, Pangburn MK et al (2001b) Binding of C4b-binding protein to porin: a molecular mechanism of serum resistance of *Neisseria gonorrhoeae*. J Exp Med 193(3):281–295

Rogers EA, Marconi RT (2007) Delineation of species-specific binding properties of the CspZ protein (BBH06) of Lyme disease spirochetes: evidence for new contributions to the pathogenesis of *Borrelia* spp. Infect Immun 75(11):5272–5281

Rogers EA, Abdunnur SV, McDowell JV, Marconi RT (2009) Comparative analysis of the properties and ligand binding characteristics of CspZ, a factor H binding protein, derived from *Borrelia burgdorferi* isolates of human origin. Infect Immun 77(10):4396–4405

Rooijakkers SH, Ruyken M, Roos A, Daha MR, Presanis JS, Sim RB, van Wamel WJ, van Kessel KP, van Strijp JA (2005) Immune evasion by a staphylococcal complement inhibitor that acts on C3 convertases. Nat Immunol 6(9):920–927

Rossmann E, Kitiratschky V, Hofmann H, Kraiczy P, Simon MM, Wallich R (2006) *Borrelia burgdorferi* complement regulator-acquiring surface protein 1 of the Lyme disease spirochetes is expressed in humans and induces antibody responses restricted to nondenatured structural determinants. Infect Immun 74(12):7024–7028

Rossmann E, Kraiczy P, Herzberger P, Skerka C, Kirschfink M, Simon MM, Zipfel PF, Wallich R (2007) Dual binding specificity of a *Borrelia hermsii*-associated complement regulator-acquiring surface protein for factor H and plasminogen discloses a putative virulence factor of relapsing fever spirochetes. J Immunol 178(11):7292–7301

Scharfstein J, Ferreira A, Gigli I, Nussenzweig V (1978) Human C4-binding protein. I. Isolation and characterization. J Exp Med 148(1):207–222

Schott M, Grosskinsky S, Brenner C, Kraiczy P, Wallich R (2010) Molecular characterization of the interaction of *Borrelia parkeri* and *Borrelia turicatae* with human complement regulators. Infect Immun 78(5):2199–2208

Schutzer SE, Fraser-Liggett CM, Casjens SR, Qiu WG, Dunn JJ, Mongodin EF, Luft BJ (2011) Whole genome sequences of thirteen isolates of *Borrelia burgdorferi*. J Bacteriol 193(4): 1018–1020

Schwalbe R, Dahlback B, Hillarp A, Nelsestuen G (1990a) Assembly of protein S and C4b-binding protein on membranes. J Biol Chem 265(27):16074–16081

Schwalbe RA, Dahlback B, Nelsestuen GL (1990b) Independent association of serum amyloid P component, protein S, and complement C4b with complement C4b-binding protein and subsequent association of the complex with membranes. J Biol Chem 265(35):21749–21757

Seling A, Siegel C, Fingerle V, Jutras BL, Brissette CA, Skerka C, Wallich R, Zipfel PF, Stevenson B, Kraiczy P (2010) Functional characterization of *Borrelia spielmanii* outer surface proteins that interact with distinct members of the human factor H protein family and with plasminogen. Infect Immun 78(1):39–48

Siegel C, Schreiber J, Haupt K, Skerka C, Brade V, Simon MM, Stevenson B, Wallich R, Zipfel PF, Kraiczy P (2008) Deciphering the ligand-binding sites in the *Borrelia burgdorferi* complement regulator-acquiring surface protein 2 required for interactions with the human immune regulators factor H and factor H-like protein 1. J Biol Chem 283(50):34855–34863

Siegel C, Hallström T, Skerka C, Eberhardt H, Uzonyi B, Beckhaus T, Karas M, Wallich R, Stevenson B, Zipfel PF et al (2010) Complement factor H-related proteins CFHR2 and CFHR5 represent novel ligands for the infection-associated CRASP proteins of *Borrelia burgdorferi*. PLoS One 5(10):e13519

Skerka C, Zipfel PF (2008) Complement factor H related proteins in immune diseases. Vaccine 26(Suppl 8):I9–I14

Skerka C, Horstmann RD, Zipfel PF (1991) Molecular cloning of a human serum protein structurally related to complement factor H. J Biol Chem 266(18):12015–12020

Skerka C, Timmann C, Horstmann RD, Zipfel PF (1992) Two additional human serum proteins structurally related to complement factor H. Evidence for a family of factor H-related genes. J Immunol 148(10):3313–3318

Stevenson B (2001) *Borrelia burgdorferi*: a (somewhat) clonal bacterial species. Trends Microbiol 9(10):471–472

Stevenson B (2002) *Borrelia burgdorferi erp* (*ospE*-related) gene sequences remain stable during mammalian infection. Infect Immun 70(9):5307–5311

Stevenson B, Miller JC (2003) Intra- and interbacterial genetic exchange of Lyme disease spirochete *erp* genes generates sequence identity amidst diversity. J Mol Evol 57(3):309–324

Stevenson B, Schwan TG, Rosa PA (1995) Temperature-related differential expression of antigens in the Lyme disease spirochete, *Borrelia burgdorferi*. Infect Immun 63(11):4535–4539

Stevenson B, Tilly K, Rosa PA (1996) A family of genes located on four separate 32-kilobase circular plasmids in *Borrelia burgdorferi* B31. J Bacteriol 178(12):3508–3516

Stevenson B, Bono JL, Schwan TG, Rosa P (1998) *Borrelia burgdorferi* Erp proteins are immunogenic in mammals infected by tick bite, and their synthesis is inducible in cultured bacteria. Infect Immun 66(6):2648–2654

Stevenson B, Zückert WR, Akins DR (2001) Repetition, conservation, and variation: the multiple cp32 plasmids of *Borrelia* species. In: Saier MH, Garcia-Lara J (eds) The spirochetes: molecular and cellular biology. Horizon Press, Oxford, pp 87–100

Stevenson B, El-Hage N, Hines MA, Miller JC, Babb K (2002) Differential binding of host complement inhibitor factor H by *Borrelia burgdorferi* Erp surface proteins: a possible mechanism underlying the expansive host range of Lyme disease spirochetes. Infect Immun 70(2):491–497

Stevenson B, Bykowski T, Cooley AE, Babb K, Miller JC, Woodman ME, von Lakum K, Riley SP (2006) The Lyme disease spirochetes Erp protein family: structure, function, and regulation of gene expression. In: Cabello FC, Godfrey HP, Hulinska D (eds) Molecular biology of spirochetes. Ios Press, Amsterdam, pp 352–372

Stubs G, Fingerle V, Wilske B, Gobel UB, Zahringer U, Schumann RR, Schroder NW (2009) Acylated cholesteryl galactosides are specific antigens of *Borrelia* causing lyme disease and frequently induce antibodies in late stages of disease. J Biol Chem 284(20):13326–13334

van Burgel ND, Kraiczy P, Schuijt TJ, Zipfel PF, van Dam AP (2010) Identification and functional characterisation of complement regulator acquiring surface protein-1 of serum resistant *Borrelia garinii* OspA serotype 4. BMC Microbiol 10:43

von Lackum K, Miller JC, Bykowski T, Riley SP, Woodman ME, Brade V, Kraiczy P, Stevenson B, Wallich R (2005) *Borrelia burgdorferi* regulates expression of complement regulator-acquiring surface protein 1 during the mammal-tick infection cycle. Infect Immun 73(11):7398–7405

Wallich R, Brenner C, Kramer MD, Simon MM (1995) Molecular cloning and immunological characterization of a novel linear-plasmid-encoded gene, pG, of *Borrelia burgdorferi* expressed only in vivo. Infect Immun 63(9):3327–3335

Wallich R, Jahraus O, Stehle T, Tran TT, Brenner C, Hofmann H, Gern L, Simon MM (2003) Artificial-infection protocols allow immunodetection of novel *Borrelia burgdorferi* antigens suitable as vaccine candidates against Lyme disease. Eur J Immunol 33(3):708–719

Wallich R, Pattathu J, Kitiratschky V, Brenner C, Zipfel PF, Brade V, Simon MM, Kraiczy P (2005) Identification and functional characterization of complement regulator-acquiring surface protein 1 of the Lyme disease spirochetes *Borrelia afzelii* and *Borrelia garinii*. Infect Immun 73(4):2351–2359

Walport MJ (2001) Complement—first of two parts. N Engl J Med 344(14):1058–1066

Wormser GP, Brisson D, Liveris D, Hanincova K, Sandigursky S, Nowakowski J, Nadelman RB, Ludin S, Schwartz I (2008) *Borrelia burgdorferi* genotype predicts the capacity for hematogenous dissemination during early Lyme disease. J Infect Dis 198(9):1358–1364

Wywial E, Haven J, Casjens SR, Hernandez YA, Singh S, Mongodin EF, Fraser-Liggett CM, Luft BJ, Schutzer SE, Qiu WG (2009) Fast, adaptive evolution at a bacterial host-resistance locus: the PFam54 gene array in *Borrelia burgdorferi*. Gene 445(1–2):26–37

Zipfel PF (2009) Complement and immune defense: from innate immunity to human diseases. Immunol Lett 126:1–7

Zipfel PF, Skerka C (2009) Complement regulators and inhibitory proteins. Nat Rev Immunol 9(10):729–740

Zipfel PF, Skerka C, Hellwage J, Jokiranta ST, Meri S, Brade V, Kraiczy P, Noris M, Remuzzi G (2002) Factor H family proteins: on complement, microbes and human diseases. Biochem Soc Trans 30(Pt 6):971–978

Chapter 5
Treponema pallidum Repeat (*tpr*) Genes and Antigenic Variation

Lorenzo Giacani and Arturo Centurion-Lara

5.1 Introduction

In the absence of treatment, syphilis is a lifelong infection with clinically manifest stages that alternate with periods of latency. The primary and secondary stages of the disease are highly infectious in that the agent of syphilis, *Treponema pallidum* subsp. *pallidum* (*T. pallidum*), actively proliferates in the lesions that stigmatize early syphilis (such as the primary "chancre", and the secondary mucosal lesions and skin rash). Within these lesions, however, the increasing bacterial burden elicits a robust immune response that culminates with the immune clearance of the great majority of the treponemes and allows spontaneous lesion resolution. Nonetheless, very few treponemes successfully evade the host's immune response and latently persist in the affected individual for decades. Eventually, in approximately one third of the patients, the disease reactivates leading to the dreaded manifestations of tertiary syphilis, such as insanity, blindness, paralysis, or death. How some *T. pallidum* cells can successfully evade the host defenses is a key question mark in syphilis pathogenesis. A variety of mechanisms were proposed over the last three decades, including residence of the pathogen within intracellular or immune privileged niches, or its ability to cloak its surface with host proteins or mucopolysaccharides to disguise as "self" and avoid immune recognition. Although not abandoned, the strength of the above theories was significantly lessened by the discovery of phase variation mechanisms for four hypothetical outer membrane proteins (OMPs) candidates encoded by the *T. pallidum tpr* (*T. pallidum* repeat) gene family, and the demonstration that the variable antigen TprK is involved in treponemal immune selection during experimental infection. Such findings strongly suggest a role for some *tpr* genes in immune evasion and persistence. After one century from the

L. Giacani, Ph.D. (✉) • A. Centurion-Lara, M.D.
University of Washington, Department of Medicine, Division of Allergy and Infectious Diseases, Harborview Medical Center, Seattle, WA 98104, USA
email: giacal@u.washington.edu; acentur@u.washington.edu

M.E. Embers (ed.), *The Pathogenic Spirochetes: Strategies for Evasion of Host Immunity and Persistence*, DOI 10.1007/978-1-4614-5404-5_5,

isolation of the first *T. pallidum* strain in 1912 by Nichols and Hough, progress to elucidate the molecular mechanisms behind *T. pallidum*'s persistence is still hindered by the inability to grow in vitro or genetically manipulate *T. pallidum*, the pathogen's limited viability outside of a host, and an uncommonly fragile cellular ultrastructure. Despite such limitations, much evidence support that immune evasion by the syphilis spirochete is based on sophisticated molecular strategies whose investigation could significantly deepen our understanding of syphilis pathogenesis and also favor the identification of broadly protective vaccines.

5.2 Historical and Epidemiological Notes

The first syphilis epidemics occurred in Europe in 1494 in the context of the invasion of Italy and the siege of Naples perpetrated by the French troops of King Charles VIII and by the army of Ludovico Sforza, Duke of Milan (Oriel 2004; Lobdell 1974). From this center, the disease is thought to have rapidly spread across Europe reaching endemic proportions by the mid-sixteenth century, after only a few decades from its identification (Diamond 1997). The outbreak among the French troops caused the disease initially named "*morbus gallicus*" (latin for "French Disease"), except among the French, that comprehensibly referred to this scourge as the "Neapolitan Disease". The term syphilis was first introduced in 1530 by Girolamo Fracastoro's epic poem in three books "Syphilis Sive Morbus Gallicus" (latin for "Syphilis or The French Disease"), where a shepherd named Syphilus (perhaps a variant spelling of Sipylus, a character in Ovid's Metamorphoses) was coursed with this disease for his blasphemy (Arrizabalaga et al. 1997). Rather than for the first use of the name syphilis, Fracastoro's work is indication that the disease was at that point widely disseminated, justifying the publication of books on the subject, that also contemplated possible remedies against syphilis such as the use of mercury. Although a precise estimate of syphilis incidence in the sixteenth century is not possible, from the nosocomial and mortality records of the time is not inaccurate to conclude that syphilis was a fairly common disease (Kevin 2004). Also, by the time of the appearance of Fracastoro's work, the majority of physicians were already familiar of the venereal nature of the disease (Temkin 1977).

The exact knowledge of when and where the first syphilis epidemic emerged in Europe strongly contrasts with the uncertainty that surrounds the origin of the disease. The most popular hypothesis among the three currently accepted is that syphilis originated in the American continent (New World) and was brought back to European soil (Old World) by Columbus' sailors (Luger 1993). Such theory is based on observations by paleopathologists that at the time of Columbus' expedition, bone damage similar to that caused by late or congenital syphilis could only be found in the New World, but not in the Old World, at least until the late fifteenth century (Rothschild 2003; Rothschild et al. 2000; Radolf and Lukehart 2006). Alternatively, the first syphilis outbreak at the end of the fifteenth century has been imputed to a change in sexual behavior or hygiene standards in the European populations of that time. This theory is however controversial, mostly because of the scarcity of historical records to support it (McGough and Erbelding 2006; Foa 1990). A third theory is that syphilis always existed in northern Europe, either as a non-venereal treponematosis that then evolved into venereal syphilis, or already as the disease we know today, that then spread to southern Europe as the consequence of the Italian war scenario (McGough and Erbelding 2006; Foa 1990). It is however beyond the scope of this chapter to provide a detailed overview of the theories on the origin of syphilis for which more exhaustive treatises can be consulted (Foa 1990). It is worth mentioning however that despite serological testing (Wasserman test, developed in 1906) and treatments (Salvarsan in 1912, and penicillin in the1940s) became available in the first half of the twentieth century, a report of the Institute of Medicine in 1997 on sexually transmitted diseases (STDs) morbidity identified syphilis and other STDs as "hidden epidemics of tremendous health and economic consequence in the United States". Of the same period is the World Health Organization (WHO) estimate of an incidence of 12 million new cases per year in adults, and a prevalence of at least 25 million cases worldwide (Gerbase et al. 1998). These data emphasize how today syphilis is still a burden for global health. The incidence and prevalence data above precede the dramatic increase in syphilis cases witnessed in the U.S., Europe, and China since 2000 (CDC 2007; Renton et al. 1998; Chen et al. 2007). In the U.S. alone the incidence of primary and secondary syphilis more than doubled from 2000 to 2008, particularly among men who have sex with men (MSM), and estimates suggest a continuing increase of ~10% per year (CDC 2007).

5.3 Pathogenesis of Syphilis

Syphilis is a venereal infection that occurs when *T. pallidum* reaches dermal abrasions or mucous membranes of a naïve host, most commonly during sexual intercourse. The infection then progresses with clinically distinct stages (primary, secondary, and tertiary syphilis) and periods of latency.

5.3.1 Primary Syphilis

At the site of entry, a typical lesion called chancre usually appears approximately 3 weeks after exposure to the pathogen, and it is generally accompanied by moderate regional lymphadenopathy (Mindel et al. 1989). The primary chancre then becomes indurated and finally ulcerates, but without any purulent discharge. In men, primary chancres commonly appear on the penis, but can be equally found in the anorectal area (more frequent in MSM) and oral cavity (Mindel et al. 1989; Hourihan et al. 2004; Fiumara and Berg 1974). The labia or cervix are the most common sites in women. Because the chancre is painless, difficult to locate, or sometimes with atypical features, a correct diagnosis of syphilis infection is not uncommonly missed until the manifestations of secondary syphilis appear (Chapel 1978, 1980; Rompalo et al. 2001). The primary chancre then heals spontaneously within 4–6 weeks, but not necessarily before the onset of the symptoms of secondary syphilis (U.S. Department of Health, Education, and Welfare 1968).

5.3.2 Secondary Syphilis

T. pallidum is highly invasive, and contemporaneously to the onset of the primary stage of the disease, the pathogen systemically disseminates from the site of initial contact to virtually every organ of the body (Baughn and Musher 2005). Secondary syphilis usually occurs within 3 months of infection, and is characterized by a protean array of symptoms, including sore throat, muscle aches, malaise, weight loss, and generalized lymphadenopathy. Disseminated mucocutaneous lesions, where *T. pallidum* is actively replicating, are however the most common manifestation of secondary syphilis (Chapel 1980; Hira et al. 1987). Lesions appear frequently on the trunk and proximal extremities, but they can extend to the totality of the body, including palms and soles. Very rarely, lesions might become necrotic, causing a condition normally referred to as "lues maligna" (DiCarlo and Martin 1997). Concomitantly to the secondary rush, about 10% of patients develop condilomata lata and mucous patches. The former are enlarged lesions that tend to appear in warm and moist areas such as the perineum or anus; while the latter are preferentially localized in the oral cavity, tongue, and genital mucosa (Mindel et al. 1989; Chapel 1980). Both condilomata and mucous patches are teeming with treponemes, and therefore highly infectious. Renal, hepatic, neurological, and ocular involvement is not uncommon during secondary syphilis (Chapel 1980; Stokes et al. 1944).

5.3.3 Neurological Involvement

Classically, neurological complications of syphilis have been associated with tertiary disease, but several studies have demonstrated that invasion of the central nervous system (CNS) by *T. pallidum* begins during earlier stages of disease (Marra et al. 2004). Because however CNS invasion by *T. pallidum* resolves in most patients, presence of the spirochete in the cerebrospinal fluid (CSF) alone does not measure the risk of a subject of developing neurosyphilis (Marra et al. 2004; Cameron 2008). Neurosyphilis is currently defined by CSF pleocytosis or elevated protein levels, or as an immunoreactive CSF when tested by VDRL (Venereal Disease Research Laboratory, a screening test for syphilis infection) (Marra et al. 2004). During the early stages of the disease, typical symptoms of neurosyphilis include fever, headache, nausea, vomiting, and stiff neck. Ocular inflammation (uveitis) might occur, as well as photophobia, blurred vision, and hearing loss as the result of cranial nerve involvement (Marra et al. 2004; Hook and Marra 1992).

5.3.4 Latent Syphilis

Manifestations of secondary syphilis usually resolve within 3 months after their appearance, leaving the untreated patient without symptoms ascribable to syphilis infection (Gjestland 1955). Approximately one fourth of the patients entering this stage experience recurrence of secondary manifestations during the first year of latency and therefore, for the first year after infection, patients are considered to have early latent syphilis. When asymptomatic infection prolongs for over 1 year, patients then enter in the late latent stage. Despite this lack of clinical manifestations, serologic testing during the latent stage is positive, but horizontal transmission is unlikely, due to the very limited number of treponemes and the absence of sores. Vertical transmission during pregnancy is however possible (Gjestland 1955). Latent syphilis ends with the appearance of tertiary syphilis manifestations or as the result of antibiotic treatment of the patient.

5.3.5 Tertiary Syphilis

Our current understanding of the clinical picture associated with tertiary syphilis comes from the historic Oslo (Gjestland 1955) and Tuskegee studies (Kampmeier 1964). In the Oslo study, conducted when no antibiotics were available, approximately one-third of individuals with latent syphilis infection developed tertiary syphilis from 20 to 40 years after the initial infection. In 15% of patients with untreated syphilis, progressive inflammation caused gumma (late benign syphilis), a localized form of tissue and bone destruction (Stokes et al. 1944; Handsfield et al. 1983). Cardiovascular syphilis, usually seen as aortic insufficiency or aneurysm, was observed in 10% of untreated patients (Kampmeier 1964; Jackman and Radolf 1989). Symptomatic late neurosyphilis was recognized in 6.5% of untreated patients. In the Tuskegee study, 408 African-American men with untreated latent syphilis in Macon County, Alabama, were followed up prospectively between 1932 and 1972 (Kampmeier 1964). This study provided valuable information about the natural history and late manifestations of the infection, but it also has been righteously questioned because of issues related to ethical standards. Informed consent was not obtained from subjects and antibiotic treatment, although available at the time, was not offered. Late neurosyphilis, presenting as general paresis (rapidly progressive dementia with psychotic features) or tabes dorsalis (a disorder of the spinal cord with sensory ataxia and bowel and bladder dysfunction), appears two to three decades after infection. Symptoms of general paresis include personality changes, emotional instability, memory impairment, hallucinations, and hyperactive reflexes (Merritt et al. 1946; Simon 1994). Tabes dorsalis presents as sensory ataxia in the lower extremities, paresthesia, and sudden-onset vomiting or abdominal pain. Wide-based gait often results from the loss of position and vibratory senses. Trophic lesions in ankles, knees, and hips, commonly known as Charcot's joints, can also occur (Simon 1994).

5.3.6 Congenital Syphilis

In pregnant women with syphilis *T. pallidum* can cross the placenta and reach the fetus causing congenital infection. Although fetal infection can happen at any stage, congenital syphilis is more frequent if pregnancy occurs within the first year of infection (Radolf et al. 1999; Sheffield et al. 2002). Antibiotic treatment of the mother is effective in preventing negative outcomes if administrated within the first two trimesters of pregnancy, but late treatment (in the third trimester) or absence of treatment may result in fetal death, fetal damage, or premature birth of an infected infant. Newborns with congenital syphilis usually are underweight (Mascola et al. 1985) and exposed to opportunistic bacterial infections, hepatitis, and subject to pulmonary hemorrhage that altogether account for approximately 4% of deaths of *T. pallidum*-infected neonates soon after birth (Chapel 1988; Fiumara 1975).

5.4 Etiologic Agent

T. pallidum was first identified as the etiologic agent of syphilis in 1905 by Schaudinn and Hoffmann (Turner and Hollander 1957), but not until 1912 however, the first *T. pallidum* strain was isolated from the cerebrospinal fluid of a patient with neurosyphilis by the U.S. Army Major H.J. Nichols (Nichols and Hough 1913). The agent of syphilis is an obligate human parasite with no known reservoirs in nature. Outside of its host, *T. pallidum* rapidly loses viability and infectious capability. Because currently there is no artificial media or cell culture-based system to support the continuous growth of this pathogen in vitro, *T. pallidum* must be propagated in rabbits to obtain enough organisms for laboratory use; for this purpose, serial passage of the syphilis spirochete in rabbit testes is the most common procedure (Turner and Hollander 1957). Furthermore, because rabbit infection with *T. pallidum* recapitulates very closely, clinically and histologically, the primary and secondary stages of the disease in humans, rabbits are the animal model of choice for syphilis (Sell et al. 1980; Baker-Zander and Sell 1980).

For simplicity, due to its double-membrane envelope, *T. pallidum* is structurally assimilated to conventional Gram-negative bacteria, such as *Escherichia coli* (*E. coli*). Nonetheless, profound structural and biochemical differences characterize the *T. pallidum* envelope when compared to *E. coli*. For example, similarly to other spirochetes, *T. pallidum* flagella are internal (and therefore called endoflagella), namely enclosed in the periplasmic space (Jepsen et al. 1968). *T. pallidum* outer membrane (OM) does not contain lipopolysaccharide (LPS) (Hardy and Levin 1983; Liang et al. 2000), and has an unusual paucity of integral membrane proteins (approximately 1% of the protein content found in the *E. coli* OM) (Radolf 1995; Radolf et al. 1989, 1995). Also, the peptidoglycan layer, which in conventional Gram-negative bacteria is structurally associated with the OM, in *T. pallidum* overlies the cytoplasmic membrane (Hovind-Hougen 1983; Radolf and Norgard 1988). Such unique localization could explain at least partially the unusual fragility of *T. pallidum* envelope. Viability of the syphilis spirochete is in fact easily compromised by mechanical manipulation (i.e., centrifugation and resuspension), as well as detergent treatment (Cox et al. 1992; Radolf et al. 1988; Cunningham et al. 1988). Not surprisingly, despite a century elapsed since the isolation of the Nichols strain, still no *T. pallidum* surface-exposed integral membrane protein has been directly visualized with the exception of Tp326, a recently identified BamA ortholog (Desrosiers et al. 2011). Genetic intractability is another limiting consequence of the inability to grow in vitro this pathogen and, unlike the cultivable spirochetes *Treponema denticola* and *Borrelia burgdorferi*, a system for genetic manipulation of *T. pallidum* is yet to be developed.

5.4.1 T. pallidum Genome Sequence

A significant contribution to our understanding of the *T. pallidum* biology and syphilis pathogenesis came from the release of the genome sequence of the Nichols strain in 1998 by Fraser et al. (Liang et al. 2000). The small size of the *T. pallidum* genome (only 1.13 Mbp coding for 1,041 putative proteins) strongly suggests that during evolution this spirochete has undergone progressive reduction of its genetic information with systematic elimination of many metabolic and biosynthetic functions resulting in a strict dependence on its host as a source of different key groups of nutrients. *T. pallidum* lacks the enzymes for both the tricarboxylic acid cycle and oxidative phosphorylation, leaving glycolysis as the principal biochemical pathway to generate energy. This is likely one of the reasons why *T. pallidum*'s generation time (GT) estimated to be 30–33 h in vivo is markedly slow compared to other Gram-negative bacteria such as *E. coli* (GT is ~20 min) (Cumberland and Turner 1949; Magnuson et al. 1948). The analysis of the *T. pallidum* genome also suggests the pathogen's inability to use alternative carbon sources for energy and to synthesize de novo enzyme cofactors, fatty acids, amino acids, and nucleotides (Liang et al. 2000). To compensate such dearth of metabolic capabilities, *T. pallidum* possesses a wide repertoire of inner membrane transporters thought to be specific for host-derived nutrients (Liang et al. 2000); the transporters in the OM are yet to be identified. Despite being metabolically crippled, *T. pallidum* is highly infectious, capable of persisting in its host despite a robust immune response, and responsible for an impressive variety of symptoms in the affected individuals. Although the *T. pallidum* genome sequence did not reveal any classical bacterial virulence factors, this pathogen has likely retained in its genetic repertoire pathogenesis-associated molecules responsible for the establishment of the infection and persistence of the pathogen in the host.

5.5 Mechanisms of Antigenic Variation

Pathogens vary surface-exposed determinants at the pathogen–host interface mainly to adapt to a particular milieu or to evade a harmful immune response (Barrett et al. 2008; Craig and Scherf 2003). This is likely true also for a highly invasive organism such as

T. pallidum, which not only can invade and adapt to a wide range of micro-environments represented by virtually every organ and tissue in the host, but also establish a chronic infection in spite of eliciting a robust immune response (Stokes et al. 1944; Nell and Hardy 1966; Uhlenhuth 1913a, b). In general, pathogens can modify their molecular make up by exploiting genetic and epigenetic mechanisms (van der Woude and Baumler 2004); genetic mechanisms encompass mutations and recombination events that induce a change in the nucleotide sequence of a surface antigen-encoding gene or its regulatory element, affecting therefore either the primary amino acid sequence of the antigen or its level of expression. For historical reasons however, genetic mechanisms which modify DNA sequences and alter the level of expression of a surface-exposed antigen in an ON-OFF manner to induce antigenic variation are often referred to as phase (or phenotypic) variation (van der Woude and Baumler 2004). On the opposite hand, epigenetic phenomena such as DNA methylation or RNA interference are also known to affect the expression of surface-exposed molecules but without altering their nucleotide sequence.

T. pallidum, initially thought to be a pathogen with a stable genome unable to diversify, has proven otherwise as an organism with remarkable genome plasticity. Several recent publications have expanded our knowledge about genetic diversity in *T. pallidum* (Arrizabalaga et al. 1997; Marra et al. 2010; Matejkova et al. 2008). Genome wide sequence changes include single nucleotide polymorphisms (SNPs) (synonymous/non-synonymous mutations and SNPs in intergenic regions) and indels of various lengths distributed both in coding and intergenic sequences. Marra et al. (2010) effectively underscored the genetic diversity among *T. pallidum* strains by showing that analysis of diversity in only six different *T. pallidum* genes can separate syphilis isolates in 22 different strain types. *T. pallidum* genetic diversity however can only be relevant to adaptation or immune evasion when it has a functional impact on a surface-exposed antigen. Classic surface components that have been well characterized in other bacterial pathogens have not been definitively identified in *T. pallidum*. *T. pallidum* OMPs candidates include the Tp0257 (glycerophosphodiester phosphodiesterase, Gpd/GlpQ) antigen (Desrosiers et al. 2011; Cameron et al. 2000) and the experimentally identified extracellular matrix (ECM) binding proteins Tp0751, Tp0136, Tp0155, and Tp0483 (Cameron et al. 2004; Cameron 2003; Brinkman et al. 2008). A fundamental concept in microbial pathogenesis is the ability to adhere to host surfaces, a necessary step in successful colonization and production of disease (Finlay and Falkow 1997). It is generally accepted that *T. pallidum* must first attach to epithelial cells and extracellular matrix components to gain the “foothold” needed to establish infection, suggesting that expression of adhesion molecules on its surface is key for survival. Despite the effect that changes in these putative surface-exposed antigens can have in the ability of *T. pallidum* to gain access to the host, the lack of sequence or antigenic changes during infection rules these molecules out as means for evading the host immune response.

Antigenic variation has however been described in *T. pallidum* within a family of paralogs called the *tpr* (*T. pallidum* repeat) genes, encoding several putative OM proteins with unknown function. For simplicity, from this point forward, we will refer to the term antigenic variation in *T. pallidum* only to describe sequence changes in the *tprK* gene. These changes arise in single cells during infection as the result of recombination of DNA segments into the expression site coding for the TprK antigen. For the *tprE, F, G, I,* and *J* genes, whose expression is modulated by varying DNA sequences within their promoters, we will use instead the term phase variation.

5.5.1 The Hub of the Antigenic Variation Theory in *T. pallidum*: The *tpr* Genes

Following the elucidation of the Nichols strain genome (Liang et al. 2000), the putative proteins encoded by the *tpr* genes quickly became a major object of investigation in virtue of their likelihood of being virulence factors. Alone, this family of 12 paralogs (named *tprA-tprL*) encompasses ~2% of the small *T. pallidum* genome, suggesting an important role for these genes in an organism that has selectively eliminated vital metabolic functions (Liang et al. 2000; Centurion-Lara et al. 1999). Furthermore, these genes show sequence homology with the major sheath protein (*msp*) gene of the oral pathogen *Treponema denticola* (*T. denticola*), which encodes an abundant and highly immunogenic surface-exposed antigen with adhesin and pore-forming proprieties (Fenno et al. 1996, 1997). Several algorithms predict Nichols' antigens encoded by *tprA, B, C, D, E, F, I, J,* and *K* to be located in the outer membrane of the microorganism (Cox et al. 2010). Also suggestive of surface exposure of some of the Tpr antigens are protection experiments with recombinant TprK, and TprF/TprI peptides showing significant attenuation of lesion development after homologous challenge, although infection was not prevented. (Centurion-Lara et al. 1999; Sun et al. 2004; Giacani et al. 2005a). Recent data obtained by our group (Molini et al. unpublished) show that immune sera raised against those Tpr antigens (Tprs) can opsonize *T. pallidum* and facilitate its ingestion by activated rabbit macrophages, supporting the in silico prediction of an OM localization of these Tprs.

Based on amino acid and nucleotide sequence homology, the antigens encoded by the *tpr* genes are currently organized into three subgroups: Subfamily I (TprC, D, F, and I), Subfamily II (TprE, G, and J), and Subfamily III (TprA, B, H, K, and L) Tprs (Centurion-Lara et al. 1999). Both Subfamily I and II antigens show sequence conservation at their amino- and carboxyl-termini, separated by unique central domains differing in sequence and length (Fig. 5.1). In the Nichols strain, TprC and TprD are identical, and the TprF COOH-terminal region and part of the central domain are absent due to a deletion and a frameshift in the gene open reading frame (ORF) (Centurion-Lara et al. 1999). Comparatively, less homology can be found among Subfamily III Tpr antigens. In the Nichols strain, also *tprA* contains a frameshift at nucleotide 712, in correspondence of a CT dinucleotide repeat deletion which truncates the ORF (Centurion-Lara et al. 1999). In *T. pallidum* strains isolated more recently than Nichols, a considerable diversity emerged among Subfamily I and II Tprs. Some treponemal isolates, for instance, carry an alternative *tprD* allele, named *tprD2*, or *tprD* genes with a small number of SNPs. Similarly, some strains show diversity at the *tprC* locus, while others, as in Nichols, carry two

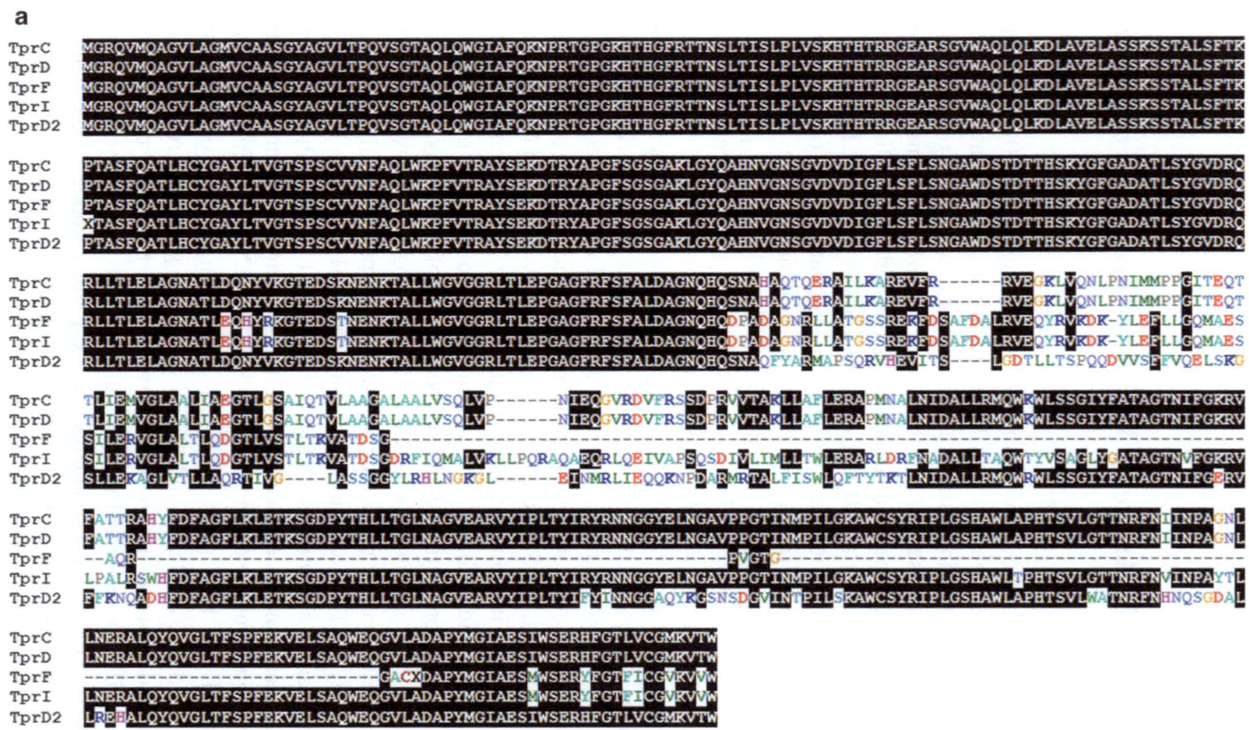

Fig. 5.1 Alignments of predicted amino acid sequences Subfamily I, II, and III *tpr* genes. *Dark shaded* areas indicate sequence conservation among the Tpr proteins. (**a**) Alignment of Subfamily I (Tpr C, D, D2, F, and I). (**b**) Alignment of Subfamily II (Tpr E, G, and J). (**c**) Alignment of Subfamily III (Tpr A1, A2, B, H, K, and L). Alignments were generated using Bioedit (http://www.mbio.ncsu.edu/bioedit/bioedit.html). The Nichols strain genome (Liang et al. 2000) (GenBank accession number AE000520) served as source for the Tpr sequences. The TprD2 sequence (Centurion-Lara et al. 2000) is also available in GenBank (accession number AAF64133)

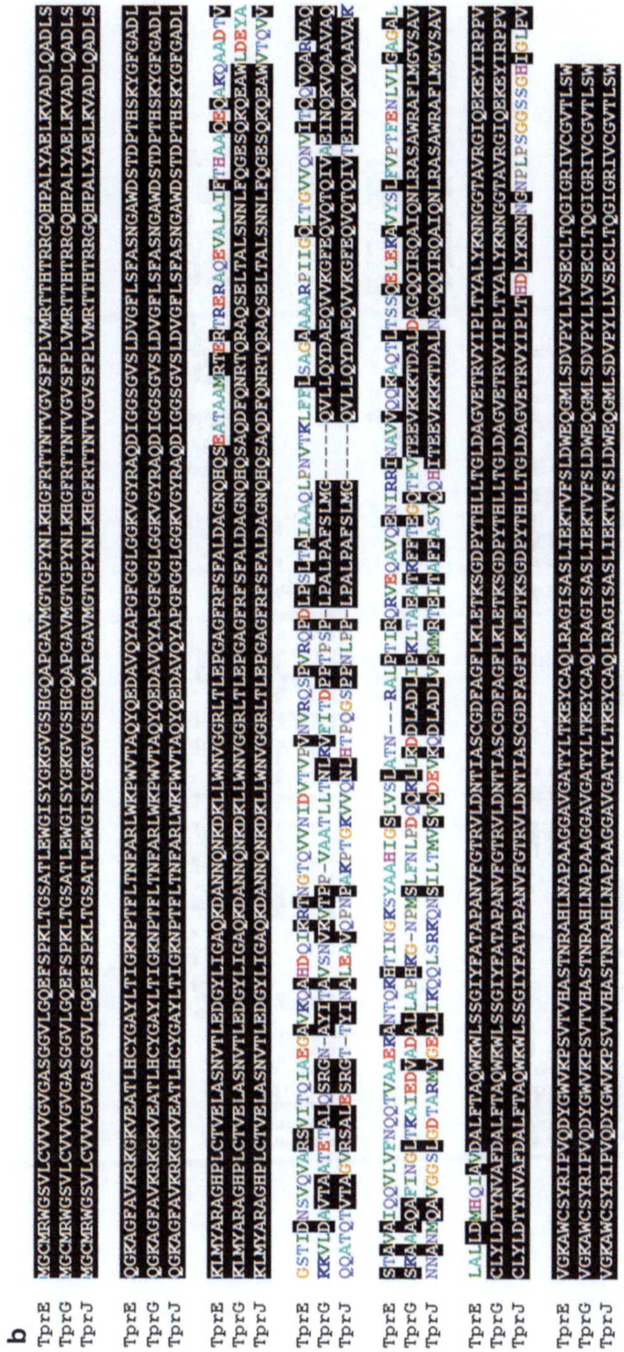

Fig. 5.1 (continued)

Fig. 5.1 (continued)

identical ORFs in the *tprC* and *tprD* loci. To date, no changes or very limited sequence variability has been detected in the *tprF* and *tprI* genes, respectively (Sun et al. 2004; Centurion-Lara et al. 2000, 2006). Although variability among Subfamily II *tprs* is practically absent in syphilis isolates, some strains, such as Sea81-4, carry a hybrid *tprG/J* allele within the *tprJ* locus, where the *tprJ* central region is replaced by the central portion of the *tprG* gene (Centurion-Lara et al. 2000; Giacani et al. 2005b), and a premature termination is introduced towards the 3′ end of the *tprG* locus (Giacani et al. 2005b). This type of variability in the *tpr* genes cannot be considered true antigenic variation, but rather antigenic drift, given that diversity in these antigens seems to be isolate-specific and does not accumulate within a strain during infection. Further investigation is needed to elucidate the full extent of sequence variability among the *tpr* genes in syphilis isolates and its significance for the pathogenesis of the disease. Recent efforts to decipher the genome sequences from other *T. pallidum* isolates (Matejkova et al. 2008; Giacani et al. 2010a) have proven very helpful in identifying genome wide sequence differences among isolates, and future genome projects will map the breath of diversity among *T. pallidum* isolates. Diversity might reflect a process of functional diversification of redundant genes within isolates, or, because many Tprs are predicted as OM proteins, sequence diversity could reflect the ability of these antigens to interact with different ligands in the host tissues, and therefore mediate *T. pallidum* preferential dissemination to certain organs. Tpr diversity could also provide an efficient way to diversify *T. pallidum* surface antigenic profile and perhaps favor re-infection of a host with a non-homologous syphilis strain. Although sequence heterogeneity is observed in several members of the Tpr antigens among isolates, to date only a subset of *tpr* genes have been associated with phase variation (*tprF, E,G,J, and I*) or antigenic variation (*tprK*) during syphilis infection. The remaining of this chapter will focus therefore on these Tpr antigens.

5.5.2 *Phase Variation of Subfamily I and II tpr Genes*

Leader et al. (Leader et al. 2003) showed that, during experimental syphilis, humoral immune responses against Subfamily I and II Tpr antigens tend to appear late (approximately 45 days postinfection) and peak at circa day 60. In the same study however, antibody titers elicited by the different Tprs greatly differed depending on the strain used for experimental infection. At day 45 postinfection, for instance, antibody response against TprC, D, E, F, I, and J was greater in rabbits infected with the Chicago strain of *T. pallidum* than in those infected with the Bal73-1 isolate. Vice versa, antibody titers against TprG were greater in Bal73-1 infected rabbits than those inoculated with Chicago. At the same time point, when reactivity could be detected, all rabbits infected with the Nichols strain exhibited lower antibody titers with respect to both Bal73-1 and Chicago infected animals (Leader et al. 2003). Different titers of specific antibodies against the Tprs likely reflect the fact that these antigens are differentially expressed within and among *T. pallidum* strains during experimental infection (Giacani et al. 2007a). Although the mechanisms and factors that regulate *tpr* gene expression are still unknown for the majority of these genes, transcription of at least four *tpr* genes (*tprE, F, G, I,* and *J*) was recently shown to be controlled by phase variation through a varying homopolymeric tract of guanosines (poly-G) located in the promoters upstream of *tprE, G,* and *J* (Giacani et al. 2007b). In bacterial pathogens, homopolymeric repeats located within or in the proximity of promoter regions are known to regulate expression of a variety of genes coding for virulence determinants and surface antigens (van der Woude and Baumler 2004; van der Ende et al. 1995; Saunders et al. 1998; Palmer et al. 2009). Variation in the length of these homopolymeric tracts is a well-recognized mechanism for controlling amount of message and ON-OFF states of transcription. During DNA replication or repair, these repeats can expand or contract via a phenomenon known as slipped-strand mispairing (SSM). Because misalignment can occur either on the leading or lagging strands, SSM can respectively result in increase or decrease of the number of repeated nucleotides in the new DNA molecule (Levinson and Gutman 1987). Classic examples of modulation of gene expression by single nucleotide repeats are the *porA* and *opc* loci of *Neisseria meningitidis* (*N. meningitidis*), in which variation in the length of the poly-G and poly-C repeats in their promoters induces high, moderate, or no expression of these genes (van der Ende et al. 1995; Sarkari et al. 1994). The analysis of the Nichols strain genome revealed the presence of poly-G repeats of varying length upstream of the ORFs encoding both Subfamily I (*tprC, D, F,* and *I*) and II (*tprE, G,* and *J*) *tprs*. Poly-G tracts associated with *tprE* were found to oscillate between 8 and 12 nucleotides in length, while those upstream of *tprG* and *J*, respectively, were found to vary from 8 to 11, and 7 to 11 nucleotides, respectively. Overall, less variability was found in the poly-G's associated with *tprF* and *tprI* (8 to 10 nucleotides) (Giacani et al. 2007b). Using a combination of in vitro and in vivo approaches, transcription from *tprE, G,* and *J* promoters was shown to be significantly enhanced by the presence of eight or fewer G's in their poly-G sequences, while the same promoters with nine or more G's gave modest or no transcriptional signal above background. In contrast, variation in the number of G's within the *tprF/I* promoter was not found to influence transcription initiation of these genes. In the Nichols strain, however, *tprF* and *tprI* are located immediately downstream of *tprG* and *tprJ*, respectively, and transcriptional analysis of these loci revealed that *tprG/F* and *tprJ/I* are co-transcribed from the promoter sequences located upstream of *tprG* and *J*, respectively (Giacani et al. 2007b). This implies that also transcription of *tprF* and *I* genes can be influenced by changes in the poly-G tracts upstream of both *tprG* and *tprJ*, as well as by the presence of promoters upstream of *tprF* and *I* able to generate sub-operonic messages (Giacani et al. 2007b).

Although homopolymeric sequences represent a common *cis*-acting regulatory mechanism in several bacterial pathogens, it is not clear how changes in the poly-G tracts can affect promoter strength of Subfamily II *tprs* in *T. pallidum* in vivo. In *N. meningitidis*, for instance, poly-G and poly-C repeats control transcription of the *porA* and *opc* genes, controlling both their ON-OFF states and the volume of transcription of these genes (van der Ende et al. 1995; Sarkari et al. 1994). In the *porA* promoter, the poly-G tract is located between the −10 and −35 signatures, while the *opc* poly-C region encompasses the −35 region. Changes in the poly-G length in the *porA* promoter regulate gene transcription by varying the optimal spacing of 17 nucleotides between the consensus sequences to 15 residues (van der Ende et al. 1995); in contrast, the homopolymeric repeats located in the −35 consensus most likely modulate transcription of the *opc* loci by influencing the binding of a regulatory protein (Sarkari et al. 1994). In the *tpr* promoters however, the poly-G repeats are located immediately upstream of their experimentally determined TSSs, and consequently variation in the length of the repeats would not alter the spacing between these signatures, ruling out a *porA*-like mechanism (Giacani et al. 2007b). Although there is evidence that the *T. pallidum* cAMP Receptor Protein (TpCRP) also influences transcription of Subfamily II *tprs*, the location of the TpCRP binding sites does not support any interaction between the *tprE, G,* and *J* poly-G's with *T. pallidum* CRP (Giacani et al. 2009). All that given, one could hypothesize that in *T. pallidum*, variation in the number of G's might affect conversion of double-stranded DNA (closed complex) into single-stranded DNA (open complex) driven by the RNA polymerase (RNAP) to initiate transcription. However, studies on DNA base pair opening rates attribute an overall higher instability of the double helix to DNA tracts rich in G nucleotide repeats. Consequently, a sequence with a high content in G repeats would alter the double helix conformation and decrease the RNAP ability to recognize and bind the DNA region in which the poly-G is contained (Dornberger et al. 1999; Bayliss et al. 2004). According to this hypothesis, it seems plausible that a lower number of G residues within the *tpr* promoter region would decrease the DNA double helix instability, perhaps facilitating the binding of the RNAP to the closed complex and the subsequent formation of the open complex, which leads to transcription initiation.

Modulation of expression of these antigens is perhaps an important piece of *T. pallidum*'s strategy to avoid clearance by opsonophagocytosis in resolving primary and secondary lesions, and perhaps increasing its chances to be transmitted to a new host.

5.5.3 The tprK Gene

Whole genome sequences of at least three *T. pallidum* isolates revealed conservation of the gene organization at the *tprK* locus and surrounding genes (Matejkova et al. 2008; Giacani et al. 2010a; Centurion-Lara et al. 2004). *tprK* is predicted to be a single transcription unit coding for a protein of approximately 53 kDa in size. Extensive sequence analysis demonstrates intra- and interstrain variations with diversity localized to seven discrete variable regions (V1–V7) flanked by highly conserved sequences. The seven V regions share a similar DNA sequence structure with two identical four-base terminal repeats defining their left and right boundaries. Diversity within V regions is due to insertions, deletions, or nucleotide changes

(Centurion-Lara et al. 2004). Sequence diversity accumulates actively during infection and is higher in some V regions than in others. In nearly all strains examined to date, V6 demonstrates the greatest capacity for variation. TprK heterogeneity occurs both spontaneously during routine propagation of *T. pallidum* in rabbits and when adaptive immunity develops in the rabbit model with rates of sequence change differing among strains. Although the degree of intrastrain diversity in the *tprK* locus is not uniform among isolates, all strains of *T. pallidum* examined to date have a certain degree of intrastrain variation in the *tprK* expression site. For example, most of the Nichols isolates maintained in different laboratories show a modest degree of intrastrain heterogeneity except for the Nichols isolate maintained at the University of North Carolina (Stamm and Bergen 2000), which has high rates of diversity resembling more recent *T. pallidum* strains such as the Chicago or Sea81-4 isolates (Centurion-Lara et al. 2004; LaFond et al. 2003). Possible mechanisms associated with the intrinsically different rates of variation among strains are currently under investigation. Due to the very high degree of sequence identity among *T. pallidum* isolates as revealed by whole genome sequencing, it is likely that minimal changes in coding sequences of key molecules or in DNA regions controlling gene expression may be responsible for the differential rates of diversity in the V regions.

Targeted sequencing of the *tprK* locus, coordinates 974319–975833 in the Nichols strain genome sequence (Liang et al. 2000), identified within isolates a large number of distinct *tprK* variants defined by their unique V regions. Initial Southern blot and genomic DNA restriction digest analyses suggested the presence of a single copy gene in the *T. pallidum* chromosome (Stamm and Bergen 2000), observations later confirmed when whole genome sequences of several *T. pallidum* strains were available. The clonal nature of newly arisen *tprK* variants and the unique sequence anatomy of the V regions prompted then the search for recombination mechanisms that could explain the impressive degree of *tprK* heterogeneity. It resulted in the development of a two-component model of non-reciprocal recombination mediated by gene conversion, which could virtually create any *tprK* variant observed to date. In this model, donor sites undergo duplicative transposition into the expression site (Fig. 5.2). The first element of this system, the donor sites, consists of >50 DNA silent DNA mini-cassettes identical to partial or nearly complete *tprK* V regions localized in the 5′ and 3′ flanking regions of *tprD*, another member of the *tpr* gene family. The second element is the expression site, the *tprK* locus. *tprK* variants are created when a donor cassette replaces an entire *tprK* V region or a segment of it, therefore called segmental gene conversion, a non-reciprocal recombination event which maintains the pool of donor sites unchanged overtime (Centurion-Lara et al. 2004).

5.5.3.1 The TprK Antigen Elicits a Strong Immune Response and Is Involved in Immune Selection During Syphilis Infection

The ability of a small number of spirochetes to avoid opsonophagocytosis and subsequent killing by activated macrophages results in the establishment of a

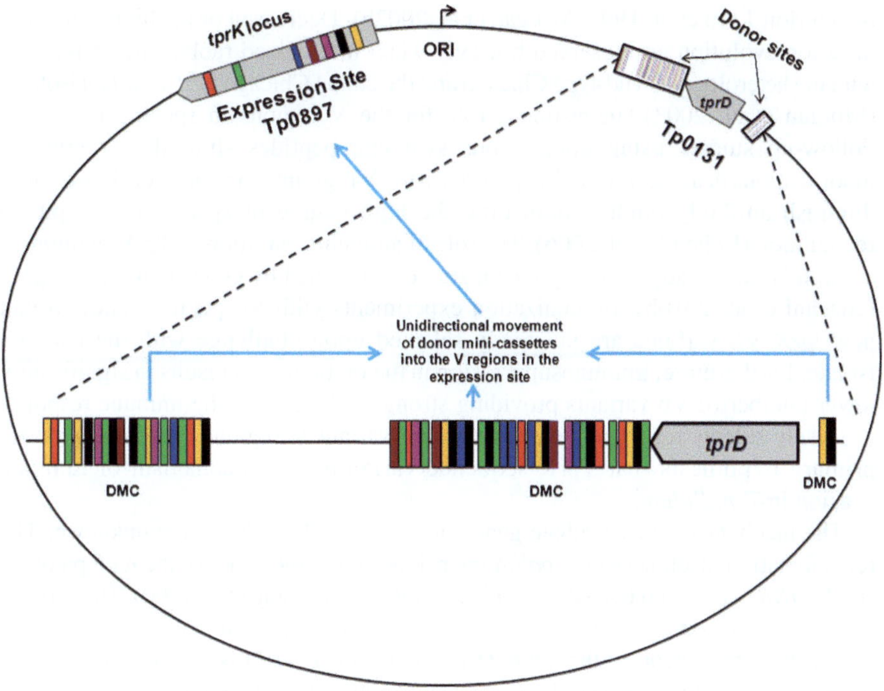

Fig. 5.2 *tprK* gene conversion model. DMC indicates donor mini-cassettes. ORI is the origin of replication. *Colored bars* highlight donor mini-cassettes specific for each V region in the *tprK* expression site. The *T. pallidum* subspecies *pallidum* circular chromosome is 1.13 Mbp long

persistent, mostly silent syphilitic infection (Lukehart et al. 1992). How these few treponemes can survive and avoid clearance by the host immune response for decades is amenable to speculation, but one likely mechanism is the modification of epitopes recognized by opsonic antibodies on the surface of the organism (LaFond et al. 2006). Although the cellular localization of TprK is still disputed, there are two pieces of information providing strong evidence that TprK is exposed on the external surface: (1) immunization with several TprK peptides consistently elicits the production of opsonic antibodies in the rabbit model, and (2) antigenic variation of TprK results in positive selection of variants when adaptive immunity develops during infection (Giacani et al. 2010b).

Upon infection, the infected host mounts a robust cellular and humoral immune response against TprK indicative that this antigen is actively expressed during infection (Centurion-Lara et al. 1999; Giacani et al. 2007a, 2010b). Morgan et al. (Morgan et al. 2002a) demonstrated that antibody responses specifically target the V regions while cellular immunity is preferentially directed to conserved regions of the protein further supporting the hypothesis of a role of TprK in immune evasion. Initial studies found that immunization of rabbits with the recombinant TprK from the Nichols strain consistently results in elicitation of opsonic antibodies and significant attenuation of lesion development against homologous challenge in the rabbit model

(Centurion-Lara et al. 1999; Morgan et al. 2002b). Decreased or negligible changes in lesion evolution are observed however when immunized rabbits are challenged with the heterologous wild-type Chicago and the clonal Chicago C *T. pallidum* isolates (Morgan et al. 2003) suggesting a role for the V regions in specific immunity. Follow-up studies using homologous synthetic peptides showed that minimal modifications in the amino acid sequences in the V regions can abolish or significantly diminish antibody binding indicating the high degree of specificity of epitope recognition (LaFond et al. 2006). The role of antigenic variation of TprK in immune evasion is further supported by recent studies published by Giacani and colleagues (Giacani et al. 2010b). Immunization experiments with V6 peptides showed that new *tprK* V6 variants are positively selected upon challenge with the parental isolate. Furthermore, immunosuppression in the rabbit model results in significantly lower number of V6 variants providing strong evidence that the immune response positively selects for treponemes expressing variant V regions. This supports our primary hypothesis that TprK sequence variation is a mechanism of immune evasion in *T. pallidum.*

The mechanisms that mediate gene conversion in *T. pallidum* are unknown. The recombination machinery of *T. pallidum* includes *recA*, the genes of the *recF* pathway (*recFGJNR*), *recX*, and two site-specific recombinases (Liang et al. 2000). The syphilis spirochete also contains *uvr* excision repair and mismatch repair genes, but reportedly it lacks the components of the *recBCD* system (Liang et al. 2000). More contemporaneous bioinformatics analysis, however, suggests that *T. pallidum* may indeed contain a *recBCD* or *addAB* system (unpublished). Disregarding the fact that the recombination system responsible for the unidirectional insertion of donor sequences into the expression site remains to be discovered, it obviously has a key role in the survival of *T. pallidum* for prolonged periods of time in the infected host.

5.6 Conclusions and Future Directions

The discovery of the *tpr* genes represents a milestone for syphilis investigators aiming to unveil the strategies developed by *T. pallidum* to persist in its host. Since their identification, a great wealth of information has been gathered for these genes regarding (1) allelic diversity among *T. pallidum* strains, (2) mechanisms of transcriptional regulation, including phase-variation to switch ON and OFF transcription of a subset of *tpr* genes, (3) the ability of the agent of syphilis to generate countless variants of TprK through non-reciprocal gene conversion, and (4) TprK involvement in treponemal selection by the host's immune response during experimental syphilis. The above mechanisms are widely used by microbial pathogens to escape the host's defenses and cause persistent infection, and their discovery in *T. pallidum* further confirms the observation of development of shared strategies among parasites, bacteria, and viruses for the avoidance of the immune response (Deitsch et al. 2009). Unlike antigenic variation in blood-borne pathogens, Palmer and colleagues (Palmer et al. 2009) have suggested that TprK antigenic variation is a mechanism important during primary and secondary syphilis for increasing the chances of transmission from an infected individual to a susceptible host by delaying clearance of *T. pallidum* cells from early lesions. It seems also plausible that people who already had syphilis would likely not have developed antibodies against all of the possible TprK variants that can be generated in *T. pallidum*, therefore facilitating re-infection.

Although the Tpr antigens have been actively investigated for over a decade, very important questions still need to be answered, including the cellular processes that these antigens mediate, their structure, and actual surface exposure. Also important will be a more comprehensive elucidation of all the *cis*- and *trans*-acting elements that regulate transcription of these genes as well as a better understanding of the actual degree of diversity of these proteins among *T. pallidum* isolates. Renewed whole-genome sequencing effort along with genome wide bacterial population analysis will expand our knowledge about the *tpr* genes and potentially identify new hotspots of variation in other coding or non-coding sequences directly or indirectly associated with the interplay at the pathogen–host interface.

References

Arrizabalaga J, Henderson J, French R (1997) The great Pox: the French disease in renaissance Europe. Yale University Press, New Haven

Baker-Zander S, Sell S (1980) A histopathologic and immunologic study of the course of syphilis in the experimentally infected rabbit. Demonstration of long-lasting cellular immunity. Am J Pathol 101(2):387–414

Barrett LG, Thrall PH, Burdon JJ, Linde CC (2008) Life history determines genetic structure and evolutionary potential of host-parasite interactions. Trends Ecol Evol 23(12):678–685

Baughn RE, Musher DM (2005) Secondary syphilitic lesions. Clin Microbiol Rev 18(1):205–216

Bayliss CD, Dixon KM, Moxon ER (2004) Simple sequence repeats (microsatellites): mutational mechanisms and contributions to bacterial pathogenesis. A meeting review. FEMS Immunol Med Microbiol 40(1):11–19

Brinkman MB, McGill MA, Pettersson J, Rogers A, Matejkova P, Smajs D, Weinstock GM, Norris SJ, Palzkill T (2008) A novel *Treponema pallidum* antigen, TP0136, is an outer membrane protein that binds human fibronectin. Infect Immun 76(5):1848–1857

Cameron CE (2003) Identification of a *Treponema pallidum* laminin-binding protein. Infect Immun 71(5):2525–2533

Cameron DJ (2008) An appraisal of "chronic Lyme disease" [comment]. N Engl J Med 358(4):429–430, author reply 430–421

Cameron CE, Lukehart SA, Castro C, Molini B, Godornes C, Van Voorhis WC (2000) Opsonic potential, protective capacity, and sequence conservation of the *Treponema pallidum* subspecies *pallidum* Tp92. J Infect Dis 181(4):1401–1413

Cameron CE, Brown EL, Kuroiwa JM, Schnapp LM, Brouwer NL (2004) *Treponema pallidum* fibronectin-binding proteins. J Bacteriol 186(20):7019–7022

CDC (2007) Sexually transmitted disease surveillance, 2006. US. Department of Health and Human Services, Centers for Disease Control and Prevention, Atlanta, GA

Centurion-Lara A, Castro C, Barrett L, Cameron C, Mostowfi M, Van Voorhis WC, Lukehart SA (1999) *Treponema pallidum* major sheath protein homologue Tpr K is a target of opsonic antibody and the protective immune response. J Exp Med 189(4):647–656

Centurion-Lara A, Sun ES, Barrett LK, Castro C, Lukehart SA, Van Voorhis WC (2000) Multiple alleles of *Treponema pallidum* repeat gene D in *Treponema pallidum* isolates. J Bacteriol 182(8):2332–2335

Centurion-Lara A, LaFond RE, Hevner K, Godornes C, Molini BJ, Van Voorhis WC, Lukehart SA (2004) Gene conversion: a mechanism for generation of heterogeneity in the *tprK* gene of *Treponema pallidum* during infection. Mol Microbiol 52(6):1579–1596

Centurion-Lara A, Molini BJ, Godornes C, Sun E, Hevner K, Van Voorhis WC, Lukehart SA (2006) Molecular differentiation of *Treponema pallidum* subspecies. J Clin Microbiol 44(9):3377–3380

Chapel TA (1978) The variability of syphilitic chancres. Sex Transm Dis 5(2):68–70

Chapel TA (1980) The signs and symptoms of secondary syphilis. Sex Transm Dis 7(4):161–164

Chapel TA (1988) Congenital syphilis. Compr Ther 14(3):25–28

Chen ZQ, Zhang GC, Gong XD, Lin C, Gao X, Liang GJ, Yue XL, Chen XS, Cohen MS (2007) Syphilis in China: results of a national surveillance programme. Lancet 369(9556):132–138

Cox DL, Chang P, McDowall AW, Radolf JD (1992) The outer membrane, not a coat of host proteins, limits antigenicity of virulent *Treponema pallidum*. Infect Immun 60(3):1076–1083

Cox DL, Luthra A, Dunham-Ems S, Desrosiers DC, Salazar JC, Caimano MJ, Radolf JD (2010) Surface immunolabeling and consensus computational framework to identify candidate rare outer membrane proteins of *Treponema pallidum*. Infect Immun 78(12):5178–5194

Craig AG, Scherf A (2003) Antigenic variation. Academic, Boston

Cumberland MC, Turner TB (1949) The rate of multiplication of *Treponema pallidum* in normal and immune rabbits. Am J Syph Gonorrhea Vener Dis 33(3):201–212

Cunningham TM, Walker EM, Miller JN, Lovett MA (1988) Selective release of the *Treponema pallidum* outer membrane and associated polypeptides with Triton X-114. J Bacteriol 170(12):5789–5796

Deitsch KW, Lukehart SA, Stringer JR (2009) Common strategies for antigenic variation by bacterial, fungal and protozoan pathogens. Nat Rev Microbiol 7:493–503

Desrosiers DC, Anand A, Luthra A, Dunham-Ems SM, Ledoyt M, Cummings MA, Eshghi A, Cameron CE, Cruz AR, Salazar JC et al (2011) TP0326, a *Treponema pallidum* beta-barrel assembly machinery A (BamA) orthologue and rare outer membrane protein. Mol Microbiol 80(6):1496–1515

Diamond J (1997) Guns, germs and steel. W.W. Norton, New York

DiCarlo RP, Martin DH (1997) The clinical diagnosis of genital ulcer disease in men. Clin Infect Dis 25(2):292–298

Dornberger U, Leijon M, Fritzsche H (1999) High base pair opening rates in tracts of GC base pairs. J Biol Chem 274(11):6957–6962

Fenno JC, Muller KH, McBride BC (1996) Sequence analysis, expression, and binding activity of recombinant major outer sheath protein (Msp) of *Treponema denticola*. J Bacteriol 178(9):2489–2497

Fenno JC, Wong GW, Hannam PM, Muller KH, Leung WK, McBride BC (1997) Conservation of *msp*, the gene encoding the major outer membrane protein of oral *Treponema* spp. J Bacteriol 179(4):1082–1089

Finlay BB, Falkow S (1997) Common themes in microbial pathogenicity revisited. Microbiol Mol Biol Rev 61(2):136–169

Fiumara NJ (1975) Syphilis in newborn children. Clin Obstet Gynecol 18(1):183–189

Fiumara NJ, Berg M (1974) Primary syphilis in the oral cavity. Br J Vener Dis 50(6):463–464

Foa A (1990) The new and the old: the spread of syphilis (1494–1530). In: Muir E, Ruggiero G (eds) Sex and gender in historical perspective. Johns Hopkins University Press, Baltimore

Gerbase AC, Rowley JT, Heymann DH, Berkely SF, Piot P (1998) Global prevalence and incidence estimates of selected curable STDs. Sex Transm Infect 74(Suppl 1):S12–S16

Giacani L, Sambri V, Marangoni A, Cavrini F, Storni E, Donati M, Corona S, Lanzarini P, Cevenini R (2005a) Immunological evaluation and cellular location analysis of the TprI antigen of *Treponema pallidum* subsp. *pallidum*. Infect Immun 73(6):3817–3822

Giacani L, Hevner K, Centurion-Lara A (2005b) Gene organization and transcriptional analysis of the *tprJ*, *tprI*, *tprG* and *tprF* loci in the Nichols and Sea 81–4 *Treponema pallidum* isolates. J Bacteriol 187(17):6084–6093

Giacani L, Molini B, Godornes C, Barrett L, Van Voorhis WC, Centurion-Lara A, Lukehart SA (2007a) Quantitative analysis of *tpr* gene expression in *Treponema pallidum* isolates: differences among isolates and correlation with T-cell responsiveness in experimental syphilis. Infect Immun 75(1):104–112

Giacani L, Lukehart S, Centurion-Lara A (2007b) Length of guanosine homopolymeric repeats modulates promoter activity of Subfamily II tpr genes of *Treponema pallidum* ssp. *pallidum*. FEMS Immunol Med Microbiol 51:289–301

Giacani L, Godornes C, Puray-Chavez M, Guerra-Giraldez C, Tompa M, Lukehart SA, Centurion-Lara A (2009) TP0262 is a modulator of promoter activity of tpr Subfamily II genes of *Treponema pallidum* ssp. *pallidum*. Mol Microbiol 72(5):1087–1099

Giacani L, Jeffrey BM, Molini BJ, Le HT, Lukehart SA, Centurion-Lara A, Rockey DD (2010a) Complete genome sequence and annotation of the *Treponema pallidum* subsp. *pallidum* Chicago strain. J Bacteriol 192(10):2645–2646

Giacani L, Molini BJ, Kim EY, Godornes BC, Leader BT, Tantalo LC, Centurion-Lara A, Lukehart SA (2010b) Antigenic variation in *Treponema pallidum*: TprK sequence diversity accumulates in response to immune pressure during experimental syphilis. J Immunol 184(7):3822–3829

Gjestland T (1955) The Oslo study of untreated syphilis. Acta Derm Venereol 35(Suppl):11–368

Handsfield HH, Lukehart SA, Sell S, Norris SJ, Holmes KK (1983) Demonstration of *Treponema pallidum* in a cutaneous gumma by indirect immunofluorescence. Arch Dermatol 119(8):677–680

Hardy PH Jr, Levin J (1983) Lack of endotoxin in *Borrelia hispanica* and *Treponema pallidum*. Proc Soc Exp Biol Med 174(1):47–52

Hira SK, Patel JS, Bhat SG, Chilikima K, Mooney N (1987) Clinical manifestations of secondary syphilis. Int J Dermatol 26(2):103–107

Hook EW III, Marra CM (1992) Acquired syphilis in adults. N Engl J Med 326(16):1060–1069

Hourihan M, Wheeler H, Houghton R, Goh BT (2004) Lessons from the syphilis outbreak in homosexual men in east London. Sex Transm Infect 80(6):509–511

Hovind-Hougen K (1983) Morpholgy. In: Schell RF, Musher DM (eds) Pathogenesis and immunology of Treponemal infection. Marcel Dekker, Inc., New York

Jackman JD Jr, Radolf JD (1989) Cardiovascular syphilis. Am J Med 87(4):425–433

Jepsen OB, Hougen KH, Birch-Andersen A (1968) Electron microscopy of *Treponema pallidum* Nichols. Acta Pathol Microbiol Scand 74(2):241–258

Kampmeier R (1964) The late manifestations of syphilis: skeletal, visceral and cardiovascular. Med Clin North Am 48:667–697

Kevin S (2004) Venereal disease, hospitals and urban poor: London's foul wards, 1600–1800. University of Rochester Press, Rochester

LaFond RE, Centurion-Lara A, Godornes C, Rompalo AM, Van Voorhis WC, Lukehart SA (2003) Sequence diversity of *Treponema pallidum* subsp. *pallidum tprK* in human syphilis lesions and rabbit-propagated isolates. J Bacteriol 185(21):6262–6268

LaFond RE, Molini BJ, Van Voorhis WC, Lukehart SA (2006) Antigenic variation of TprK V regions abrogates specific antibody binding in syphilis. Infect Immun 74(11):6244–6251

Leader BT, Hevner K, Molini BJ, Barrett LK, Van Voorhis WC, Lukehart SA (2003) Antibody responses elicited against the *Treponema pallidum* repeat proteins differ during infection with different isolates of *Treponema pallidum* subsp. *pallidum*. Infect Immun 71(10):6054–6057

Levinson G, Gutman GA (1987) Slipped-strand mispairing: a major mechanism for DNA sequence evolution. Mol Biol Evol 4(3):203–221

Liang FT, Jacobson RH, Straubinger RK, Grooters A, Philipp MT (2000) Characterization of a *Borrelia burgdorferi* VlsE invariable region useful in canine Lyme disease serodiagnosis by enzyme-linked immunosorbent assay. J Clin Microbiol 38(11):4160–4166

Lobdell J (1974) OD: the origin of syphilis. J Sex Res 10(1):4

Luger A (1993) The origin of syphilis. Clinical and epidemiologic considerations on the Columbian theory. Sex Transm Dis 20(2):110–117

Lukehart SA, Shaffer JM, Baker-Zander SA (1992) A subpopulation of *Treponema pallidum* is resistant to phagocytosis: possible mechanism of persistence. J Infect Dis 166(6):1449–1453

Magnuson HJ, Eagle H, Fleischman R (1948) The minimal infectious inoculum of *Spirochaeta pallida* (Nichols Strain), and a consideration of its rate of multiplication *in vivo*. Am J Syphilis 32(1):1–18

Marra CM, Maxwell CL, Smith SL, Lukehart SA, Rompalo AM, Eaton M, Stoner BP, Augenbraun M, Barker DE, Corbett JJ et al (2004) Cerebrospinal fluid abnormalities in patients with syphilis: association with clinical and laboratory features. J Infect Dis 189(3):369–376

Marra CM, Sahi SK, Tantalo LC, Godornes C, Reid T, Behets F, Rompalo A, Klausner JD, Yin YP, Mulcahy F et al (2010) Enhanced molecular typing of *Treponema pallidum*: geographical distribution of strain types and association with neurosyphilis. J Infect Dis 202(9):1380–1388

Mascola L, Pelosi R, Blount JH, Alexander CE, Cates W Jr (1985) Congenital syphilis revisited. Am J Dis Child 139(6):575–580

Matejkova P, Strouhal M, Smajs D, Norris SJ, Palzkill T, Petrosino JF, Sodergren E, Norton JE, Singh J, Richmond TA et al (2008) Complete genome sequence of *Treponema pallidum* ssp. *pallidum* strain SS14 determined with oligonucleotide arrays. BMC Microbiol 8:76

McGough LJ, Erbelding E (2006) Historical evidence of syphilis and other treponematosis. In: Radolf JD, Lukahart SA (eds) Pathogenic treponema—molecular and cellular biology. Caister Academic Press, Norfolk, England

Merritt HH, Adams RD, Solomon HC (1946) Neurosyphilis. Oxford University Press, New York

Mindel A, Tovey SJ, Timmins DJ, Williams P (1989) Primary and secondary syphilis, 20 years' experience. 2. Clinical features. Genitourin Med 65(1):1–3

Morgan CA, Molini BJ, Lukehart SA, Van Voorhis WC (2002a) Segregation of B and T cell epitopes of *Treponema pallidum* repeat protein K to variable and conserved regions during experimental syphilis infection. J Immunol 169(2):952–957

Morgan CA, Lukehart SA, Van Voorhis WC (2002b) Immunization with the N-terminal portion of *Treponema pallidum* repeat protein K attenuates syphilitic lesion development in the rabbit model. Infect Immun 70(12):6811–6816

Morgan CA, Lukehart SA, Van Voorhis WC (2003) Protection against syphilis correlates with specificity of antibodies to the variable regions of *Treponema pallidum* repeat protein K. Infect Immun 71(10):5605–5612

Nell EE, Hardy PH (1966) Studies on the chemical composition and immunologic properties of a polysaccharide from the Reiter treponeme. Immunochemistry 3(3):233–245

Nichols HJ, Hough WH (1913) Demonstration of *Spirochaeta pallida* in the cerebrospinal fluid. JAMA 60:108–110

Oriel JD (2004) The scars of venus: a history of venerealogy. Springer, London

Palmer GH, Bankhead T, Lukehart SA (2009) 'Nothing is permanent but change'—antigenic variation in persistent bacterial pathogens. Cell Microbiol 11(12):1697–1705

Radolf JD (1995) *Treponema pallidum* and the quest for outer membrane proteins. Mol Microbiol 16(6):1067–1073

Radolf JD, Lukehart SA (eds) (2006) Pathogenic *Treponema*: molecular and cellular biology. Caister Academic Press, Norfolk, England.

Radolf JD, Norgard MV (1988) Pathogen specificity of *Treponema pallidum* subsp. *pallidum* integral membrane proteins identified by phase partitioning with Triton X-114. Infect Immun 56(7):1825–1828

Radolf JD, Chamberlain NR, Clausell A, Norgard MV (1988) Identification and localization of integral membrane proteins of virulent *Treponema pallidum* subsp. *pallidum* by phase partitioning with the nonionic detergent triton X-114. Infect Immun 56(2):490–498

Radolf JD, Norgard MV, Schulz WW (1989) Outer membrane ultrastructure explains the limited antigenicity of virulent *Treponema pallidum*. Proc Natl Acad Sci U S A 86(6):2051–2055

Radolf JD, Robinson EJ, Bourell KW, Akins DR, Porcella SF, Weigel LM, Jones JD, Norgard MV (1995) Characterization of outer membranes isolated from *Treponema pallidum*, the syphilis spirochete. Infect Immun 63(11):4244–4252

Radolf JD, Sanchez PJ, Schultz KF, Murphy KM (1999) Congenital syphilis, chapter 84. In: Holmes KK, Sparling PF, Mardh PA, Lemon SM, Stamm WE, Piot P, Wasserheit JN (eds) Sexually transmitted diseases, 3rd edn. McGraw-Hill, New York

Renton AM, Borisenko KK, Meheus A, Gromyko A (1998) Epidemics of syphilis in the newly independent states of the former Soviet Union. Sex Transm Infect 74(3):165–166

Rompalo AM, Joesoef MR, O'Donnell JA, Augenbraun M, Brady W, Radolf JD, Johnson R, Rolfs RT (2001) Clinical manifestations of early syphilis by HIV status and gender: results of the syphilis and HIV study. Sex Transm Dis 28(3):158–165

Rothschild B (2003) Infectious processes around the dawn of civilization. In: Greenblatt CL, Spigelman M (eds) Emerging pathogens: the archaeology, ecology, and evolution of infectious disease. Oxford University Press, London

Rothschild BM, Calderon FL, Coppa A, Rothschild C (2000) First European exposure to syphilis: the Dominican Republic at the time of Columbian contact. Clin Infect Dis 31(4):936–941
Sarkari J, Pandit N, Moxon ER, Achtman M (1994) Variable expression of the Opc outer membrane protein in *Neisseria meningitidis* is caused by size variation of a promoter containing poly-cytidine. Mol Microbiol 13(2):207–217
Saunders NJ, Peden JF, Hood DW, Moxon ER (1998) Simple sequence repeats in the *Helicobacter pylori* genome. Mol Microbiol 27(6):1091–1098
Sell S, Gamboa D, Baker-Zander SA, Lukehart SA, Miller JN (1980) Host response to *Treponema pallidum* in intradermally-infected rabbits: evidence for persistence of infection at local and distant sites. J Invest Dermatol 75(6):470–475
Sheffield JS, Sanchez PJ, Morris G, Maberry M, Zeray F, McIntire DD, Wendel GD Jr (2002) Congenital syphilis after maternal treatment for syphilis during pregnancy. Am J Obstet Gynecol 186(3):569–573
Simon RP (1994) Neurosyphilis. Neurology 44(12):2228–2230
Stamm LV, Bergen HL (2000) The sequence-variable, single-copy *tprK* gene of *Treponema pallidum* Nichols strain UNC and street strain 14 encodes heterogeneous TprK proteins. Infect Immun 68(11):6482–6486
Stokes JH, Beerman H, Ingram NR (1944) Modern clinical syphilology. W.B. Saunders, Philadelphia
Sun ES, Molini BJ, Barrett LK, Centurion-Lara A, Lukehart SA, Van Voorhis WC (2004) Subfamily I *Treponema pallidum* repeat protein family: sequence variation and immunity. Microbes Infect 6(8):725–737
Temkin O (1977) The double face of janus and other essays on the history of medicine. Johns Hopkins University Press, Baltimore
Turner TB, Hollander DH (1957) Biology of the Treponematoses. World Health Organization, Geneva
U.S. Department of Health, Education, and Welfare (1968) Syphilis—a synopsis. Center for Disease Control, Atlanta, GA
Uhlenhuth PT (1913a) Weitere mitteilungen über die infectiosität des blutes und anderer körperflüssigkeiten syphilitischer menschen für das kaninchen. Berl klin Wchnschr 50:769–775
Uhlenhuth PT (1913b) Weitere mitteilungen über ergebnisse der experimetnellen syphilisforschung. Berl klin Wchnschr 50:2031–2035
van der Ende A, Hopman CT, Zaat S, Essink BB, Berkhout B, Dankert J (1995) Variable expression of class I outer membrane protein in *Neisseria meningitidis* is caused by variation in the spacing between the −10 and −35 regions of the promoter. J Bacteriol 177(9):2475–2480
van der Woude MW, Baumler AJ (2004) Phase and antigenic variation in bacteria. Clin Microbiol Rev 17(3):581–611

Chapter 6
Antigenic Variation of VlsE in *Borrelia burgdorferi*

Troy Bankhead

6.1 Introduction

The multisystem disease known as Lyme disease (or Lyme borreliosis) is the most prevalent vector-borne infection affecting humans in North America and Europe (Barbour 2001; Steere et al. 2004). Causative bacterial agents of the disease, the spirochetes *Borrelia burgdorferi*, *Borrelia garinii*, and *Borrelia afzelii*, are transmitted by hard-bodied ticks of the genus *Ixodes*. Infected ticks transmit Lyme disease *Borreliae* to humans during feeding, which can result in a localized infection (erythema migrans) at the site of the tick bite. Disseminated and chronic stages of infection follow; these are characterized by neurological, cardiac, and arthritic manifestations of disease. Infection with Lyme disease *Borrelia* can last from months to years due to avoidance of the host immune response, and key to its successful evasion tactics is recombination at the *vls* locus located at the telomeric end of a 28-kb linear plasmid (Norris 2006; Palmer et al. 2009; Zhang et al. 1997). Recombinational switching at the *vls* locus results in sequence variation of the surface protein, VlsE, thus altering its antigenic properties and allowing the spirochete to evade the host's antibody-mediated response (Zhang et al. 1997; Zhang and Norris 1998a, b).

A schematic of the *vls* locus is shown in Fig. 6.1. The locus consists of the *vlsE* expression site that encodes the 35 kDa lipoprotein VlsE, located 82 bp from the right telomere end of the linear plasmid, lp28-1 (Norris 2006; Zhang et al. 1997; Zhang and Norris 1998a). Also included in the *vls* locus is a tandem array of 15 silent cassettes (*vls2-16*; each approximately 500 bp in length) that are oriented in the opposite direction of the *vlsE* gene. The *vlsE* locus and the silent cassettes are separated by a short intergenic region (~160 bp) that contains a near-perfect inverted

T. Bankhead, Ph.D. (✉)
Department of Veterinary Microbiology and Pathology

Paul G. Allen School for Global Animal Health,
Washington State University, Pullman, Washington, USA
e-mail: tbankhead@vetmed.wsu.edu

M.E. Embers (ed.), *The Pathogenic Spirochetes: Strategies for Evasion of Host Immunity and Persistence*, DOI 10.1007/978-1-4614-5404-5_6,

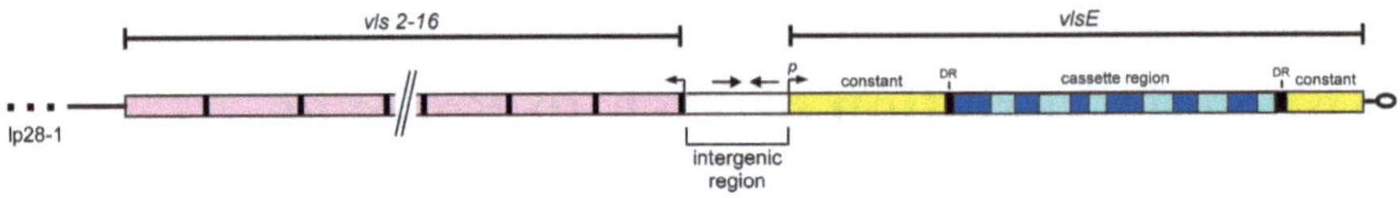

Fig. 6.1 The *vls* locus of *B. burgdorferi*. Illustration showing the arrangement of the *vls* expression site, *vlsE*, and the contiguous array of 15 silent cassettes comprising the *vls* locus on the right telomeric end of lp28-1. The six variable regions of the central *vlsE* cassette are colored *light blue*, while the six invariant regions are colored *dark blue*. The *black bars* flanking the *vlsE* cassette region and silent cassettes represent the 17 bp direct repeats. The silent cassettes (*vls2-16*) are not drawn to scale. *Arrows* positioned at the beginning of *vlsE* and silent cassettes indicate the respective orientations. The *arrows* located within the intergenic region denote the inverted DNA repeat. DR, 17 bp direct repeat; *p*, *vlsE* promoter

repeat of a 51 bp sequence capable of forming a highly stable DNA stem loop (Hudson et al. 2001). In addition, a portion of the promoter sequence required for *vlsE* expression is located within this inverted repeat (Hudson et al. 2001). The *vlsE* expression region is comprised of a central variable cassette (Fig. 6.1, blue region) that is flanked by constant regions (yellow). At the junction of the variable and constant regions are 17 bp direct repeats. With the exception of the final cassette, 17 bp direct repeat sequences are also found at either end of the silent cassettes. The *vlsE* cassette region exhibits roughly 90 % sequence identity with each of the silent cassettes (Zhang et al. 1997), and most of the sequence differences reside in six variable regions (Fig. 6.1, light blue) that are flanked by six highly conserved, or invariant, sequences (dark blue).

6.2 Evidence for the Role of *vls* Locus in Immune Evasion and Persistence

A number of studies over the years have provided strong evidence implicating the importance of VlsE antigenic variation for *B. burgdorferi* persistence. Assessing the rate of variation has proven difficult, primarily because it cannot be detected in vitro and *B. burgdorferi* is not readily sampled from the blood. However, recombination events have been detected as early as 4 days after infection of mice, and continue to occur throughout infection (Zhang and Norris 1998b). The sequence changes primarily occur within the six variable regions, and VlsE variation occurs to such a degree that each clone examined from a single mouse skin biopsy can have a different *vlsE* sequence after only 28 days post infection. Early work also demonstrated that antibodies specific for the variable regions of VlsE were produced during experimental infection of mice (McDowell et al. 2002). Interestingly, *vlsE* antigenic switching in *B. burgdorferi* only occurs during mammalian infections, suggesting that host factors may be required to activate the antigenic variation process (Indest et al. 2001).

More evidence for the role of the *vls* system in immune avoidance came from studies involving the *vls*-resident plasmid, lp28-1 (Labandeira-Rey and Skare 2001;

Purser and Norris 2000). Clones lacking lp28-1 were shown to exhibit an infectivity phenotype whereby these spirochetes were able to disseminate to tissue sites but were unable to persist in the murine host. Notably, these same clones are capable of long-term survival in severe-combined immunodeficient (SCID) mice which lack an effective antibody response (Labandeira-Rey et al. 2003; Purser et al. 2003). Lp28-1-deficient isolates also grow normally in a dialysis membrane chamber implanted in the peritoneal cavity of rats, where exposure to either antibodies or immune cells is restricted (Purser et al. 2003). Finally, immunocompetent mice infected with an lp28-1 minus strain complemented with only the *vlsE* gene (*sans* the *vls* silent cassettes) are able to clear infection, demonstrating that it is not the mere presence of VlsE that provides the capacity for persistent infection, but rather the ability to undergo *vls* antigenic variation to produce VlsE variants (Lawrenz et al. 2004). Together, these studies provided a strong indication that lp28-1, and presumably the *vls* locus, is required for persistence only in the presence of an effective humoral immune response. Nevertheless, definitive evidence for the role of VlsE antigenic variation in persistence remained elusive, prior to the recent generation of a genetic knockout of the *vls* locus.

The construction of a *vls* knockout mutant in *B. burgdorferi* was made possible by utilizing a newly developed technology involving targeted deletion (see Fig. 6.2a). This method involves integration of a plasmid containing a replicated telomere within a chosen linear plasmid of *B. burgdorferi* (Bankhead and Chaconas 2007; Beaurepaire and Chaconas 2005; Chaconas 2005). The end result of the internal placement of this DNA site is deletion of DNA through the action of the *Borrelia* telomere resolvase, ResT (Chaconas 2005; Kobryn and Chaconas 2002). Spirochetes that lost the *vls* locus due to telomere-mediated removal were completely cleared from immunocompetent C3H mice by 21 days post infection (Bankhead and Chaconas 2007), matching the phenotype observed with the complete loss of lp28-1. The *vls* knockout clones exhibited long-term survival in SCID mice, consistent with the findings that the lp28-1 plasmid is not required for persistent infection in the absence of an adaptive immune response (Labandeira-Rey et al. 2003; Purser et al. 2003; Lawrenz et al. 2004), confirming the hypothesis that *vls* recombination functions to evade the humoral immune response in the mouse host (Norris 2006; Zhang et al. 1997).

Targeted deletion was also utilized to obtain lp28-1 mutants containing mainly the *vls* locus and the necessary genes for autonomous replication of the plasmid (Fig. 6.2b). The only other potential genes retained on the lp28-1 mutant plasmid were three small open reading frames predicted to encode for proteins consisting of 56, 43, and 82 amino acids (*bbf27*, *bbf28*, and *bbf30*, respectively). Infectivity experiments showed that these *B. burgdorferi* clones are fully infectious and persistent in immunocompetent mice, providing further evidence that the silent *vls* cassettes and *vlsE* are involved in spirochete persistence (Bankhead and Chaconas 2007). Moreover, these mutant clones carried out *vlsE* recombination in immunocompetent C3H mice, indicating that protein factors required for antigenic switching are likely not carried on lp28-1, but encoded elsewhere. Although the truncated lp28-1 plasmid in these mutant clones contained the *bbf27*, *bbf28*, and *bbf30* open reading frames in addition to *vlsE* and the silent cassettes, the respective protein

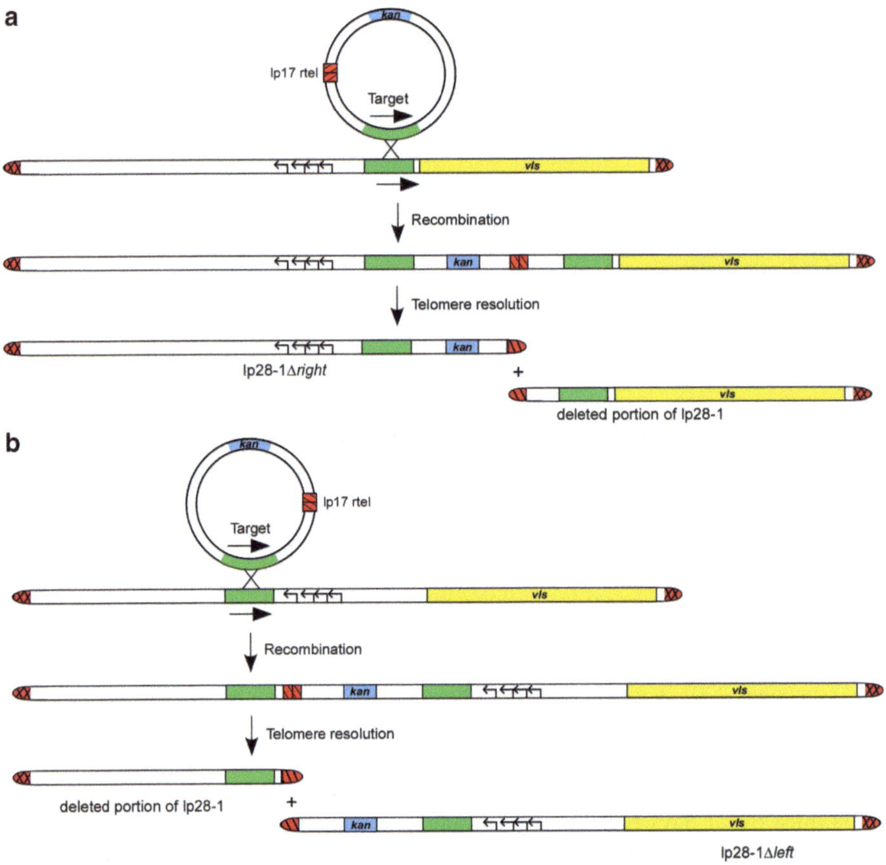

Fig. 6.2 Targeted deletion of the *vls* locus by insertion of a replicated telomere. (**a**) Schematic of the construction strategy for the lp28-1Δ*right* plasmid. The target sequence for insertion (*green*) is chosen so that only the *vls* locus would be deleted from lp28-1. The genes encoding plasmid maintenance proteins that have been previously shown to allow autonomous replication are shown as *black-lined arrows*. (**b**) Schematic of the construction strategy for the lp28-1Δ*left* plasmid. The target sequence for insertion (*green*) was chosen so that only the genes encoding proteins that allow autonomous replication of lp28-1 (shown as *arrows* arranged from left to right), and the right side of lp28-1 would remain. Hairpin telomeres are shown as *red* hatched regions. Figure adapted from (Bankhead and Chaconas 2007)

products are likely too small to have any role in *vlsE* switching. However, a recent study found that one or more of these putative genes may have a role in persistence through involvement in the downregulation of the immunodominant outer surface protein C (OspC) (Embers et al. 2008). Previous studies have shown that OspC is downregulated in *B. burgdorferi* shortly after establishing infection in the animal host, and it has been suggested that this provides the spirochete with a mechanism to avoid clearance mediated by anti-OspC antibodies (Fung et al. 1994; Liang et al. 2004a; Xu et al. 2006). Nevertheless, the generation of the *vls* deletion mutant

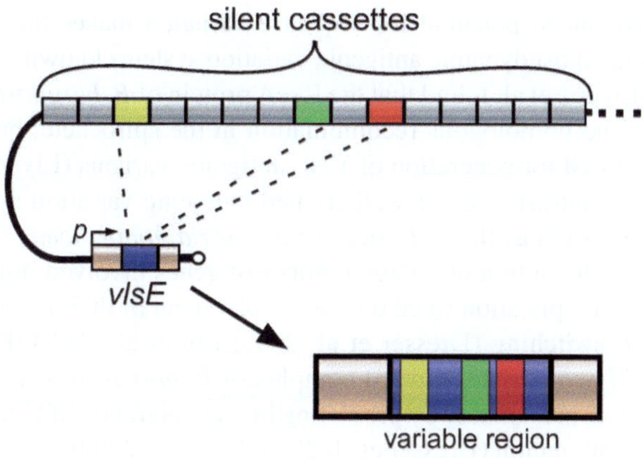

Fig. 6.3 Overview of *vlsE* antigenic switching in *B. burgdorferi*. Variant-specific segments act as a source of DNA for non-reciprocal recombination events with the *vlsE* expression locus. Through this process, segments of the variable region (*blue*) are replaced by sections of varied length and location from the donor sequences. In the example shown, three sequential gene conversion events (represented by *dashed lines*) occur within each expression site through recombination with the colored donor sections (*yellow*, *green*, or *red*) to generate a new expression site sequence with a mosaic structure. Figure adapted from (Palmer et al. 2009)

offered the first definitive evidence that VlsE antigenic variation is absolutely crucial for persistent infection by *B. burgdorferi*.

6.3 Mechanism of *vlsE* Recombination

Although gene conversion has been implicated in *vlsE* antigenic variation, little is known about the exact mechanism of recombination and the proteins involved in this process. Since the discovery of the *vls* system just over a decade ago, progress towards elucidating the mechanistic details of recombinational switching has been impeded by two factors. First, switching does not occur in culture and instead requires passage through a mammalian host (Norris 2006), and second, the *vls* locus is genetically unstable when cloned in *E. coli* (Zhang et al. 1997; Casjens et al. 2000).

The proposed model for *vlsE* antigenic switching involves a non-reciprocal gene conversion mechanism whereby segments within the *vlsE* central cassette region are replaced by sections of varied length and location from the silent cassettes (Fig. 6.3). Only the cassette region of *vlsE* displays sequence variation resulting from gene conversion; the sequence and organization of the silent cassettes remain unaltered (Zhang and Norris 1998b). An elaborate study by Norris and colleagues found that the *vlsE* antigenic variation system promotes both short and long recombination events within each cassette region (Coutte et al. 2009). In addition to the gene conversion events, template-independent changes lead to important sequence variation of *vlsE*. The end result of these accumulated changes is a new *vlsE* sequence with a mosaic structure, and the total combined potential for sequence variation makes the *vlsE* switching process one of the most dynamic antigenic variation systems known.

A study by Liveris et al. found that the RecA protein of *B. burgdorferi* is involved in DNA repair and homologous recombination in the spirochete, but surprisingly, RecA is not required for generation of VlsE antigenic variants (Liveris et al. 2008). This is in sharp contrast to other well-studied antigenic variation systems that are RecA-dependent, such as the *pilE* locus of *Neisseria gonorrhoeae* (Koomey et al. 1987). Similarly, disruption of a large number of genes involved in DNA recombination, repair, and replication ruled out the involvement of their respective encoded proteins in *vlsE* switching (Dresser et al. 2009; Lin et al. 2009). However, these same studies did find that the RuvAB complex of *B. burgdorferi* is required for *vls* antigenic variation, presumably by promoting branch migration of Holliday junctions during *vlsE* recombination (Dresser et al. 2009; Lin et al. 2009). Interestingly, while *B. burgdorferi* contains genes that encode the branch-migration complex RuvAB, these spirochetes are noticeably lacking the gene for the Holliday junction (HJ) resolvase, RuvC. In fact, no recognizable HJ resolvase exists in the *B. burgdorferi* genomic sequence, which suggests that a specialized HJ resolvase may be expressed in the spirochete.

6.4 The VlsE Lipoprotein

The crystal structure of the recombinant-variant VlsE outer surface lipoprotein has been solved, and the resulting three-dimensional structure presents a protein fold very different from most other solved protein structures (Eicken et al. 2002). However, crystallography of the VlsE protein did reveal a structure similar to the variable surface protein (Vsp) of *B. hermsii*, as well as the OspC protein of *B. burgdorferi*. The overall primary structure of VlsE consists of N-terminal and C-terminal constant domains that flank the central cassette region. The C terminus is positioned close to the N terminus, and the loops containing the variable regions are localized at the distal surface of the protein. Gel filtration chromatography of VlsE showed that the protein is primarily monomeric in solution, although the observed interface in the crystal structure raises the possibility that VlsE could exist as a dimer on the bacterial cell surface.

The observed position of the variable regions at the distal surface of VlsE confirms their accessibility to antibodies. In contrast to the variant regions, the invariant regions of the central cassette have very limited surface exposure. Despite the apparent inaccessibility to antibodies, a strong humoral response is mounted against the conserved regions of VlsE (most notably invariant region 6). In fact, VlsE-based recombinant proteins or synthetic peptides are used for the immunodiagnosis of Lyme disease (Bacon et al. 2003; Embers et al. 2007a, b; Liang et al. 1999; Schulte-Spechtel et al. 2003). The function of VlsE independent of its antigenic variation properties is not currently known. There have been previous indications that VlsE may be involved in tissue tropism, including reports that VlsE production is elevated during mammalian infection (Liang et al. 2004b) with higher levels observed in spirochetes recovered from joint and skin tissues than from heart tissue (Crother et al. 2003). However, a recent study found no obvious differences in the amino acid sequence of VlsE variants recovered from different tissue sites, suggesting the absence of any VlsE role in tissue tropism (Coutte et al. 2009).

6.5 Future Directions

Past research involving *B. burgdorferi* antigenic variation has focused on either analyzing *vlsE* sequence changes, the dynamics of those changes, or the role of *vlsE* switching in *B. burgdorferi* pathogenesis. However, studies that address the fundamental question of how antigenic variation occurs in *B. burgdorferi* have been seriously lacking. In addition, the use of molecular approaches to study *vls* recombination has been limited. In fact, despite the widespread nature of this effective pathogenic strategy, few molecular details of the recombinational switching process have been reported for any organism.

6.5.1 Mechanistic Elements

Future work investigating the molecular details of the *vlsE* switching process will likely focus on several currently unknown aspects. For example, the subtelomeric positioning of the *vls* locus is a common feature associated with many antigenic variation systems found in both eubacterial and protozoal pathogens (Barry et al. 2003; Kraemer and Smith 2006). The reasons behind this phenomenon are still unclear and of great interest to those studying antigenic variation. The role of the 17 bp direct repeats that demarcate the *vlsE* variable region and flank the silent cassettes also remains unknown. Although it was originally thought that these sequences might play a role in recombinational switching (Zhang et al. 1997), sequence analysis showed that these repeats are not well conserved across *Borrelia* species and even within some species, casting doubt on a proposed role in recombination (Wang et al. 2003). Nevertheless, any potential role of these direct repeat elements for *vlsE* recombination has not been directly investigated. Also remaining to be determined is whether the *vlsE*-silent cassette intergenic region is required for *vls* antigenic variation. Finally, a recent study demonstrated that antigenic switching does not occur with a *trans* copy of *vlsE* during infection, suggesting a possible *cis* requirement for the *vls* recombination mechanism (Lawrenz et al. 2004). However, the *trans* copy of *vlsE* was carried on a circular shuttle vector, as opposed to the linear lp28-1 plasmid in which *vlsE* normally resides, and whether this may have inhibited recombination between the shuttle plasmid and lp28-1 remains to be tested.

6.5.2 Biological Role of VlsE

A role for the VlsE protein other than providing an antigenic disguise is not currently known, but it has been proposed that the protein might function in other forms of immune evasion (Liang et al. 2002; Philipp et al. 2001). Although a number of other surface proteins exist that are immunogenic, VlsE is the only known *B. burgdorferi* antigen that exhibits variation of its surface epitope. A daunting question regarding *B. burgdorferi* immune escape has been how such a feat is accomplished through sequence variation of this single lipoprotein, despite the presence of a substantial number of additional antigens residing on the bacterial surface. Many pathogens utilize their antigenically variable proteins in a number of ways as an evasion strategy, and several models have been suggested for how VlsE might become the primary target for the host immune response. In other pathogen systems, such as those found in *Plasmodium falciparum* and *Trypanosoma brucei*, the variable surface protein hides other antigens by essentially coating the surface of the organism (Dzikowski et al. 2006; Taylor and Rudenko 2006). It is possible that VlsE may act as a shield to obscure the epitopes of other surface antigens. In fact, crystallography data suggest that the binding of VlsE to other proteins on the membrane surface may block antibody binding to the lateral surface of VlsE (Eicken et al. 2002). This in turn may protect other surface antigens from antibody recognition. A precedent for this type of interaction has been demonstrated in studies with the *Borrelia* protein P66, in which the protein is protected from proteolytic cleavage in Lyme Borrelia expressing high levels of the outer surface protein OspA (Bunikis and Barbour 1999). Unfortunately, results from a recent study investigating the possibility that VlsE directly masks other surface antigens were inconclusive (Bankhead and Chaconas 2007).

It has also been proposed that VlsE might be a T-cell-independent antigen that could directly stimulate B cells (Bankhead and Chaconas 2007; Philipp et al. 2001). The resulting humoral response generated by VlsE may serve to override antibody production against other potential surface antigens. In support of this, *B. burgdorferi* clones expressing both variable and static forms of VlsE on the cell surface are not cleared in immunocompetent mice, suggesting that the immune response may have focused preferentially on the multiple variable forms (Lawrenz et al. 2004). Moreover, it has been observed that *vls* mutant *B. burgdorferi* clones complemented with a non-switchable form of *vlsE* are cleared faster in immunocompetent mice relative to non-VlsE-complemented mutants (Bankhead and Chaconas 2007). It is conceivable that this outcome could be the result of direct stimulation of B cells by VlsE, and this modulating ability could result in more effective clearance of these spirochetes due to the absence of VlsE antigenic variation in those clones. Fortunately, the generation of a *vls* knockout mutant of *B. burgdorferi* now provides experimental opportunities to begin answering questions relating to the protective role that VlsE may play during mammalian infection.

References

Bacon RM, Biggerstaff BJ, Schriefer ME, Gilmore RD Jr, Philipp MT, Steere AC, Wormser GP, Marques AR, Johnson BJ (2003) Serodiagnosis of Lyme disease by kinetic enzyme-linked immunosorbent assay using recombinant VlsE1 or peptide antigens of Borrelia burgdorferi compared with 2-tiered testing using whole-cell lysates. J Infect Dis 187(8):1187–1199

Bankhead T, Chaconas G (2007) The role of VlsE antigenic variation in the Lyme disease spirochete: persistence through a mechanism that differs from other pathogens. Mol Microbiol 65(6):1547–1558

Barbour AG (2001) Borrelia: a diverse and ubiquitous genus of tick-borne pathogens. In: Scheld MW, Craig WA, Hughes JM (eds) Emerging Infections 5. American Society for Microbiology, Washington, D.C, pp 153–173

Barry JD, Ginger ML, Burton P, McCulloch R (2003) Why are parasite contingency genes often associated with telomeres? Int J Parasitol 33(1):29–45

Beaurepaire C, Chaconas G (2005) Mapping of essential replication functions of the linear plasmid lp17 of B. burgdorferi by targeted deletion walking. Mol Microbiol 57(1):132–142

Bunikis J, Barbour AG (1999) Access of antibody or trypsin to an integral outer membrane protein (P66) of *Borrelia burgdorferi* is hindered by Osp lipoproteins. Infect Immun 67(6):2874–2883

Casjens S, Palmer N, Van Vugt R, Huang WH, Stevenson B, Rosa P, Lathigra R, Sutton G, Peterson J, Dodson RJ et al (2000) A bacterial genome in flux: the twelve linear and nine circular extra-chromosomal DNAs in an infectious isolate of the Lyme disease spirochete *Borrelia burgdorferi*. Mol Microbiol 35(3):490–516

Chaconas G (2005) Hairpin telomere and genome plasticity in *Borrelia*: all mixed up in the end. Mol Microbiol 58:625–635

Coutte L, Botkin DJ, Gao L, Norris SJ (2009) Detailed analysis of sequence changes occurring during vlsE antigenic variation in the mouse model of Borrelia burgdorferi infection. PLoS Pathog 5(2):e1000293

Crother TR, Champion CI, Wu XY, Blanco DR, Miller JN, Lovett MA (2003) Antigenic composition of Borrelia burgdorferi during infection of SCID mice. Infect Immun 71(6):3419–3428

Dresser AR, Hardy PO, Chaconas G (2009) Investigation of the genes involved in antigenic switching at the vlsE locus in Borrelia burgdorferi: an essential role for the RuvAB branch migrase. PLoS Pathog 5(12):e1000680

Dzikowski R, Templeton TJ, Deitsch K (2006) Variant antigen gene expression in malaria. Cell Microbiol 8(9):1371–1381

Eicken C, Sharma V, Klabunde T, Lawrenz MB, Hardham JM, Norris SJ, Sacchettini JC (2002) Crystal structure of Lyme disease variable surface antigen VlsE of Borrelia burgdorferi. J Biol Chem 277(24):21691–21696

Embers ME, Jacobs MB, Johnson BJ, Philipp MT (2007a) Dominant epitopes of the C6 diagnostic peptide of Borrelia burgdorferi are largely inaccessible to antibody on the parent VlsE molecule. Clin Vaccine Immunol 14(8):931–936

Embers ME, Wormser GP, Schwartz I, Martin DS, Philipp MT (2007b) Borrelia burgdorferi spirochetes that harbor only a portion of the lp28-1 plasmid elicit antibody responses detectable with the C6 test for Lyme disease. Clin Vaccine Immunol 14(1):90–93

Embers ME, Alvarez X, Ooms T, Philipp MT (2008) The failure of immune response evasion by linear plasmid 28-1-deficient Borrelia burgdorferi is attributable to persistent expression of an outer surface protein. Infect Immun 76(9):3984–3991

Fung BP, McHugh GL, Leong JM, Steere AC (1994) Humoral immune response to outer surface protein C of Borrelia burgdorferi in Lyme disease: role of the immunoglobulin M response in the serodiagnosis of early infection. Infect Immun 62(8):3213–3221

Hudson CR, Frye JG, Quinn FD, Gherardini FC (2001) Increased expression of *Borrelia burgdorferi* vlsE in response to human endothelial cell membranes. Mol Microbiol 41(1):229–239

Indest KJ, Howell JK, Jacobs MB, Scholl-Meeker D, Norris SJ, Philipp MT (2001) Analysis of Borrelia burgdorferi vlsE gene expression and recombination in the tick vector. Infect Immun 69(11):7083–7090

Kobryn K, Chaconas G (2002) ResT, a telomere resolvase encoded by the Lyme disease spirochete. Mol Cell 9(1):195–201

Koomey M, Gotschlich EC, Robbins K, Bergstrom S, Swanson J (1987) Effects of recA mutations on pilus antigenic variation and phase transitions in Neisseria gonorrhoeae. Genetics 117(3):391–398

Kraemer SM, Smith JD (2006) A family affair: var genes, PfEMP1 binding, and malaria disease. Curr Opin Microbiol 9(4):374–380

Labandeira-Rey M, Skare JT (2001) Decreased infectivity in *Borrelia burgdorferi* strain B31 is associated with loss of linear plasmid 25 or 28–1. Infect Immun 69(1):446–455

Labandeira-Rey M, Seshu J, Skare JT (2003) The Absence of Linear Plasmid 25 or 28–1 of *Borrelia burgdorferi* Dramatically Alters the Kinetics of Experimental Infection via Distinct Mechanisms. Infect Immun 71(8):4608–4613

Lawrenz MB, Wooten RM, Norris SJ (2004) Effects of *vlsE* complementation on the infectivity of *Borrelia burgdorferi* lacking the linear plasmid lp28-1. Infect Immun 72(11):6577–6585

Liang FT, Steere AC, Marques AR, Johnson BJ, Miller JN, Philipp MT (1999) Sensitive and specific serodiagnosis of Lyme disease by enzyme-linked immunosorbent assay with a peptide based on an immunodominant conserved region of Borrelia burgdorferi vlsE. J Clin Microbiol 37(12):3990–3996

Liang FT, Jacobs MB, Bowers LC, Philipp MT (2002) An immune evasion mechanism for spirochetal persistence in Lyme borreliosis. J Exp Med 195(4):415–422

Liang FT, Brown EL, Wang T, Iozzo RV, Fikrig E (2004a) Protective niche for Borrelia burgdorferi to evade humoral immunity. Am J Pathol 165(3):977–985

Liang FT, Yan J, Mbow ML, Sviat SL, Gilmore RD, Mamula M, Fikrig E (2004b) *Borrelia burgdorferi* changes its surface antigenic expression in response to host immune responses. Infect Immun 72(10):5759–5767

Lin T, Gao L, Edmondson DG, Jacobs MB, Philipp MT, Norris SJ (2009) Central role of the holliday junction helicase RuvAB in vlsE recombination and infectivity of Borrelia burgdorferi. PLoS Pathog 5(12):e1000679

Liveris D, Mulay V, Sandigursky S, Schwartz I (2008) Borrelia burgdorferi vlsE antigenic variation is not mediated by RecA. Infect Immun 76(9):4009–4018

McDowell JV, Sung SY, Hu LT, Marconi RT (2002) Evidence That the Variable Regions of the Central Domain of VlsE Are Antigenic during Infection with Lyme Disease Spirochetes. Infect Immun 70(8):4196–4203

Norris SJ (2006) Antigenic variation with a twist - the *Borrelia* story. Mol Microbiol 60(6): 1319–1322

Palmer GH, Bankhead T, Lukehart SA (2009) 'Nothing is permanent but change'- antigenic variation in persistent bacterial pathogens. Cell Microbiol 11(12):1697–1705

Philipp MT, Bowers LC, Fawcett PT, Jacobs MB, Liang FT, Marques AR, Mitchell PD, Purcell JE, Ratterree MS, Straubinger RK (2001) Antibody response to IR6, a conserved immunodominant region of the VlsE lipoprotein, wanes rapidly after antibiotic treatment of *Borrelia burgdorferi* infection in experimental animals and in humans. J Infect Dis 184(7):870–878

Purser JE, Norris SJ (2000) Correlation between plasmid content and infectivity in *Borrelia burgdorferi*. Proc Natl Acad Sci USA 97(25):13865–13870

Purser JE, Lawrenz MB, Caimano MJ, Howell JK, Radolf JD, Norris SJ (2003) A plasmid-encoded nicotinamidase (PncA) is essential for infectivity of *Borrelia burgdorferi* in a mammalian host. Mol Microbiol 48(3):753–764

Schulte-Spechtel U, Lehnert G, Liegl G, Fingerle V, Heimerl C, Johnson BJ, Wilske B (2003) Significant improvement of the recombinant Borrelia-specific immunoglobulin G immunoblot test by addition of VlsE and a DbpA homologue derived from Borrelia garinii for diagnosis of early neuroborreliosis. J Clin Microbiol 41(3):1299–1303

Steere AC, Coburn J, Glickstein L (2004) The emergence of Lyme disease. J Clin Invest 113(8):1093–1101

Taylor JE, Rudenko G (2006) Switching trypanosome coats: what's in the wardrobe? Trends Genet 22(11):614–620

Wang D, Botkin DJ, Norris SJ (2003) Characterization of the vls antigenic variation loci of the Lyme disease spirochaetes Borrelia garinii Ip90 and Borrelia afzelii ACAI. Mol Microbiol 47(5):1407–1417

Xu Q, Seemanapalli SV, McShan K, Liang FT (2006) Constitutive expression of outer surface protein C diminishes the ability of Borrelia burgdorferi to evade specific humoral immunity. Infect Immun 74(9):5177–5184

Zhang JR, Norris SJ (1998a) Genetic variation of the *Borrelia burgdorferi* gene *vlsE* involves cassette-specific, segmental gene conversion. Infect Immun 66(8):3698–3704

Zhang JR, Norris SJ (1998b) Kinetics and in vivo induction of genetic variation of *vlsE* in *Borrelia burgdorferi*. Infect Immun 66(8):3689–3697

Zhang JR, Hardham JM, Barbour AG, Norris SJ (1997) Antigenic variation in Lyme disease borreliae by promiscuous recombination of VMP-like sequence cassettes. Cell 89(2):275–285

Chapter 7
Modulation of Cytokine Signaling by *B. burgdorferi*

Vida A. Dennis, Aarti Gautam, and Saurabh Dixit

7.1 Introduction

Lyme disease, caused by the spirochete *Borrelia burgdorferi sensu lato,* is a multisystem disease and the most prevalent tick-borne disease world-wide. The spirochete infects humans and other mammals via the bite of infected *Ixodes* ticks (Burgdorfer et al. 1982). Following infection, spirochetes disperse to multiple organs including the skin, central and peripheral nervous systems, bladder, joints, eyes, and heart and persist in them for a long time causing many of the diverse manifestations of Lyme disease. Manifestations of the disease include, among others, acute or chronic arthritis, carditis, and neuroborreliosis (Steere 2001; Steere et al. 2004)

Lyme disease is thought to ensue largely as a consequence of both acute and chronic inflammatory responses, induced either by the spirochete or by spirochetal antigens left in tissues after bacterial demise. Invasion of the mammalian host with spirochetes is accompanied by the activation of inflammatory pathways that lead to the release of inflammatory mediators and an influx of inflammatory cells (Glickstein et al. 2001; Guerau-de-Arellano et al. 2005) that result in many of the ensuing clinical manifestations of Lyme disease. The potent stimulatory properties of the spirochete, its lipoproteins and/or spirochete antigens left in tissues after bacterial demise, are thought to be responsible for inflammatory foci in tissues, as well as other host responses against the infection, all of which have been linked to Lyme disease pathogenesis.

V.A. Dennis, Ph.D. (✉) • S. Dixit, Ph.D.
Department of Biological Sciences, Center for Nanobiotechnology and Life Science Research, Alabama State University, 1627 Hall Street, Montgomery, AL 36104, USA
e-mail: vdennis@alasu.edu; sdixit5@gmail.com

A. Gautam, Ph.D.
Systems and Integrative Biology, US Army Medical Command, USACEHR
568 Doughten Drive, Fort Detrick, MD 21702, USA

M.E. Embers (ed.), *The Pathogenic Spirochetes: Strategies for Evasion of Host Immunity and Persistence*, DOI 10.1007/978-1-4614-5404-5_7,

7.2 Cytokines and Chemokines in Lyme Disease

Cytokines and chemokines are regulators of early innate immune responses; they link innate and adaptive immunity, and are thought to play key roles in the pathogenesis of Lyme disease. *B. burgdorferi*-induced inflammatory foci are typically infiltrated with a plethora of pro-inflammatory mediators. The spirochetes along with their lipoproteins have been shown to induce cytokines (G-CSF, IL-10, interleukin (IL)-17, IL-1β, IL-6, IL-12, TNF, IFN-γ, GM-CSF), as well as T-cell (CXCL9, CXCL10), monocyte (CCL2, CCL3, CCL4, CCL5), and neutrophil (IL-8, CXCL1, CXCL2, CXCL5) chemoattractants in several cells and tissue types or body fluids from humans or animal models, e.g., synoviocytes (Shin et al. 2007; Singh et al. 2005), monocytes/macrophages (Giambartolomei et al. 2002; Murthy et al. 2000; Diterich et al. 2003; Dennis et al. 2009), microglia (Myers et al. 2009), lymph node cells (Ganapamo et al. 2000, 2001), erythema migrans skin lesions (Jones et al. 2008; Mullegger et al. 2007), and cerebrospinal fluid (Ramesh et al. 2009). These pro-inflammatory mediators have potent effector functions that overlap extensively to bring about the many components of inflammation, e.g., tissue necrosis (Waage et al. 1987; Casey et al. 1993), chemotaxis of cellular infiltrates, induction of collagenase and prostaglandin secretion (Dayer et al. 1986), bone resorption and cartilage destruction (Arend and Dayer 1995), a plethora of microbicidal effector mechanisms (Nussler et al. 1999), as well as amplification of immune responses. IFN-γ, for example, reportedly can promote inflammation by synergizing with *B. burgdorferi* to potentiate expression of chemokines and cytokines in human endothelial cells (Dame et al. 2007) and mouse macrophages (Sabino et al. 2011), along with enhancing macrophages effector functions for spirochete clearance (Olson et al. 2009). Humans injected intradermally with synthetic analogs for spirochetal lipoproteins (lipopeptides derived from the N termini of the *B. burgdorferi* outer surface protein C) elicited cellular infiltrates comprising monocytes/macrophages, dendritic cells, neutrophils and T cells and marked up-regulation of IL-12, IL-10, IFN-γ, and IL-6 (Salazar et al. 2005). Inflammatory immune responses are of crucial importance for the early containment of infectious pathogen but at the same time have the potential to result in immunopathological consequences. Thus cytokine-induced signal transduction pathways must be tightly regulated to avoid the detrimental consequences of excessive stimulation.

7.3 *B. burgdorferi* Induces the Production of IL-10

Exaggerated, unchecked inflammation in pathogen-induced diseases, or of any origin, can be deleterious to the host and, consequently, several regulatory mechanisms have evolved to control the magnitude and duration of inflammation. One such pivotal regulator of inflammation is the anti-inflammatory cytokine, IL-10. IL-10 plays a pivotal role in limiting the inflammatory response and preventing tissue damage. This is mainly achieved by down-regulating the expression of pro-inflammatory cytokines and chemokines as well as inhibiting effector functions of T cells and mononuclear phagocytes (Donnelly et al. 1999). In addition to these activities, IL-10 anti-inflammatory effects are also directed at regulating the effector functions of B cells, NK cells, cytotoxic cells, mast cells, granulocytes, dendritic cells, keratinocytes, and endothelial cells (Moore et al. 2001).

B. burgdorferi not only triggers the production of pro-inflammatory mediators, but also production of the anti-inflammatory cytokine, IL-10. IL-10 is produced in response to stimulation by *B. burgdorferi* in different cells and tissue types of animal models as reported in mice (Dennis et al. 2006; Miller et al. 2008; Brown et al. 1999; Lazarus et al. 2008), rhesus macaques (Giambartolomei et al. 1998), and human patients (Jones et al. 2008; Strle et al. 2009). IL-10 production is not exclusively linked to outer surface protein A (OspA) and its lipid moiety (Giambartolomei et al. 2002) but can also be produced in response to stimulation with sonicated, heat killed, live or freeze thawed *B. burgdorferi* spirochetes (Dennis et al. 2006). IL-10 expression has been detected in joint tissues (Miller et al. 2008), skin tissues (Lazarus et al. 2006), lymph node cells (Ganapamo et al. 2001), splenocytes (Anguita et al. 1997), glial cells (Rasley et al. 2006), and macrophages (Dennis et al. 2006; Brown et al. 1999; Lazarus et al. 2008) in the murine model of Lyme disease. Macrophages from C57BL/6 J (C57) mice, which are Lyme-disease resistant, produce higher levels of IL-10 than do macrophages from the disease-susceptible C3H/HeN (C3H) mice in response to *B. burgdorferi* or its lipoproteins (Brown et al. 1999). There are numerous reports that *B. burgdorferi* induces the production of IL-10 *in vitro* in mononuclear cells present in peripheral blood (Giambartolomei et al. 2002; Diterich et al. 2001), monocytes/macrophages, dendritic cells (Sjowall et al. 2005), lymphocytes (Pohl-Koppe et al. 1998), CSF (Cepok et al. 2003), synovium of *B. burgdorferi*-infected patients (Yin et al. 1997), microglia (Cassiani-Ingoni et al. 2006), skin (Salazar et al. 2005), and erythema-migrans skin lesions of *B. burgdorferi*-infected patients (Jones et al. 2008). A study by Yin et al. (1997) showed that IL-10 is produced concomitantly with Th1 cytokines when synoviocytes are incubated with proteins of *B. burgdorferi*. *B. burgdorferi* infection induces a novel population of IL-12-induced IFN-gamma/IL-10-secreting T cells in patients with Lyme arthritis. Not only bacterial antigen, but the tick saliva reportedly also up-regulates IL-10 production that consequently inhibits the production of pro-inflammatory cytokines (Kopecky et al. 1999).

7.4 *In Vitro* Effects IL-10 on *B. burgdorferi*-Induced Inflammation

IL-10 has emerged to be a key cytokine in regulating inflammatory responses elicited by *B. burgdorferi* by controlling the production and function of various pro-inflammatory cytokines. Experiments *in vitro* (Giambartolomei et al. 2002; Diterich et al. 2003; Miller et al. 2008; Lazarus et al. 2008; Pohl-Koppe et al. 1998; Ganapamo et al. 2003; Lisinski and Furie 2002) indicate that in response to *B. burgdorferi* or its lipoproteins, IL-10 dampens pro-inflammatory responses of cells that are involved in innate and acquired immunity. IL-10 limits/potentiates the production of proinflammatory cytokines (including IL-1α, IL-1β, IL-6, IL-12, IL-9, IFN-γ, and TNF-α) and chemokines (CCL3, CCL5, CXCL8, CXCL10, MCP1, G-CSF, CXCL2, CXCL5, CCl2, and CCL3) (Gautam et al. 2011). The Yin et al. (1997) study showed that endogenously produced IL-10 triggered by *B. burgdorferi* can inhibit *B. burgdorferi*-specific lymphocyte proliferation as well as TNF-α and IFN-γ production. By using the human monocytic THP-1 cells, it was demonstrated that endogenously synthesized IL-10 down-modulated the production of IL-6 and IL-12, whereas exogenous IL-10 dampened the production of IL-1β, TNFα, IL-12, and IL-6, as elicited by spirochetal lipoproteins (Murthy et al. 2000). IL-10 was identified as the main mediator involved in *B. burgdorferi* induction of tolerance in human PBMCs (Diterich et al. 2003). Studies by Lisinski and Furie (2002) showed that IL-10 decreases production of CXCL8 in *B. burgdorferi*-stimulated endothelial cells. Interleukin 10 produced by synovial cells inhibits antigen-mediated lympho-proliferation, production of TNF and IFN-γ, and reduces transendothelial migration of monocytes and lymphocytes, thus leading to prevention of tissue lesions (Lisinski and Furie 2002).

7.5 *In Vivo* Effects of IL-10 on *B. burgdorferi*-Induced Inflammation

Earlier studies provided evidence for the *in vivo* anti-inflammatory effects of IL-10 during a *B. burgdorferi* infection in mice. IL-10 was suggested to be a key mediator by down-regulation of Th1 immune responses in mice exposed to UV irradiation followed by *B. burgdorferi* immunization (Brown et al. 1995). Administration of anti-IL-10 antibody to UV-irradiated *B. burgdorferi*-infected mice exacerbated Lyme arthritis and abrogated the suppression of cellular and antibody responses in mice (Brown et al. 2001). Observations of high levels of IL-10 produced by macrophages of C57 mice may, in part, help to explain why these mice are better able to regulate inflammation after infection with *B. burgdorferi* than C3H mice, resulting in less intense Lyme arthritis. The fact that C57 (Lazarus et al. 2008) and C3H (Brown et al. 2008) mice deficient in IL-10 and infected with *B. burgdorferi* develop more severe arthritis and harbor fewer spirochetes in joints than wild-type mice underscores the significance of IL-10 in controlling joint inflammation in Lyme disease. Moreover, IL10–/– mice are significantly better at clearing *B. burgdorferi* spirochetes at target tissue sites than are wild-type mice (Lazarus et al. 2006). The above studies clearly indicate the ability of IL-10 to repress inflammation, and are congruent with results by Lazarus et al. (2006), where TNF-α and IL-6 production by macrophages were greatly enhanced in the C57 IL-10–/– mouse as compared to wild-type controls macrophages. Other investigators have also shown that several cytokine and chemokine genes are up-regulated in joints of *B. burgdorferi*-infected IL-10 deficient mice (Dennis et al. 2006) indicating that the expression of these genes is tightly controlled by IL-10. Sonderegger and colleagues (2012) further demonstrated that elevated numbers of NK cells, NKT cells, α/β T cells, and macrophages, including inflammatory cytokines and chemokines and IFN-γ-producing cells in joint tissues of infected C57 IL-10–/–, contributed to their arthritis development, substantiating IFN-γ synergy with spirochetes to promote inflammation (Dame et al. 2007; Sabino et al. 2011). These studies suggest that IL-10 induced by *B. burgdorferi* and/or its lipoproteins contributes to the control of inflammation in Lyme disease, such that spirochetes would be endowed with a means of controlling the inflammation that they themselves are able to induce. Thus, IL-10 could have an effect both on the innate and the acquired immune response to *B. burgdorferi* and influence the magnitude of the infectious burden as well as disease outcome.

7.6 Mechanism of IL-10 Control of *B. burgdorferi*-Induced Inflammation

Initially shown to regulate T-cell responses, many of the effects of IL-10 on T cell, neutrophils, dendritic cells, and NK cell function are now known to be indirect, being mediated via a direct effect of IL-10 on monocytes/macrophages (Couper et al. 2008). Monocytes/macrophages are the primary IL-10 responders to pathogen-associated molecular patterns (PAMPS), and their activation initiates most inflammatory responses. Apparently, monocytes/macrophages are the main target cells of IL-10 inhibitory effects, and an intact macrophage function is critical for combating *B. burgdorferi* infection or any bacterial infection. Recently, transcriptome-profiling studies were conducted to identify genes induced by stimulation of a mouse J774 macrophage cell line with live spirochetes, and which were perturbed by addition of IL-10 to spirochete cultures (Gautam et al. 2011). IL-10 specifically inhibited the *B. burgdorferi*-induced transcriptional activation of TLR1 (Toll-like receptor1) and TLR2 (Toll-like receptor 2), along with several of their down-stream signaling components, such as IRAK2, IRAK3, TRAF1, and TNFAIP3, leading to activation of NF-κB, an important regulator of cellular events (Bourteele et al. 2007) and MAP kinase signaling pathways (Kobayashi et al. 2002). Proteome analysis further revealed the IL-10 modulation and/or potentiation of RANTES/CCL5, MIP-2/CXCL2, IP-10/CXCL10, MIP1-α/CCL3, G-CSF/CSF3, CXCL1, CXCL5, CCL2, CCL4, IL-6, TNF-α, IL-1α, IL-1β, IFN-γ, and IL-9, all of which have been recognized as components of the *B. burgdorferi*-induced inflammation in cells and tissues (Gautam et al. 2011). TLR-stimulated macrophages induce effectors of the adaptive immune system, such as CD40, CD80, and CD86, to drive T-cell activation and proliferation (Medzhitov and Janeway 1998), as well as IFN-α/IFN-β inducible genes, all of which were inhibited by IL-10 (Gautam et al. 2011). This study documented IL-10-mediated inhibition of effectors of the TLR pathway thus providing some insight into the role of this cytokine in the control of Lyme disease inflammation. Studies have also shown that the mechanism of IL-10-mediated inhibition of pro-inflammatory gene expression involves inhibition of NF-κB or P38 MAPK pathways, destabilization of RNA message (Dokter et al. 1996; Kishore et al. 1999) or other intracellular regulatory mechanisms. Recently, it has become clear that at least three different classes of inhibitors contribute to the negative regulation of cytokine signaling: protein tyrosine phosphatases such as SHP-1 and CD45, the protein inhibitors of activated STATs (PIAS), and the suppressor of cytokine signaling (SOCS) proteins (Hilton 1999; Irie-Sasaki et al. 2001). The focus here will be on SOCS as a mediator in regulating cytokine signaling in macrophages after their exposure to *B. burgdorferi* alone or combined with IL-10.

7.7 Suppressor of Cytokine Signaling Protein

SOCS proteins have been identified as negative feedback inhibitors for various cytokines (Alexander and Hilton 2004). To date, eight members (SOCS1 to SOCS7 and CIS) have been identified in this protein family, all sharing a central Src homology 2 (SH2) domain and a C-terminal conserved domain called the SOCS box (Alexander and Hilton 2004; Greenhalgh and Hilton 2001). SOCS inhibitory effects are derived from the direct interaction of SOCS proteins with cytokine receptors and/or Janus kinases (JAKs), thereby preventing recruitment of signal transducers and activators of transcription (STATs) to the signaling complex (Kinjyo et al. 2002). SOCS induction and action can also be caused by a much broader variety of stimuli and might even act on signaling pathways distinct from JAK/STAT. Recent studies show that SOCS proteins can be induced by TLR-mediated stimuli, and in turn regulate TLR signaling in innate immune cells (Bode et al. 1999; Stoiber et al. 2001; Nakagawa et al. 2002). Besides TLRs, other proteins recognized as being important in activating pro-inflammatory responses within innate immune cells are C-type lectin receptors (CLRs), NOD-like receptors (NLRs), and RIG-l-like receptors (RLRs), which either recognizes microbial patterns extracellularly or in cytosolic compartments (Kumar et al. 2011; Creagh and O'Neill 2006; Berende et al. 2010). These receptors use adaptor proteins like MyD88/TRIF (for TLRs), caspase recruitment domain-containing protein 9 (CARD9) (for CLRs), and interferon-beta promoter stimulator 1(IPS1) (for RLRs) to trigger NF-kB or Interferon regulatory factor (IRF) family members of transcription factors. Together, these pathways culminate in the regulation of gene expression in the nucleus, leading to target cell differentiation, proliferation, survival, apoptosis, or activation (Nicola 1994) and SOCS induction to regulate innate cells inflammatory mediators.

7.8 Induction of SOCS Proteins by *B. burgdorferi*

The ability of SOCS1 to suppress cytokine signaling in immune cells is well substantiated *in vivo* in $SOCS1^{-/-}$ mice which die early because of the development of a systemic disease that results from increased sensitivity to IFN-γ and elevated serum levels of this cytokine (Alexander 2002). In rheumatoid arthritis, where proinflammatory cytokines play an important role in the pathogenesis of the disease, STATs are activated together with elevated SOCS3 expression in joint tissues of patients (Suzuki et al. 2001). The use of *SOCS3* gene therapy suppressed proliferation of synovial fibroblasts, as well as the production of IL-6 *in vitro,* and significantly reduced the onset and progression of collagen- and antigen-induced arthritis in mice (Shouda et al. 2001). SOCS1 and SOCS3 (Alexander 2002), both of which are key physiological regulators of macrophages, play significant roles in the regulation of inflammation. Reportedly, *B. burgdorferi* can induce the expression of SOCS1 and SOCS3 in mouse macrophages (Miller et al. 2008). Stimulation of mouse J774 macrophages with either live or freeze/thawed *B. burgdorferi* spirochetes, L-OspA or the triacylated lipopeptide Pam3Cys augmented SOCS1 and SOCS3 expression levels (Miller et al. 2008). Neither de novo protein synthesis nor endogenously produced IL-10 was required for SOCS1 and SOCS3 to be induced in macrophages by these stimulants. The observation that *B. burgdorferi* or its lipoproteins can alone induce SOCS1 and SOCS3 expression in macrophages indicates that the spirochete may control its own induced inflammation. SOCS expression has been similarly induced in macrophages by other bacteria such as *Listeria monocytogenes* (Stoiber et al. 2001), *Mycobacterium bovis* (Imai et al. 2003), *Mycobacterium tuberculosis* (Pai et al. 2003), *Salmonella enterica* (Uchiya and Nikai 2005), and *Staphylococcus aureus* (Murthy et al. 2000), or bacteria-derived substances such as LPS and CpG-DNA. The expression of SOCS as induced by the above pathogens yielded functional SOCS proteins, as demonstrated by SOCS' ability to inhibit inflammatory cytokine signaling in macrophages via the JAK/STAT pathway. Recent studies have revealed that SOCS induced by TLR stimulation in macrophages inhibited IL-6, GM-CSF, type I IFN, and IFN-γ signaling (Stoiber et al. 2001; Suzuki et al. 2001; Imai et al. 2003). Thus, SOCS1/SOCS3 induction by these pathogens mediates a variety of signaling pathways to suppress macrophage activation and subsequently the inflammatory potential of these cells. *B. burgdorferi* may similarly regulate cytokine signaling pathways in macrophages via induction of SOCS. Consistent with a role for SOCS1 and SOCS3 in *B. burgdorferi*-induced inflammation was the observation that CD14 receptor-mediated recognition of *B. burgdorferi* triggers p38-dependent SOCS, and that reduced SOCS expression in cells resulted in greater expression of cytokines through diminished regulation of the TLR2 pathway (Sahay et al. 2009).

7.9 Induction of SOCS Proteins by IL-10 and *B. burgdorferi*

IL-10 initiates its actions on target cells by binding with high affinity to its tetrameric receptors (Asadullah et al. 2003). Upon increase of IL-10R expression, IL-10 mediates signal transduction and immune responses. The IL-10/IL-10R interaction results in transactivation of the receptor-associated Janus kinases, JAK1 and TYK2, which result in the phosphorylation and stimulation of JAK1 and the subsequent activation of a signaling cascade involving phosphorylation of STAT1 and STAT3 (Heeg and Dalpke 2003). STAT3 has been shown to be essential for the IL-10-mediated inhibition of macrophage inflammatory mediator production (Moore et al. 2001). The phosphorylation of STATs allows them to dimerize and translocate to the nucleus where they stimulate the transcription of cytokine-responsive genes including those that encode the SOCS proteins.

SOCS3, in particular, has been shown to be a major player in the IL-10-mediated inhibition of LPS-induced proinflammatory actions in mouse macrophages. SOCS3 was shown to be essential for IL-10-mediated inhibition of LPS-stimulated production of TNF-α, IL-6, GM-CSF, and nitric oxide (NO) in mouse macrophages (Qasimi et al. 2006). The same study observed that the SH2 domain, SOCS box, and both Tyr^{204} and Tyr^{221} were required for IL-10 inhibition of TNF-α mRNA and protein expression in mouse macrophages (Berlato et al. 2002). Stimulation of mouse J774 macrophages with recombinant mouse IL-10 resulted in up-regulation of SOCS1 and SOCS3 expression (Miller et al. 2008). Further co-stimulation of macrophages with IL-10 and *B. burgdorferi* stimuli (live or sonicated spirochetes, L-OspA) additively augmented the induced SOCS1 and SOCS3 expression. Enhanced expression of SOCS1 and SOCS3 by co-stimulation temporally correlated with the ability of IL-10 to down-modulate the transcription and production of IL-6, IL-12p40, IL-1β, IL-18, and TNF-α. Both SOCS1 and SOCS3 proteins are known to regulate JAK/STAT and non-JAK/STAT cytokines in innate immune cells. As such, both JAK/STAT (IL-6) and non-JAK/STAT cytokines (TNF-α, IL-1β) were inhibited by IL-10 in the presence of augmented SOCS1/SOCS3 (Miller et al. 2008), suggesting that expression of SOCS may be a part of the mechanism of IL-10-mediated inhibition of inflammatory cytokines elicited by *B. burgdorferi* in macrophages. The up-regulation of SOCS1 and SOCS3 expression in macrophages in response to IL-10, together with spirochetes or its lipoproteins, strongly suggests a potential role of SOCS1/SOCS3 involvement in the IL-10-mediated anti-inflammatory activity produced in macrophages. The capacity of *B. burgdorferi* alone to induce SOCS1 and SOCS3 expression in macrophages indicates that the spirochete can control inflammation via a mechanism independent from that involving IL-10, even though both involve SOCS.

7.10 SOCS3 Mediates IL-10 Inhibition of Induced Inflammatory Mediators

The significance of the up-regulation of SOCS by IL-10 combined with *B. burgdorferi* (Bb) was further explored by employing murine whole-genome microarrays to investigate differences in gene expression profiling before and after siRNA silencing of the *Socs3* gene in macrophages, in response to sonicated (Son) Bb and recombinant IL-10 (Dennis et al. 2007). Microarray analyses identified over 1,500 genes that were differentially expressed in macrophages following co-stimulation with sonicated Bb and IL-10 for 24 h. Ingenuity Pathways Analysis (IPA®) identified several canonical pathways and gene ontologies that were overrepresented in the differentially expressed genes at 24 h. Further Gene Ontology analysis revealed amongst genes with altered profiles, the overrepresentation of genes encoding mediators from several pathways, chief amongst which were TLR, JAK/STAT, and p38 MAPK. A comparison of the microarray data before and after SOCS3 knockdown in macrophages identified several molecular mediators from TLR (Figs. 7.1a, b, and 7.2) pathways that may be regulated by SOCS3 expression as induced by sonicated Bb and rIL-10 in these cells (Table 7.1). Those genes which were either (1) down-regulated (green) with Bb + IL-10 and up-regulated in the SOCS3-inhibited cells, or (2) unaffected with Bb + IL-10 and up-regulated in the presence of SOCS3 inhibition are those genes likely affected by the IL-10/SOCS3 pathway for suppression of inflammation. These findings suggest the convergence of mediators from different signaling pathways for optimal SOCS3 expression in macrophages after co-stimulation with sonicated Bb and IL-10. The data demonstrate that silencing of the *Socs3* gene in macrophages aborted the IL-10 anti-inflammatory actions against several inflammatory mediators in response to sonicated Bb and IL-10. This indicates that SOCS3 may play a key role in the IL-10-mediated down-regulation of inflammation induced by sonicated Bb in macrophages.

A proposed model for *B. burgdorferi* and IL-10 control of inflammation in Lyme disease is depicted in Fig. 7.3. This model proposes that both *B. burgdorferi* and IL-10 are involved in regulation of inflammation during the early stage of infection. SOCS1 and SOCS3 would act either as mediators of the IL-10 anti-inflammatory activity or as direct modulators of inflammatory cytokine signaling in macrophages. IL-10 would regulate *B. burgdorferi*-induced inflammation by inhibiting multiple signaling pathways to control inflammation. The IL-10 anti-inflammatory effect may be mediated by SOCS1 and SOCS3. On the other hand, *B. burgdorferi* spirochetes may regulate their own induced inflammation by serving as mediators of a variety of cytokine signaling pathway in macrophages via SOCS1 and SOCS3 induction to suppress macrophage activation and subsequently, the inflammatory potential of these cells. The full activation of macrophages enhances their capacity to become microbicidal, synthesize membrane receptors, and secrete molecules that function in a proinflammatory immune response. Thus, SOCS1 and SOCS3, as induced by *B. burgdoferi,* affect signal transduction pathways to reduce production of inflammatory mediators by macrophages. Future studies should be directed at subjecting the proposed model to experimental investigations both *in vitro* and in the mouse model of Lyme disease.

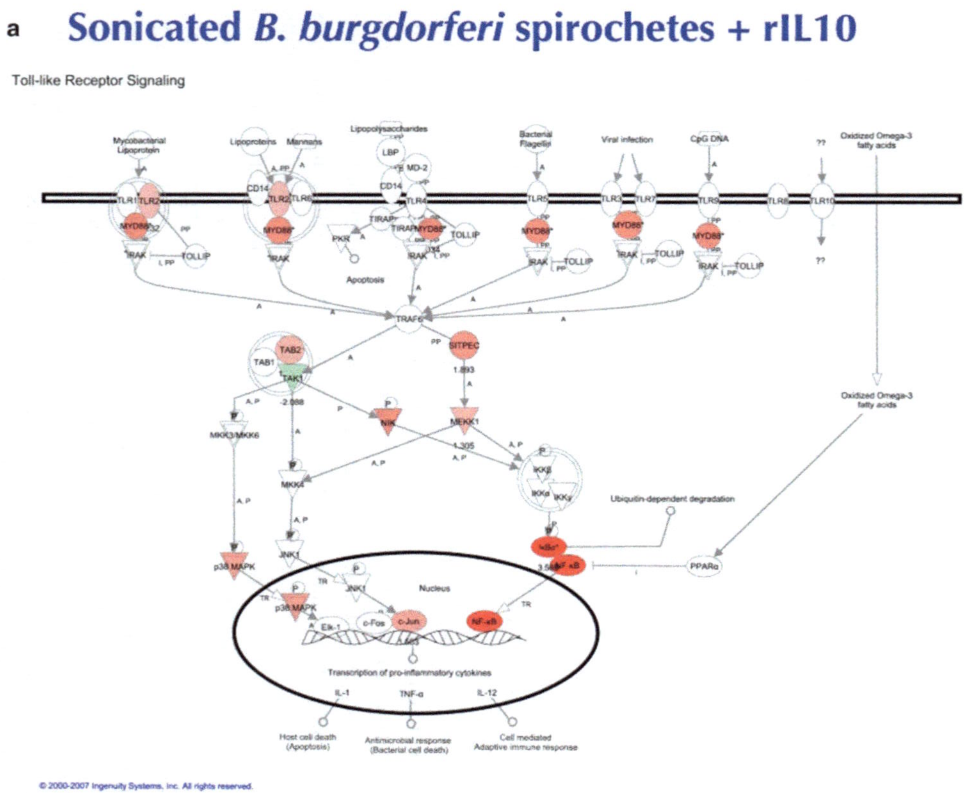

Fig. 7.1 Perturbation of the Toll-like receptor signaling pathway in response to co-stimulation with sonicated *B. burgdorferi* spirochetes + rIL-10 (**a**) or sonicated *B. burgdorferi* spirochetes + rIL-10 after SOCS3 silencing (**b**). Microarrays were performed after 24 h of co-stimulation. Shown are genes significantly down-regulated (*green*) and up-regulated (*red*). The Ingenuity Pathways® legend indicates the type of gene product and interaction of products in the pathway

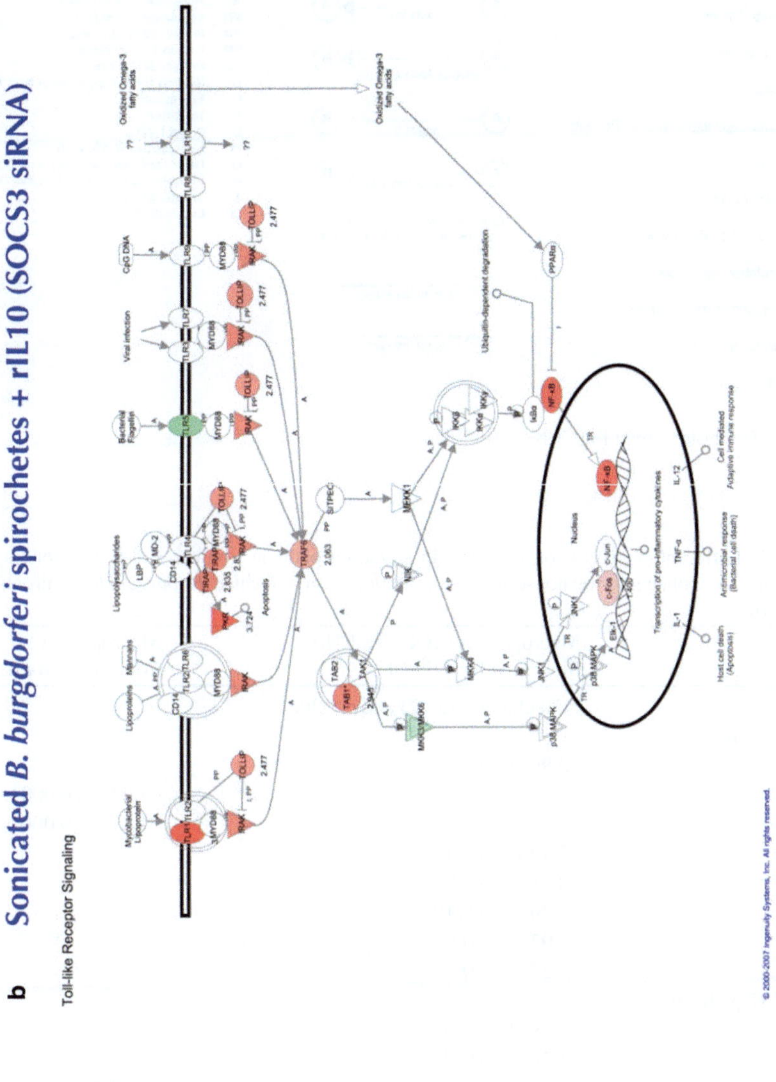

Fig. 7.1 (continued)

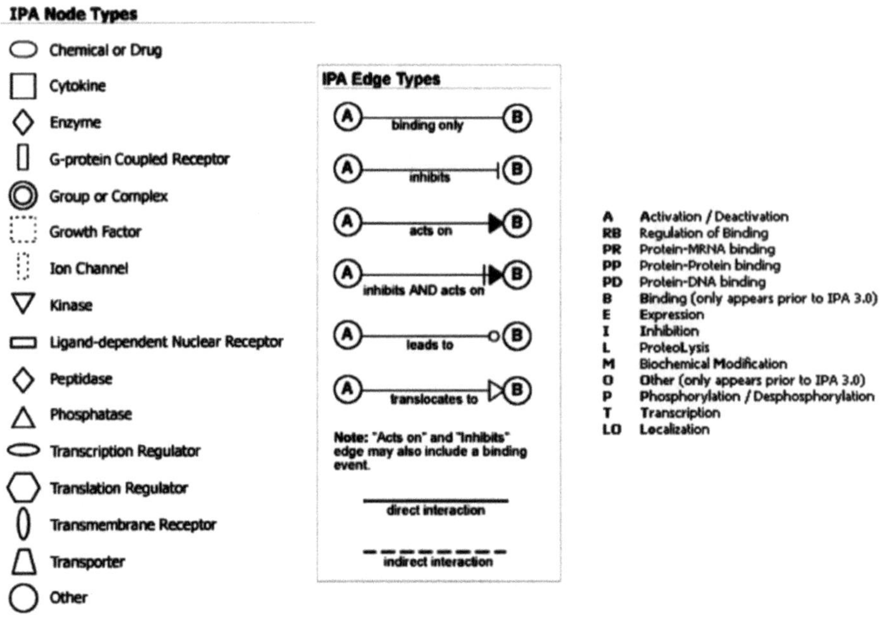

Fig. 7.2 Symbols to read Fig. 7.1

Table 7.1 Differential expression of genes in signaling pathways before and after silencing of SOCS3 in macrophages in response to co-stimulation with sonicated *B. burgdorferi* spirochetes and rIL-10

	Macrophages (SOCS3 siRNA) Sonicated Bb+IL-10	Macrophages Sonicated Bb+IL-10
Genes	Fold change (Log2) and *p*-value	
TLR pathway		
TLR1	**3.90** $P<0.001$	
TLR2	–	**1.23** $P<0.0001$
MyD88	–	**1.21** $P<0.0001$
IRAK	**2.90** $P<0.0001$	–
TOLLIP	**2.55** $P<0.001$	–
TRAF6	**2.00** $P<0.0001$	–
TIRAP	**3.02** $P<0.001$	–
TLR5	**2.27** $P<0.001$	–
JAK/STAT pathway		
JAK2	**2.49** $P<0.001$	–
Rassf4	**1.30** $P<0.001$	–
AKT	–	**1.90** $P<0.001$
STAT2	**1.56** $P<0.001$	–
STAT1	**2.71** $P<0.0001$	**1.45** $P<0.001$
STAT4	–	**1.60** $P<0.001$
STAT3	–	**2.80** $P<0.0010$

(continued)

Table 7.1 (continued)

	Macrophages (SOCS3 siRNA) Sonicated Bb+IL-10	Macrophages Sonicated Bb+IL-10
Genes	Fold change (Log2) and *p*-value	
SOCS3	–	**6.8** $P<0.001$
P21cip	–	**1.97** $P<0.001$
ERK1/2	–	**1.40** $P<0.001$
p38 MAPK pathway		
TNF/FASL	**1.04** $P<0.001$	–
TNAFaip3	**3.62** $P<0.001$	**2.13** $P<0.001$
TNFrsf1b	**5.20** $P<0.0001$	–
TNFaip2	**2.50** $P<0.001$	–
TNFrs22	**2.50** $P<0.001$	–
FADD	**2.67** $P<0.001$	–
HPK1	–	**2.34** $P<0.001$
IRAK	**2.90** $P<0.0001$	–
TRAF 6	**2.00** $P<0.0001$	–
STAT 1	**2.71** $P<0.0001$	–
P38MAP3	–	**1.40** $P<0.001$
P38MAP14	–	**1.35** $P<0.0001$
P38MAP13	–	**1.68** $P<0.0001$
P38MAP9	–	**1.22** $P<0.0001$
CREB2	–	**1.37** $P<0.0001$
Histone H3	–	**2.26** $P<0.0001$

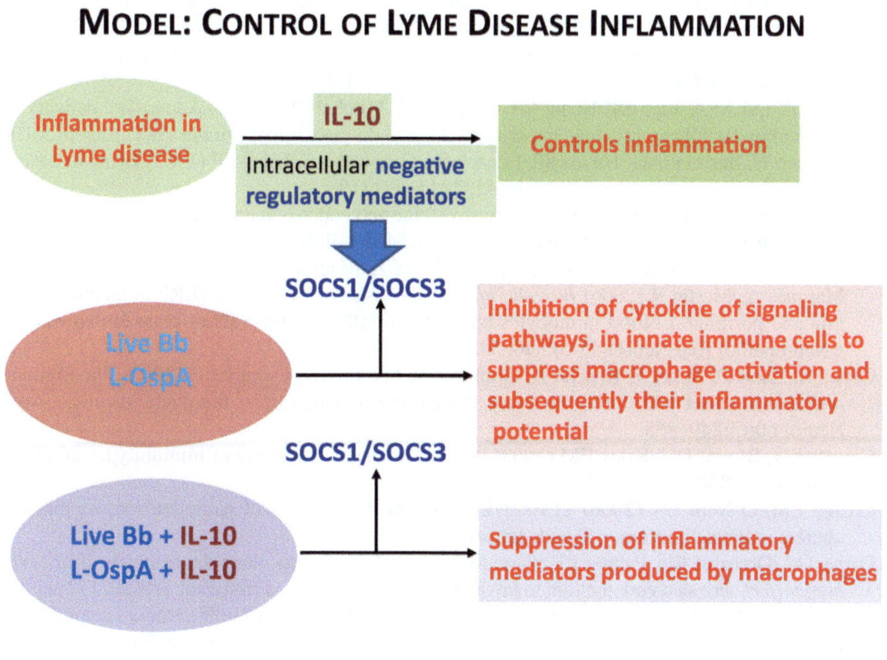

Fig. 7.3 Proposed model for the control of inflammation by the Lyme disease spirochetes, *Borrelia burgdorferi*

References

Alexander WS (2002) Suppressors of cytokine signalling (SOCS) in the immune system. Nat Rev Immunol 2(6):410–416

Alexander WS, Hilton DJ (2004) The role of suppressors of cytokine signaling (SOCS) proteins in regulation of the immune response. Annu Rev Immunol 22:503–529

Anguita J, Roth R, Samanta S, Gee RJ, Barthold SW, Mamula M, Fikrig E (1997) B7-1 and B7-2 monoclonal antibodies modulate the severity of murine Lyme arthritis. Infect Immun 65(8):3037–3041

Arend WP, Dayer JM (1995) Inhibition of the production and effects of interleukin-1 and tumor necrosis factor alpha in rheumatoid arthritis. Arthritis Rheum 38(2):151–160

Asadullah K, Sterry W, Volk HD (2003) Interleukin-10 therapy–review of a new approach. Pharmacol Rev 55(2):241–269

Berende A, Oosting M, Kullberg BJ, Netea MG, Joosten LA (2010) Activation of innate host defense mechanisms by Borrelia. Eur Cytokine Netw 21(1):7–18

Berlato C, Cassatella MA, Kinjyo I, Gatto L, Yoshimura A, Bazzoni F (2002) Involvement of suppressor of cytokine signaling-3 as a mediator of the inhibitory effects of IL-10 on lipopolysaccharide-induced macrophage activation. J Immunol 168(12):6404–6411

Bode JG, Nimmesgern A, Schmitz J, Schaper F, Schmitt M, Frisch W, Haussinger D, Heinrich PC, Graeve L (1999) LPS and TNFalpha induce SOCS3 mRNA and inhibit IL-6-induced activation of STAT3 in macrophages. FEBS Lett 463(3):365–370

Bourteele S, Oesterle K, Weinzierl AO, Paxian S, Riemann M, Schmid RM, Planz O (2007) Alteration of NFkB activity leads to mitochondrial apoptosis after infection with pathological prion protein. Cell Microbiol 9(9):2202–2217

Brown EL, Rivas JM, Ullrich SE, Young CR, Norris SJ, Kripke ML (1995) Modulation of immunity to *Borrelia burgdorferi* by ultraviolet irradiation: differential effect on Th1 and Th2 immune responses. Eur J Immunol 25(11):3017–3022

Brown JP, Zachary JF, Teuscher C, Weis JJ, Wooten RM (1999) Dual role of interleukin-10 in murine Lyme disease: regulation of arthritis severity and host defense. Infect Immun 67(10):5142–5150

Brown EL, Ullrich SE, Pride M, Kripke ML (2001) The effect of UV irradiation on infection of mice with *Borrelia burgdorferi*. Photochem Photobiol 73(5):537–544

Brown CR, Lai AY, Callen ST, Blaho VA, Hughes JM, Mitchell WJ (2008) Adenoviral delivery of interleukin-10 fails to attenuate experimental Lyme disease. Infect Immun 76(12):5500–5507

Burgdorfer W, Barbour AG, Hayes SF, Benach JL, Grunwaldt E, Davis JP (1982) Lyme disease-a tick-borne spirochetosis? Science 216(4552):1317–1319

Casey LC, Balk RA, Bone RC (1993) Plasma cytokine and endotoxin levels correlate with survival in patients with the sepsis syndrome. Ann Intern Med 119(8):771–778

Cassiani-Ingoni R, Cabral ES, Lunemann JD, Garza Z, Magnus T, Gelderblom H, Munson PJ, Marques A, Martin R (2006) *Borrelia burgdorferi* induces TLR1 and TLR2 in human microglia and peripheral blood monocytes but differentially regulates HLA-class II expression. J Neuropathol Exp Neurol 65(6):540–548

Cepok S, Zhou D, Vogel F, Rosche B, Grummel V, Sommer N, Hemmer B (2003) The immune response at onset and during recovery from *Borrelia burgdorferi* meningoradiculitis. Arch Neurol 60(6):849–855

Couper KN, Blount DG, Riley EM (2008) IL-10: The master regulator of immunity to infection. J Immunol 180(9):5771–5777

Creagh EM, O'Neill LA (2006) TLRs, NLRs and RLRs: a trinity of pathogen sensors that co-operate in innate immunity. Trends Immunol 27(8):352–357

Dame TM, Orenzoff BL, Palmer LE, Furie MB (2007) IFN-gamma alters the response of *Borrelia burgdorferi*-activated endothelium to favor chronic inflammation. J Immunol 178(2):1172–1179

Dayer JM, de Rochemonteix B, Burrus B, Demczuk S, Dinarello CA (1986) Human recombinant interleukin 1 stimulates collagenase and prostaglandin E2 production by human synovial cells. J Clin Invest 77(2):645–648

Dennis VA, Jefferson A, Singh SR, Ganapamo F, Philipp MT (2006) Interleukin-10 anti-inflammatory response to *Borrelia burgdorferi*, the agent of Lyme disease: a possible role for suppressors of cytokine signaling 1 and 3. Infect Immun 74(10):5780–5789

Dennis VA, Kaushal D, Singh SR, Bernardino A, Philipp MT: Transcriptomics of murine macrophages activated by *Borrelia burgdorferi* and interleukin-10 (poster D-042). *ASM General 107th Meeting* 2007, Toronto, Canada, May 21–25.

Dennis VA, Dixit S, O'Brien SM, Alvarez X, Pahar B, Philipp MT (2009) Live *Borrelia burgdorferi* spirochetes elicit inflammatory mediators from human monocytes via the Toll-like receptor signaling pathway. Infect Immun 77(3):1238–1245

Diterich I, Harter L, Hassler D, Wendel A, Hartung T (2001) Modulation of cytokine release in ex vivo-stimulated blood from borreliosis patients. Infect Immun 69(2):687–694

Diterich I, Rauter C, Kirschning CJ, Hartung T (2003) *Borrelia burgdorferi*-induced tolerance as a model of persistence via immunosuppression. Infect Immun 71(7):3979–3987

Dokter WH, Koopmans SB, Vellenga E (1996) Effects of IL-10 and IL-4 on LPS-induced transcription factors (AP-1, NF-IL6 and NF-kappa B) which are involved in IL-6 regulation. Leukemia 10(8):1308–1316

Donnelly RP, Dickensheets H, Finbloom DS (1999) The interleukin-10 signal transduction pathway and regulation of gene expression in mononuclear phagocytes. J Interferon Cytokine Res 19(6):563–573

Ganapamo F, Dennis VA, Philipp MT (2000) Early induction of gamma interferon and interleukin-10 production in draining lymph nodes from mice infected with *Borrelia burgdorferi*. Infect Immun 68(12):7162–7165

Ganapamo F, Dennis VA, Philipp MT (2001) CD19(+) cells produce IFN-gamma in mice infected with *Borrelia burgdorferi*. Eur J Immunol 31(12):3460–3468

Ganapamo F, Dennis VA, Philipp MT (2003) Differential acquired immune responsiveness to bacterial lipoproteins in Lyme disease-resistant and -susceptible mouse strains. Eur J Immunol 33(7):1934–1940

Gautam A, Dixit S, Philipp MT, Singh SR, Morici LA, Kaushal D, Dennis VA (2011) Interleukin-10 alters effector functions of multiple genes induced by *Borrelia burgdorferi* in macrophages to regulate Lyme disease inflammation. Infect Immun 79(12):4876–4892

Giambartolomei GH, Dennis VA, Philipp MT (1998) *Borrelia burgdorferi* stimulates the production of interleukin-10 in peripheral blood mononuclear cells from uninfected humans and rhesus monkeys. Infect Immun 66(6):2691–2697

Giambartolomei GH, Dennis VA, Lasater BL, Murthy PK, Philipp MT (2002) Autocrine and exocrine regulation of interleukin-10 production in THP-1 cells stimulated with *Borrelia burgdorferi* lipoproteins. Infect Immun 70(4):1881–1888

Glickstein L, Edelstein M, Dong JZ (2001) Gamma interferon is not required for arthritis resistance in the murine Lyme disease model. Infect Immun 69(6):3737–3743

Greenhalgh CJ, Hilton DJ (2001) Negative regulation of cytokine signaling. J Leukoc Biol 70(3):348–356

Guerau-de-Arellano M, Alroy J, Huber BT (2005) Beta2 integrins control the severity of murine Lyme carditis. Infect Immun 73(6):3242–3250

Heeg K, Dalpke A (2003) TLR-induced negative regulatory circuits: role of suppressor of cytokine signaling (SOCS) proteins in innate immunity. Vaccine 21(Suppl 2):S61–S67

Hilton DJ (1999) Negative regulators of cytokine signal transduction. Cell Mol Life Sci 55(12):1568–1577

Imai K, Kurita-Ochiai T, Ochiai K (2003) Mycobacterium bovis bacillus Calmette-Guerin infection promotes SOCS induction and inhibits IFN-gamma-stimulated JAK/STAT signaling in J774 macrophages. FEMS Immunol Med Microbiol 39(2):173–180

Irie-Sasaki J, Sasaki T, Matsumoto W, Opavsky A, Cheng M, Welstead G, Griffiths E, Krawczyk C, Richardson CD, Aitken K et al (2001) CD45 is a JAK phosphatase and negatively regulates cytokine receptor signalling. Nature 409(6818):349–354

Jones KL, Muellegger RR, Means TK, Lee M, Glickstein LJ, Damle N, Sikand VK, Luster AD, Steere AC (2008) Higher mRNA levels of chemokines and cytokines associated with macrophage

activation in erythema migrans skin lesions in patients from the United States than in patients from Austria with Lyme borreliosis. Clin Infect Dis 46(1):85–92
Kinjyo I, Hanada T, Inagaki-Ohara K, Mori H, Aki D, Ohishi M, Yoshida H, Kubo M, Yoshimura A (2002) SOCS1/JAB is a negative regulator of LPS-induced macrophage activation. Immunity 17(5):583–591
Kishore R, Tebo JM, Kolosov M, Hamilton TA (1999) Cutting edge: clustered AU-rich elements are the target of IL-10-mediated mRNA destabilization in mouse macrophages. J Immunol 162(5):2457–2461
Kobayashi K, Hernandez LD, Galán JE, Janeway CA Jr, Medzhitov R, Flavell RA (2002) IRAK-M Is a negative regulator of Toll-like receptor signaling. Cell 110(2):191–202
Kopecky J, Kuthejlova M, Pechova J (1999) Salivary gland extract from *Ixodes ricinus* ticks inhibits production of interferon-gamma by the upregulation of interleukin-10. Parasite Immunol 21(7):351–356
Kumar H, Kawai T, Akira S (2011) Pathogen recognition by the innate immune system. Int Rev Immunol 30(1):16–34
Lazarus JJ, Meadows MJ, Lintner RE, Wooten RM (2006) IL-10 deficiency promotes increased *Borrelia burgdorferi* clearance predominantly through enhanced innate immune responses. J Immunol 177(10):7076–7085
Lazarus JJ, Kay MA, McCarter AL, Wooten RM (2008) Viable *Borrelia burgdorferi* enhances interleukin-10 production and suppresses activation of murine macrophages. Infect Immun 76(3):1153–1162
Lisinski TJ, Furie MB (2002) Interleukin-10 inhibits proinflammatory activation of endothelium in response to *Borrelia burgdorferi* or lipopolysaccharide but not interleukin-1beta or tumor necrosis factor alpha. J Leukoc Biol 72(3):503–511
Medzhitov R, Janeway CA Jr (1998) Innate immune recognition and control of adaptive immune responses. Semi Immunol 10(5):351–353
Miller JC, Ma Y, Crandall H, Wang X, Weis JJ (2008) Gene expression profiling provides insights into the pathways involved in inflammatory arthritis development: murine model of Lyme disease. Exp Mol Pathol 85(1):20–27
Moore KW, de Waal Malefyt R, Coffman RL, O'Garra A (2001) Interleukin-10 and the interleukin-10 receptor. Annu Rev Immunol 19:683–765
Mullegger RR, Means TK, Shin JJ, Lee M, Jones KL, Glickstein LJ, Luster AD, Steere AC (2007) Chemokine signatures in the skin disorders of Lyme borreliosis in Europe: predominance of CXCL9 and CXCL10 in erythema migrans and acrodermatitis and CXCL13 in lymphocytoma. Infect Immun 75(9):4621–4628
Murthy PK, Dennis VA, Lasater BL, Philipp MT (2000) Interleukin-10 modulates proinflammatory cytokines in the human monocytic cell line THP-1 stimulated with *Borrelia burgdorferi* lipoproteins. Infect Immun 68(12):6663–6669
Myers TA, Kaushal D, Philipp MT (2009) Microglia are mediators of *Borrelia burgdorferi*-induced apoptosis in SH-SY5Y neuronal cells. PLoS Pathog 5(11):e1000659
Nakagawa R, Naka T, Tsutsui H, Fujimoto M, Kimura A, Abe T, Seki E, Sato S, Takeuchi O, Takeda K et al (2002) SOCS-1 participates in negative regulation of LPS responses. Immunity 17(5):677–687
Nicola N (1994) Guidebook to cytokines and their receptors. Oxford University Press, Oxford
Nussler AK, Wittel UA, Nussler NC, Beger HG (1999) Leukocytes, the Janus cells in inflammatory disease. Langenbecks Arch Surg 384(2):222–232
Olson CM Jr, Bates TC, Izadi H, Radolf JD, Huber SA, Boyson JE, Anguita J (2009) Local production of IFN-gamma by invariant NKT cells modulates acute Lyme carditis. J Immunol 182(6):3728–3734
Pai RK, Convery M, Hamilton TA, Boom WH, Harding CV (2003) Inhibition of IFN-gamma-induced class II transactivator expression by a 19-kDa lipoprotein from *Mycobacterium tuberculosis*: a potential mechanism for immune evasion. J Immunol 171(1):175–184
Pohl-Koppe A, Balashov KE, Steere AC, Logigian EL, Hafler DA (1998) Identification of a T cell subset capable of both IFN-gamma and IL-10 secretion in patients with chronic *Borrelia burgdorferi* infection. J Immunol 160(4):1804–1810

Qasimi P, Ming-Lum A, Ghanipour A, Ong CJ, Cox ME, Ihle J, Cacalano N, Yoshimura A, Mui AL (2006) Divergent mechanisms utilized by SOCS3 to mediate interleukin-10 inhibition of tumor necrosis factor alpha and nitric oxide production by macrophages. J Biol Chem 281(10):6316–6324

Ramesh G, Borda JT, Gill A, Ribka EP, Morici LA, Mottram P, Martin DS, Jacobs MB, Didier PJ, Philipp MT (2009) Possible role of glial cells in the onset and progression of Lyme neuroborreliosis. J Neuroinflammation 6:23

Rasley A, Tranguch SL, Rati DM, Marriott I (2006) Murine glia express the immunosuppressive cytokine, interleukin-10, following exposure to *Borrelia burgdorferi* or *Neisseria meningitidis*. Glia 53(6):583–592

Sabino GJ, Hwang SJ, McAllister SC, Mena P, Furie MB (2011) Interferon-gamma influences the composition of leukocytic infiltrates in murine lyme carditis. Am J Pathol 179(4):1917–1928

Sahay B, Patsey RL, Eggers CH, Salazar JC, Radolf JD, Sellati TJ (2009) CD14 signaling restrains chronic inflammation through induction of p38-MAPK/SOCS-dependent tolerance. PLoS Pathog 5(12):e1000687

Salazar JC, Pope CD, Moore MW, Pope J, Kiely TG, Radolf JD (2005) Lipoprotein-dependent and -independent immune responses to spirochetal infection. Clin Diagn Lab Immunol 12(8):949–958

Shin JJ, Glickstein LJ, Steere AC (2007) High levels of inflammatory chemokines and cytokines in joint fluid and synovial tissue throughout the course of antibiotic-refractory lyme arthritis. Arthritis Rheum 56(4):1325–1335

Shouda T, Yoshida T, Hanada T, Wakioka T, Oishi M, Miyoshi K, Komiya S, Kosai K, Hanakawa Y, Hashimoto K et al (2001) Induction of the cytokine signal regulator SOCS3/CIS3 as a therapeutic strategy for treating inflammatory arthritis. J Clin Invest 108(12):1781–1788

Singh SK, Morbach H, Nanki T, Girschick HJ (2005) Differential expression of chemokines in synovial cells exposed to different *Borrelia burgdorferi* isolates. Clin Exp Rheumatol 23(3):311–322

Sjowall J, Carlsson A, Vaarala O, Bergstrom S, Ernerudh J, Forsberg P, Ekerfelt C (2005) Innate immune responses in Lyme borreliosis: enhanced tumour necrosis factor-alpha and interleukin-12 in asymptomatic individuals in response to live spirochetes. Clin Exp Immunol 141(1):89–98

Sonderegger FL, Ma Y, Maylor-Hagan H, Brewster J, Huang X, Spangrude GJ, Zachary JF, Weis JH, Weis JJ (2012) Localized production of IL-10 suppresses early inflammatory cell infiltration and subsequent development of IFN-gamma-mediated Lyme arthritis. J Immunol 188(3): 1381–1393

Steere AC (2001) Lyme disease. N Engl J Med 345(2):115–125

Steere AC, Coburn J, Glickstein L (2004) The emergence of Lyme disease. J Clin Invest 113(8):1093–1101

Stoiber D, Stockinger S, Steinlein P, Kovarik J, Decker T (2001) Listeria monocytogenes modulates macrophage cytokine responses through STAT serine phosphorylation and the induction of suppressor of cytokine signaling 3. J Immunol 166(1):466–472

Strle K, Drouin EE, Shen S, Khoury JE, McHugh G, Ruzic-Sabljic E, Strle F, Steere AC (2009) *Borrelia burgdorferi* stimulates macrophages to secrete higher levels of cytokines and chemokines than *Borrelia afzelii* or *Borrelia garinii*. J Infect Dis 200(12):1936–1943

Suzuki A, Hanada T, Mitsuyama K, Yoshida T, Kamizono S, Hoshino T, Kubo M, Yamashita A, Okabe M, Takeda K et al (2001) CIS3/SOCS3/SSI3 plays a negative regulatory role in STAT3 activation and intestinal inflammation. J Exp Med 193(4):471–481

Uchiya K, Nikai T (2005) Salmonella pathogenicity island 2-dependent expression of suppressor of cytokine signaling 3 in macrophages. Infect Immun 73(9):5587–5594

Waage A, Halstensen A, Espevik T (1987) Association between tumor necrosis factor in serum and fatal outcome in patients with meningococcal disease. Lancet 1:355–357

Yin Z, Braun J, Neure L, Wu P, Eggens U, Krause A, Kamradt T, Sieper J (1997) T cell cytokine pattern in the joints of patients with Lyme arthritis and its regulation by cytokines and anticytokines. Arthritis Rheum 40(1):69–79

Chapter 8
Leptospira: Invasion, Pathogenesis and Persistence

Syed M. Faisal, Sean P. McDonough, and Yung-Fu Chang

8.1 Introduction

Leptospirosis, a zoonosis of worldwide distribution and recognized as an emerging infectious disease, is caused by Gram-negative bacteria belonging to the genus *Leptospira* (Bharti et al. 2003; Levett 2001). Occupational exposure, recreational activity, flooding and household environments in close contact with animals increase the risk of transmission to humans. The clinical syndromes of leptospirosis vary from a subclinical, self-limiting, anicteric, febrile illness to severe and potentially fatal disease. Leptospires are widely distributed in both urban and rural areas of tropical, sub-tropical, and temperate regions (Riley et al. 2007; Reis et al. 2008). Over the past several decades, changes in epidemiological patterns underscore the importance of leptospirosis as a major public health problem. More than 500,000 cases of severe leptospirosis are reported each year, with case fatality rates exceeding 10%. This review focuses on recent advances made in understanding leptospiral pathogenesis, highlighting strategies for invasion, persistence and immune evasion in the susceptible host. The progress made in understanding host immune responses and successes in developing more effective vaccines are also discussed.

S.M. Faisal, Ph.D. • Y.-F. Chang, D.V.M., Ph.D. (✉)
Department of Population Medicine and Diagnostic Sciences, College of Veterinary Medicine, Cornell University, Ithaca, NY, USA
e-mail: yc42@cornell.edu

S.P. McDonough, D.V.M., Ph.D.
Department of Biomedical Sciences, College of Veterinary Medicine, Cornell University, Ithaca, NY 14853, USA

M.E. Embers (ed.), *The Pathogenic Spirochetes: Strategies for Evasion of Host Immunity and Persistence*, DOI 10.1007/978-1-4614-5404-5_8,

8.1.1 Microbiology and Taxonomy

Leptospires are thin, helically coiled, motile spirochetes 0.2 μm in diameter by 6–20 μm in length. They are obligate aerobes with an optimum growth temperature of 28–30°C. The most widely used medium in current practice is based on the oleic acid-albumin medium, EMJH, supplemented with bovine serum albumin, vitamin B12, long-chain fatty acids, and ammonium salts (Levett 2001). The genus includes 13 pathogenic species namely *L. alexanderi, L. alstonii* , *L. borgpetersenii, L. inadai, L. interrogans, L. fainei, L. kirschneri, L. licerasiae, L. noguchi, L. santarosai, L. terpstrae, L. weilii, L. wolffii* and six saprophytic species namely *L. biflexa, L. meyeri, L. yanagawae, L. kmetyi, L. vanthielii* and *L. wolbachii*. Based on structural heterogeneity of the carbohydrate component of their lipopolysaccharide (LPS), pathogenic leptospires are categorized into 260 serovars while saprophytic species have over 60 serovars (Adler and de la Pena Moctezuma 2010; Ahmed et al. 2006; Palaniappan et al. 2007).

8.1.2 Molecular Biology and Genetic Tools

8.1.2.1 Molecular Biology

The leptospiral genome is approximately 5,000 kb and consists of one large (4,400 kb) and one small (350 kb) circular chromosomes (Ren et al. 2003). The leptospiral genome is comparatively larger than related spirochetes like *Borrelia* and *Treponema*, reflecting the ability of leptospires to live both inside the host and freely in the environment. The large chromosome carries genes for replicative proteins, such as dnaA, dnaN, GyrA, and GyrB as well as most of the essential housekeeping genes. Six genomes of three different species have been sequenced (Ren et al. 2003; Nascimento et al. 2004; Picardeau et al. 2008; Bulach et al. 2006). The genomes encode 3,085–3,600 putative coding sequences, of which approximately 40% have no assigned function. Furthermore, there is high similarity in gene content among different serovars of *L. interrogans* as revealed by DNA hybridization (He et al. 2007). *L. interrogans* and *L. biflexa* contain a large repertoire of regulatory genes that allow for survival in diverse environments, but have been lost in *L. borgpetersenii* limiting that strain's transmission potential outside the host (Bulach et al. 2006). *Leptospira* species have around 80 genes involved in motility and chemotaxis that are conserved between both pathogenic and saprophytic species (Xue et al. 2009).

8.1.2.2 Genetic Tools

Attempts to perform genetic manipulations in pathogenic *Leptospira* spp. have been hampered due to the lack of naturally occurring plasmids and their slow growth in artificial media. However, with the development of an *E. coli-Leptospira biflexa* shuttle vector that can replicate autonomously in both bacteria, DNA can be successfully introduced (Girons et al. 2000). A new genetic tool involving conjugation between *E. coli* and *Leptospira* spp. utilizing the RP4 plasmid was developed recently (Picardeau 2008). Several genes including *flab*, *recA*, *trpE*, *hemH*, and *metX* have been successfully inactivated by employing site-directed mutagenesis through allelic exchange (Picardeau et al. 2001; Tchamedeu Kameni et al. 2002; Bauby 2003; Guegan et al. 2003; Louvel and Picardeau 2007). In pathogenic leptospires, *ligB* was the first gene to be successfully targeted by site-directed homologous recombination (Croda et al. 2008). Subsequently, the replacement of *lipL32* and *fliY* with antibiotic resistance genes through homologous recombination has also been achieved (Liao et al. 2009; Murray et al. 2009a). A system for generating transposon insertion mutants was developed using HimarI mariner elements (Bourhy et al. 2005; Louvel et al. 2005) and has been used to generate a large number of mutants in various serovars of *Leptospira* spp., including Pomona, Manile, Canicola, and Kito (Murray et al. 2009b). LipL32 and Lao22 mutants were also generated by HimarI mutagenesis. A large number of mutant genes coding for hypothetical functions are in the early stages of development as they are not yet phenotypically characterized (Murray et al. 2009b; Ko et al. 2009). Further improvements in the methodology and identification of more readily transformable *L. interrogans* strains may allow the generation of a mutant library for high-throughput screening for specific processes that are known to be involved in pathogenesis.

8.1.3 Epidemiology

Human leptospirosis has a wide geographic distribution; however, factors such as climatic changes, flooding, a poor level of sanitation and high population density of maintenance hosts are the major determinants that govern infection and transmission rates (Pappas et al. 2008). Transmission of leptospirosis requires continuous enzootic circulation of the pathogen among animal reservoirs or maintenance hosts. Leptospires colonize and are shed from the renal tubules of a wide range of animals. After excretion in the urine, the bacteria survive for several weeks in moist soil and water, possibly by aggregating and forming biofilms (Ristow et al. 2008; Trueba et al. 2004). The source of infection for humans is usually either direct or indirect contact with the urine of an infected animal. The usual portal of entry is through abrasions or cuts in the skin or via the conjunctiva (Fig. 8.1). The epidemiologic pattern of human leptospirosis has changed in the last decade. Outbreaks have occurred at sporting events and recreational activities in regions where disease is not common, perhaps as a consequence of changing climatic patterns (Victoriano et al. 2009; McCurry 2009). The change in epidemiology from rural to urban areas has also been reported in countries like Brazil and Israel which are geographically distant (Ko et al. 1999; Kariv et al. 2001). Leptospirosis remains a major health problem in areas where the disease is endemic but is also now recognized as an emerging infectious disease in non-endemic regions due to recreational activities and tourism.

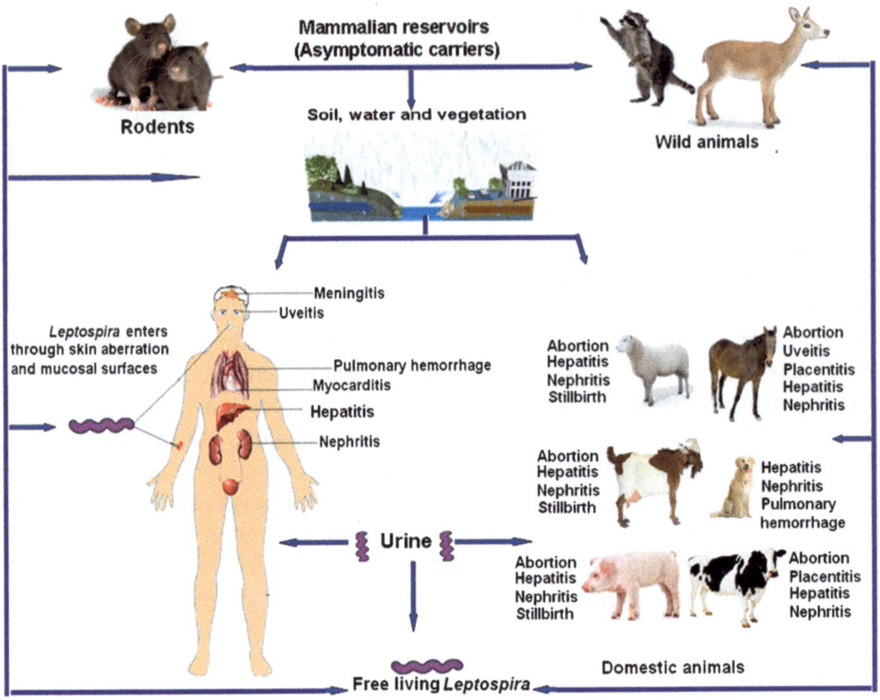

Fig. 8.1 *Leptospira* transmission cycle. Wild animals and rodents, which act as maintenance hosts and asymptomatic carriers, serve as reservoirs for *Leptospira*. They contaminate water, soil or vegetation through their urine. Humans and domestic animals get infected through direct contact with reservoir or contaminated soil. In humans, *leptospira* enter through skin abrasions and mucosal surfaces and disseminate via the blood stream to various vital organs. Involvement of the heart, liver, lung, kidneys or brain results in serious manifestations such as myocarditis, hepatitis, pulmonary hemorrhage, nephritis and meningitis. Immune responses eliminate Leptospires from all organs except the kidneys where they are shed in the urine. *Leptospira* cause serious manifestations in domestic animals like abortion, nephritis, hepatitis, uveitis and placentitis. Humans and domestic animals infect each other through direct contact or infected urine

8.1.4 Clinical Symptoms and Pathology of Leptospirosis

The spectrum of human disease caused by leptospires is extremely wide, ranging from subclinical infection to a severe syndrome of multi-organ failure with high mortality. The incubation period for leptospirosis is 5–14 days on average, with a range of 2–30 days (Levett 2001). The great majority of infections caused by leptospires in humans are anicteric (subclinical or very mild without signs of jaundice), and most patients do not seek medical attention. Symptoms include chills, headache, abdominal pain, conjunctival suffusion and less often a skin rash which usually lasts for about a week and its resolution coincides with the appearance of antibodies. Mortality is almost nil in anicteric leptospirosis. Icteric leptospirosis, which presents itself in humans as Weil's disease, is much more severe. The clinical course is often very rapidly progressive, and is characterized by jaundice, acute renal failure and bleeding with a mortality rate ranging between 5 and 15% (McBride et al. 2005). The onset of disease correlates with the appearance of agglutinating antibodies and the clearance of leptospires by antibody-mediated opsonization and lysis. The pathologic manifestations in leptospirosis are characterized by the development of vasculitis, endothelial damage and inflammatory infiltrates composed of monocytes, plasma cells and neutrophils. In almost 50% of patients, disseminated intravascular coagulation is due to activation of the coagulation cascade and organs become jaundiced. Leptospires infect almost all vital organs including kidney, liver, lung, heart and occasionally spleen and brain. In the liver, leptospirosis causes hypertrophy and hyperplasia of Kupffer cells accompanied by intrahepatic cholestasis. In the kidneys, tubulointerstitial nephritis is the major finding with an intense cellular infiltrate composed of neutrophils and monocytes and localization of leptospires in renal tubules. Minor changes in glomeruli are also evident. Impaired reabsorption of Na^+, which may be due to inhibition of renal Na^+K^+ ATPase by non-esterified unsaturated fatty acids derived from leptospires, causes a characteristic hypokalemic, nonoliguric acute renal failure (Seguro et al. 1990; Burth et al. 2005). Interstitial myocarditis with infiltration of lymphocytes, petechial hemorrhages, pericardial effusion and coronary arteritis are key pathological findings in the heart. Lungs may show pulmonary hemorrhage and congestion with infiltration of alveolar spaces by alveolar macrophages and neutrophils. The formation of hyaline membranes and the presence of leptospires in interalveolar septa are also evident.

Leptospiral components that are released after immune killing stimulate the production of pro-inflammatory cytokines and mediate inflammation and damage of end-organ tissues. LPS and outer membrane proteins activate TLR-dependent pathways (TLR4 and TLR2), which lead to the activation of nuclear factor-κb, mitogen-activated kinases and cytokines, further contributing to organ damage. Recently, a much more severe form of the disease called Leptospirosis-Associated Pulmonary Hemorrhage Syndrome (LPHS) has emerged for which the case fatality rate is more than 50% (Gouveia et al. 2008). Current research suggests that this condition might be brought about by linear deposition of immunoglobulins and complement on the alveolar surface (Croda et al. 2010). Our group demonstrated that *leptospira* may bind to fibrinogen through its surface protein LigB, thereby reducing clot formation and regulating thrombosis and fibrinolysis which might also contribute to pulmonary hemorrhage (Lin et al. 2011). In our laboratory, we have inoculated *L. interrogans* serovar Pomona into hamsters and induced serious pulmonary hemorrhages (Fig. 8.2).

8.1.5 Diagnosis

The diagnosis of leptospirosis in humans is challenging due to the similarity of its clinical symptoms with other diseases such as malaria, typhoid fever, and Dengue fever. Most cases of leptospirosis are diagnosed by serology. Antibodies are

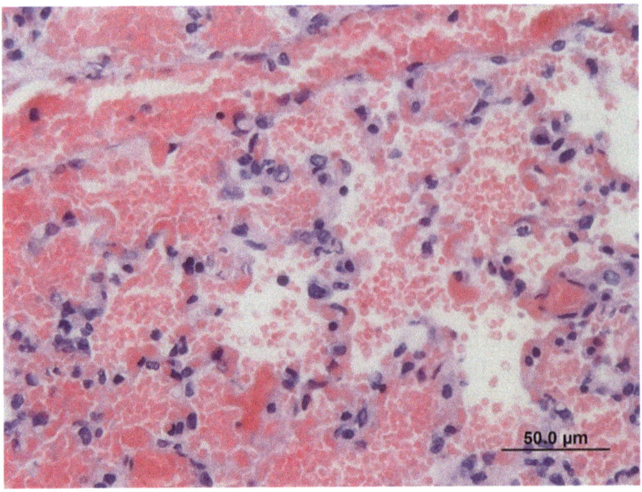

Fig. 8.2 Laboratory induced pulmonary hemorrhage in hamster. Alveoli are flooded with erythrocytes but no necrosis or hyaline membranes are noted. Pulmonary hemorrhage in the hamster model of leptospirosis resembles the highly fatal Leptospirosis-Associated Pulmonary Hemorrhage Syndrome (LPHS) seen in humans and dogs

detectable in the blood approximately 5–7 days after the onset of symptoms. The currently available Microscopic Agglutination Test (MAT) is considered the gold standard. However, the inability of MAT to differentiate between vaccination and infection is a major limitation in using this assay (Levett 2001). Moreover, MAT is unable to identify the infecting serovar due to paradoxical reactions (cross-reaction of antibodies with different serovars), particularly in specimens collected from patients with acute infections. IgM-specific ELISAs or dot blot rapid dipstick tests are currently available in leptospirosis-endemic areas and provide reliable diagnoses during potential outbreaks (Bajani et al. 2003). Attempts have also been made to use antigens that are only expressed by *Leptospira* spp. during infection for diagnosis. Although several recombinant protein antigens like LipL32 and OmpL1 have been evaluated for their potential for the diagnosis of *Leptospira* spp. infection, only Lig proteins are successful in differentiating between vaccination and infection (Boonyod et al. 2005; Mariya et al. 2006; Okuda et al. 2005; Palaniappan et al. 2004). Isolating the organism from urine or blood through culture is the most definitive diagnostic test; however, leptospiral culture is slow and insensitive (Palaniappan et al. 2007). Another diagnostic method used is a PCR assay for detecting leptospires in various biological fluids (Merien et al. 1993; Kee et al. 1994). A new multiplex PCR has been developed to differentiate between pathogenic and saprophytic leptospires and use of real-time and quantitative Taqman PCR have also been reported for use in the detection of leptospires (Kositanont et al. 2007; Palaniappan et al. 2005). However, a major limitation in using PCR for diagnosis is the inability to identify the infecting serovars. Several assays, such as single strand confirmation analysis, low stringency single-specific PCR and multi-locus variable number tandem repeat analysis have also been evaluated for their diagnostic efficacy (Ahmed et al. 2006; Wangroongsarb et al. 2007). MLST provides straightforward characterization of *Leptospira* spp. isolates, which is amenable to standardization through an online database (Ahmed et al. 2006). Use of a combination of these methods could considerably improve the ability to establish the diagnosis.

8.1.6 Antibiotic Treatment

Treatment of leptospirosis with antibiotics differs depending on the severity and duration of symptoms at the time of presentation. A wide range of antibiotics are available that are active against this disease. Doxycycline (100 mg orally twice daily) is the drug of choice and is effective in treating patients with mild disease, although it is not safe for pregnant women (Pappas and Cascio 2006). β-Lactams like penicillin and amoxicillin are useful in treating early infections (Watt et al. 1988). However, intravenous administration of penicillin G is recommended in a person with severe disease until oral tolerance of the drug improves (Katz et al. 2001). Penicillin can cause an inflammatory response known as the Jarisch-Herxheimer reaction, but this should not preclude its use (Leblebicioglu et al. 2003). Antibiotic treatment is effective during the early course of infection but often not during the late phase. Although there has been controversy over treatment with antibiotics in patients with severe disease, the consensus is that physicians should not withhold antibiotic treatment even for suspect cases. Many cases have been reported in which late administration of antibiotics have proved efficacious with reduced mortality rates (Pappas and Cascio 2006). Prophylaxis with doxycycline in persons travelling to endemic regions has shown a substantial decrease in the severity of the clinical course of disease, although infection with leptospires was not prevented (Sehgal et al. 2000). Antibiotic treatment should be administered in parallel with support of aggressive fluid intake and maintenance of electrolyte balance. Leptospires are susceptible in vitro to newer classes of antibiotics like ceftriaxone, cefotaxime, azithromycin and ciprofloxacin but outcomes for these antibiotics have not been assessed in clinical trials (Hospenthal and Murray 2003).

8.2 Animal Models

Animal models are valuable for understanding host immune responses and the pathogenesis of *Leptospira* spp. infection. Of various established animal models, hamsters are most commonly used to model acute human leptospirosis (Haake 2006; Silva et al. 2008). Guinea pigs also exhibit fulminant leptospirosis making them another appropriate model for severe disease (Nally et al. 2004). Mice and gerbils have been used for studying genetics of the immune response and vaccination trials (Pereira et al. 1998; Koizumi and Watanabe 2004). However, mice are relatively resistant and require high inocula to produce infection except for young C3H/HeJ mice (TLR4-deficient), which are highly susceptible to infection (Viriyakosol et al. 2006). The number of bacteria required to initiate infection in an animal model depends on several factors, such as the serovar or strain used, number of in vitro passages and age of the animals. The challenge dose or median lethal dose (MLD_{50}) varies and could be as low as 10 bacteria and as high as 10^8. Although rats and mice are regarded as maintenance hosts and these models do not reflect disease progression in humans, they do provide useful information regarding the mechanisms of leptospiral colonization and host resistance (Athanazio et al. 2008; Tucunduva de Faria et al. 2007). Zebrafish embryos have been recently evaluated as a model for studying leptospiral pathogenesis (Davis et al. 2009). This model may provide clues in the initial events of leptospiral infection, but its relevance to human leptospirosis is yet to be established. Dogs and monkeys have proved to be ideal large animal models for studying human leptospirosis (Faine 1957; Minette and Shaffer 1968). Recently, a monkey model has been developed for studying pulmonary hemorrhage syndrome and a horse model has been successfully established that could be useful in studying both human and equine leptospirosis (Pereira et al. 2005; Yan et al. 2010a). However, the lack of tools for genetic manipulation and the high cost of experimentation have limited the use of large animal models. No ideal animal model exists to study all the varied aspects of leptospiral pathogenesis.

8.3 Pathogenesis

Pathogenic *Leptospira* spp. produce a systemic infection after environmental exposure, establish persistent renal carriage with urinary shedding in reservoir animals, and cause tissue damage in multiple organs of susceptible hosts. The pathogenesis of leptospiral infection and establishment of infection in a particular host involve invasion of host cells with the help of various surface molecules (virulence factors), persistence or colonization, and immune evasion. The various steps involved in leptospiral pathogenesis are discussed below.

8.3.1 Invasion

The ability of *Leptospira* to attach and form pores in order to invade host cells is the initial step in infection. Attachment is mediated by various molecules (virulence factors) present on the surface or secreted by bacteria. The virulence factors of *Leptospira* spp. discovered to date are either surface-exposed proteins that are primarily involved in attachment and invasion of host cells or are secretory proteins that either form pores or cause damage/lysis of host cells. Surface proteins of *Leptospira* spp. play important roles in attachment to host fibroblasts, microglia, endothelial and epithelial cells (Cinco et al. 2006; Ito and Yanagawa 1987).

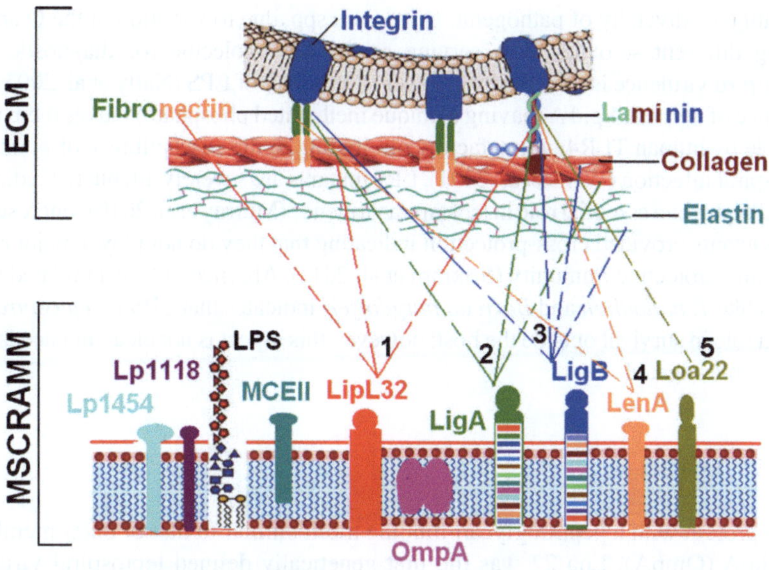

Fig. 8.3 Interaction of *Leptospiral* Microbial Surface Component Recognizing Adhesive Matrix Molecules (MSCRAMM) with host Extracellular Matrix Molecules (ECM). Various *Leptospira* MSCRAMM, which are mainly comprised of surface-exposed proteins, interact with different ECM of the host in order to attach, invade, and initiate infection. (1) LipL32 interacts with fibronectin, collagen, and laminin; (2) LigA interacts with fibronectin, collagen, and laminin; (3) LigB interacts with fibronectin, collagen, laminin, and elastin; (4) LenA interacts with fibronectin and laminin; (5) Loa22 interacts with laminin

Several surface proteins of *Leptospira* spp. have been identified that have been shown to bind to various components of extracellular matrix (ECM) and thereby mediate host-pathogen interaction (Stevenson et al. 2007; Merien et al. 2000; Atzingen et al. 2008; Hauk et al. 2008; Hoke et al. 2008; Choy et al. 2007). The various surface proteins of *leptospira* that bind to host ECM are shown schematically in Fig. 8.3. *Leptospira* also bind to host plasminogen through various surface proteins that could further contribute to invasion (Vieira et al. 2009, 2010a). Consistent with the predicted ability of *Leptospira* spp. to migrate through host tissues, the genomes of these organisms encode a wide range of putative hemolysins and proteases that may facilitate this process. Some of the important virulence factors involved in leptospiral invasion and pathogenesis are discussed below.

8.3.1.1 Lipopolysaccharide (LPS)

LPS is considered to be a general virulence factor of Gram-negative bacteria. It is a major component of the outer membrane and contributes to its structural integrity. Leptospiral LPS has a composition similar to that of other Gram-negative bacteria but has lower endotoxic activity (Vinh et al. 1986; Shimizu et al. 1987). LPS is responsible for antigenic diversity of pathogenic *Leptospira* spp. due to variation in the O antigen among different serovars, thus serving as the key molecule for diagnostic tests. *Leptospira* virulence is affected by changes in structure of LPS (Nally et al. 2005). The presence of atypical lipid A (having a unique methylated phosphate) makes them undetectable by human TLR4, a key factor contributing to the susceptibility of humans to leptospiral infection (Werts et al. 2001). LPS mutants are severely attenuated indicating that LPS plays an essential role in leptospira virulence (Murray et al. 2010). Interestingly, LPS mutants provided cross-protection indicating that they do not play a major role in imparting protective immunity (Srikram et al. 2011). Absence of LPS in related spirochetes like *T. palladium* and *Borrelia burgdorferi* indicates that LPS in *leptospira* may have a role in survival outside the host; however, this aspect is not clear and needs to be further investigated.

8.3.1.2 Loa 22

A lipoprotein with a peptidoglycan-binding motif similar to that of outer membrane protein A (OmpA), Loa 22 was the first genetically defined leptospiral virulence factor. This lipoprotein is up-regulated during leptospiral infection (Ristow et al. 2007; Koizumi and Watanabe 2003; Nally et al. 2007). Loa 22 is highly conserved among pathogenic leptospires, the mutant of which is avirulent in both the guinea pig and hamster models of leptospirosis. Complementation tests were found to restore virulence. Loa 22 is a surface-exposed protein that is recognized by sera from human leptospirosis patients (Gamberini et al. 2005). Although Loa 22 binds to laminin, the interaction is weak. The saprophytic *L. biflexa* species shares an ortholog of Loa22, but differential expression and variation in sialylation of this molecule in pathogenic leptospires could be responsible for its role as a virulence determinant (Picardeau et al. 2008).

8.3.1.3 LipL32

Also known as hemolysis-associated protein I (Hap I), LipL32 accounts for 75% of the outer membrane proteome of leptospires (Haake et al. 2000; Lee et al. 2000; Cullen et al. 2002). This lipoprotein is surface exposed and highly conserved in pathogenic *Leptospira* spp. The C-terminus portion of LipL32 binds to various ECM components such as laminin, fibronectin, and collagen (Hauk et al. 2008; Hoke et al. 2008). LipL32 is a Ca^{++}-binding protein whose structure has been determined by X-ray crystallography (Vivian et al. 2009). LipL32 induces tubulointerstitial nephritis through a TLR2-dependent pathway followed by activation of nuclear transcription factor kappa B, mitogen-activated protein kinases, and differential induction of chemokines and cytokines relevant to tubular inflammation. LipL32 also induces activation of the transforming growth factor-β/Smad-associated fibrosis pathway through TLR2 leading to accumulation of fibronectin (Tian et al. 2010). Due to its abundance and relatively higher level of expression during infection as compared to in vitro culture, LipL32 was thought to be a putative virulence factor (Nally et al. 2007). Surprisingly however, a LipL32 mutant obtained by HimarI insertion mutagenesis was not attenuated and was able to cause chronic infection in hamsters and colonization of kidneys in rats (Murray et al. 2009a). In light of the relative abundance and high level of conservation of LipL32 across a broad range of pathogenic *Leptospira* spp., it is hard to believe that this protein is not an important virulence factor. Alternative approaches need to be employed in order to confirm that LipL32 is not involved in *Leptospira* pathogenesis. Moreover, further research is required to determine why leptospires produce this protein in such a large amount. Based on our current knowledge, we speculate that LipL32 might contribute to the leptospiral life cycle outside the mammalian host.

8.3.1.4 LenA

LenA, also known as Lsa24, is a 24-kDa protein that binds to laminin as well as complement factor H, fibrinogen and fibronectin (Stevenson et al. 2007; Barbosa et al. 2006; Verma et al. 2006). LenA is member of the leptospiral endostatin-like protein family that includes LenB, LenC, LenD, LenE, and LenF, all of which bind to fibronectin. Other laminin-binding leptospiral proteins include Lsa21, Lsa27, Lsa63, and a 36 kDa membrane protein (Merien et al. 2000; Atzingen et al. 2008; Longhi et al. 2009; Vieira et al. 2010b).

8.3.1.5 Hemolysins

The genome of pathogenic *Leptospira* spp. encodes various putative hemolysins and proteases that may facilitate the entry of *Leptospira* spp. into host cells and rapid translocation into various tissues. These include pore forming proteins and sphingomylinases (Lee et al. 2002). Hemolysins have the ability to lyse erythrocytes and other cell membranes, which make them potential virulence factors. SphA from *L. interrogans* serovar Pomona causes lysis of sheep erythrocytes (Bernheimer and Bey 1986; Segers et al. 1992). SphH, a homolog of SphA, found in *L. interrogans* serovar Lai has the ability to form pores in host cells without exhibiting sphingomylinase and phospholipase activity (Gamberini et al. 2005; Barbosa et al. 2006). Sph2 found in some pathogenic leptospires causes endothelial cell and erythrocyte membrane damage. Both SphH and Sph2 are expressed during infection and exhibit cytotoxic properties (Artiushin et al. 2004; Carvalho et al. 2010; Zhang et al. 2008). However, a different research group has reported contradictory results regarding hemolytic properties of SphH and Sph2, which could be due to differences in folding or other properties of the recombinant proteins used in their assays (Carvalho et al. 2010). The absence of sphingomyelinase activity in saprophytic leptospires signals its role in virulence or survival in the host where key nutrients are limited; however, the exact role of sphingomyelinase in the pathogenesis of leptospirosis is still unclear (Adler and de la Pena Moctezuma 2010; Bulach et al. 2006).

8.3.1.6 Leptospiral Immunoglobulin-Like Proteins

The Lig proteins are members of the family of bacterial surface proteins that contain 12–13 tandem immunoglobulin-like repeat domains which serve as adhesins and invasion-mediating determinants in other bacterial pathogens such as intimin in *Escherichia coli* and invasin in *Yersinia pseudotuberculosis* (Palaniappan et al. 2002). This family contains three copies of high molecular mass proteins, namely LigA, LigB, and LigC. The distribution of copies of Lig genes in each serovar may vary. For example, serovar Pomona contains all three of these genes, serovar Lai contains only LigB, whereas serovar Grippotyphosa contains LigA and LigB but LigC is absent (McBride et al. 2009).

The Lig proteins are expressed only on the surface of pathogenic leptospires isolated from infected animals and not in saprophytic *Leptospira* species (Matsunaga et al. 2003). They are present in low-passage bacteria but are absent in high passage, culture-attenuated leptospires that have lost virulence, as tested in a hamster model. These proteins are strongly recognized by sera from patients with leptospirosis (Croda et al. 2007). Moreover, their expression can be induced by a change in the osmolarity of the culture medium, mimicking the conditions that bacteria might encounter in a mammalian host (Matsunaga et al. 2007). Recombinant Lig proteins (LigA and LigB) bind in vitro to host extracellular matrix proteins, including fibronectin (Fn), fibrinogen, collagen, and laminin (Choy et al. 2007). LigB binds to fibronectin through the C-terminal region with high affinity similar to FnBPA of *S. aureus* (Lin and Chang 2008). A high-affinity Fn-binding region of LigB has been localized to LigBCen2, which contains the partial 11th and full 12th Ig-like repeats (LigBCen2R) and 47 amino acids of the non-repeat region (LigBCen2NR) of LigB. Binding of LigB to fibronectin is through both the N-terminal domain (NTD) and Gelatin-binding domain (GBD). However, interaction with the GBD is stronger, involving a multi-domain interaction, thereby creating a unique compact globular structure (Lin et al. 2009a, b, 2010). Furthermore, fibronectin binding is modulated by Ca^{++}, and this interaction is mediated by three Ca^{++}-binding motifs which are harbored in Big domains of LigB (Lin et al. 2008). Binding of fibronectin to LigB induces a conformational change in its disordered region. LigB also binds to elastin and tropoelastin with high avidity using Ig-like domains, which is possibly mediated by salt bridges (Lin et al. 2009c). The difference in the binding mechanisms of LigB to fibronectin and elastin suggests that each Ig-like domain has unique functional characteristics. LigB also binds to fibrinogen (Fg). A conformational change in LigBCen2R when bound to the C-terminal fibrinogen αC domain (FgαCC), arginine-glycine-aspartate (RGD) motif blocked further binding to integrin $\alpha_{IIb}\beta3$ on platelets, thus preventing their aggregation. LigBCen2R binding to FgαCC reduced clot formation but did not affect plasminogen (PLG) and tissue-type plasminogen activator (tPA) interactions with FgαCC. This indicates that LigB binds to the C-terminal αC domain of Fg, which regulates thrombosis and fibrinolysis, and might contribute to pulmonary hemorrhage and thrombocytopenia seen in clinical cases of leptospirosis (Lin et al. 2011). Lig proteins also interact with human complement regulators like Factor H and C4BP indicating that these proteins are utilized by the pathogen to escape complement-mediated killing (Castiblanco-Valencia et al. 2012). Lig proteins are considered to be putative virulence factors, as well as the most promising vaccine candidates identified to date. However, a recent study showed that a *ligB* mutant was not attenuated and these bacteria retain the ability to cause acute leptospirosis in hamsters or persistent renal colonization in rats (Croda et al. 2008). It is likely that the function of LigB was taken over by LigA or other similar putative adhesins with potentially overlapping function, indicating redundancy in this bacterial pathogen.

8.3.1.7 HemO

Like other Gram-negative bacteria, *Leptospira* spp. also contain several iron-uptake systems, including TonB-dependent outer membrane receptors, indicating that iron acquisition is important for virulence (Louvel et al. 2006). *Leptospira* spp. also possess a Heme oxygenase, encoded by *hemO*, which degrades the tetrapyrrole ring of the Heme molecule, releasing ferrous iron. Disruption of the *hemO* gene severely attenuates the virulence of *L. interrogans* in the hamster model of leptospirosis, suggesting that Heme is the principal source of iron during infection (Murray et al. 2009c).

8.3.1.8 Other Potential Virulence Factors

Disruption of the *flaB* gene makes *leptospira* nonmotile due to lack of a functional endoflagellum; this indicates that the gene could be a possible virulence factor since *flaB* mutants would have a reduced ability to disseminate in the host and could be severely attenuated (Picardeau et al. 2001). This hypothesis is supported by the fact that freshly isolated pathogenic leptospires from infected animals are more motile than those cultured in vitro (Ellis et al. 1983). Proteins responsible for maintaining the cytoskeletal structure of bacteria such as Mreb, MreC, and MreD are encoded in the leptospiral genome and could be potential virulence factors that need to be tested. Multiple methyl-accepting chemotaxis proteins have been identified which could be potential virulence factors as chemotaxis of leptospires to hemoglobin is correlated with virulence (Yuri et al. 1993). Recently, mutations in two hypothetical proteins LA1641 and LA0615 were found to attenuate *Leptospira* indicating that they could be potential virulence factors; however, this discovery needs confirmation through complementation (Murray et al. 2009b). Microarray analysis has also revealed up-regulation of several hypothetical proteins in response to temperature and osmolarity, mimicking the conditions in the host. These proteins could be environment-sensing regulatory proteins involved in survival or persistence in an infected host, but their potential as virulence factors needs to be investigated by performing further studies (Matsunaga et al. 2007; Lo et al. 2006). Recently discovered proteins such as OmpL37, which binds to skin and vascular elastin, are also potential virulence factors (Pinne et al. 2010).

8.4 Leptospiral Persistence and Immune Evasion

Leptospira spp. are not facultative intracellular pathogens like *Salmonella* or *Mycobacterium* but reside within host cells only transiently as they rapidly translocate through cell monolayers in vitro (Merien et al. 1997; Barocchi et al. 2002; Liu et al. 2007). Within non-phagocytic host cells they reside either in the cytoplasm or phagosomal compartments. Further, both human and murine macrophages phagocytose leptospires with the same efficiency, but strikingly, the intracellular fate of this pathogen varies markedly. In murine macrophages, leptospires reside in lysosomes where they are degraded by lysosome hydrolases. In contrast, those spirochetes that are phagocytosed by human macrophages escape from phagosomes to the cytosol where they proliferate and induce apoptosis (Tu et al. 1982; Li et al. 2010). Thus, leptospires have acquired phagosomal escape as a mechanism to evade immune attack and quickly disseminate in various host tissues to establish infection.

There are several factors that contribute to leptospiral persistence and immune evasion in the host. Leptospires cause systemic infection and are subsequently cleared from all organs except renal tubules, where they give rise to persistent renal carriage in reservoir animals. In fact, the kidney is the primary target of *Leptospira* spp. during both acute and chronic infections, where conditions in the renal tubule favor its survival (Yang 2007). This organ seems to be an immuno-privileged site that contributes to high grade persistence. The mechanism that facilitates the survival of *Leptospira* spp. in the kidney is difficult to ascertain, however it can be speculated that anti-leptospiral antibodies are unable to kill the bacteria possibly because of the absence of complement (Monahan et al. 2008). Leptospires are resistant to both the classical and alternate pathways of complement activation. They acquire complement factor H by binding through surface-exposed Len proteins, thereby escaping the alternate pathway of complement mediated killing (Stevenson et al. 2007; Verma et al. 2006; Meri et al. 2005). Leptospires also bind to host complement fragment C4b-binding protein alpha chain (C4BPA) through the recently discovered receptor LcpA (leptospiral complement regulator-acquiring protein A) and secure protection against the classical pathway of complement activation (Barbosa et al. 2009, 2010). Further, cell aggregation and biofilm formation as observed in rat kidney tubules correlate to persistence and high concentration of urinary shedding (Ristow et al. 2008; Sterling and Thiermann 1981). The substantially higher amount of expression of LPS O antigen in leptospires isolated from the liver of hamsters with acute infection indicates that expression of O antigen facilitates the persistence of bacteria in the kidney; however, the specific mechanism involved needs to be determined (Nally et al. 2005).

Recent proteomic analysis of leptospires shed in urine as compared to those cultured in vitro reveals that antibodies present in serum of chronically infected rats react with relatively more antigens from in vitro-derived leptospira as compared to those isolated from urine (Monahan et al. 2008). These results indicate that differential protein expression and down-regulation of specific antigens by *Leptospira* spp. facilitate renal colonization, persistence, and immune evasion (Xue et al. 2010).

Evaluation of the global gene response of *Leptospira* spp. to innate immunity by microarray analysis reveals down-regulation of major OMPs, which could represent an immune evasion strategy. The development of leptospirosis and disease progression is influenced by host susceptibility factors, the dose of the infecting inoculum and the virulence characteristics of the infecting strain. Certain *Leptospira* spp. species and serovars cause severe disease in humans more frequently than others. Genetic predisposition to acute leptospiral infection in humans has been shown to be associated with expression of HLA-DQ6 (MHCII) (Lingappa et al. 2004). The various strategies adopted by *leptospira* in order to evade host immune attack are schematically presented in Fig. 8.4.

8.5 Host Immune Response

It was long thought that immunity to leptospirosis is mediated predominantly by the humoral immune system (Adler and Faine 1977). This response appears to be primarily targeted against LPS. Evidence supporting this conclusion includes: (a) passive transfer of immunity, correlated to the level of anti-LPS antibodies in transferred sera; and (b) protection of various animals like guinea pigs, hamsters, dogs, and monkeys against leptospiral infection by anti-LPS monoclonal antibodies (Adler and Faine 1978; Jost et al. 1986). However, inability of anti-LPS antibodies to confer complete protection against all serovars argues that factors other than LPS might be involved in immunity against leptospiral infection. Earlier work demonstrated that the use of protein extract could confer protection against both homologous and heterologous challenge (Sonrier et al. 2000). Following this, several outer membrane proteins have been identified that confer protection against challenge in animal models. Further studies confirm that B cells play an important role in protection, though there is support from various mediators like IFN-γ and iNOS secreted by T cells (Fig. 8.5). Although leptospires are not strict extracellular microbes, their invasive ability allows them to translocate and transiently reside in various phagocytic and non-phagocytic cells, which could help them escape from agglutinating antibodies and thus evade immune attack (Merien et al. 1997; Barocchi et al. 2002).

Other studies in cattle changed the paradigm of the host immune response following leptospiral infection. Importantly, the protective efficacy of a killed *L. borgpetersenii* serovar Hardjo vaccine in cattle correlates with its ability to stimulate a Th1 response and IFN-γ release (Bolin and Alt 2001; Brown et al. 2003; Naiman et al. 2001, 2002). In spite of inducing high antibody levels, vaccines that fail to elicit a Th1 response do not impart protection. The requirement of a Th1 response for protection is not restricted to a single serovar (Hardjo) or host animal (cattle). Several reports in recent years have demonstrated a significantly enhanced level of protection against challenge with *L. interrogans* serovar Pomona that correlates with the level of the Th1 response in hamsters (Faisal et al. 2008, 2009a). Live pathogenic leptospires also can stimulate production of type 1 cytokines

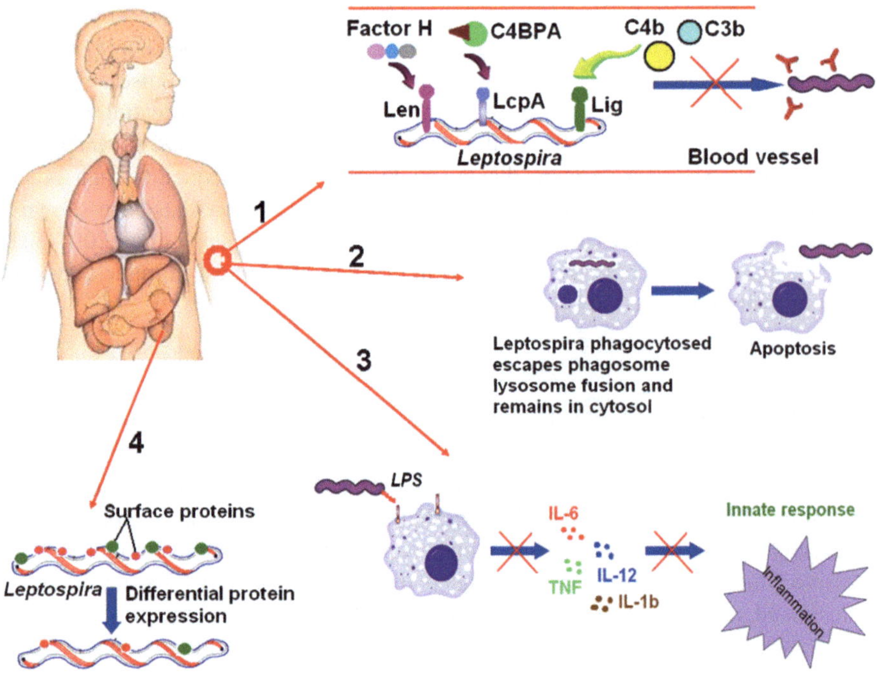

Fig. 8.4 Strategies for *Leptospiral* immune evasion. (1) In blood, *Leptospira* escape complement mediated killing by binding various key molecules of the cascade like Factor H, C3b and C4b through surface proteins like Lig, Len, and LcpA. (2) After phagocytosis, *Leptospira* escape phagosome-lysosome fusion, remain in the cytosol and induce apoptosis when conditions are favorable. (3) *Leptospira* cause variation in O antigen of LPS which prevents their recognition by TLR4, thereby inhibiting the MyD88/MAP kinase pathway, leading to blockage of innate immune responses and subsequent clearance from the host. (4) Differential protein expression facilitates renal colonization, persistence and immune evasion of *Leptospira* spp.

(Vernel-Pauillac and Merien 2006). Several other lines of evidence suggest a possible role for cell-mediated immunity, specifically CD4 and γδ-T cells with the generation of IFN-γ against *Leptospira* infection. Leptospires stimulate expansion of γδ-T cells in naïve PBMCs and leptospirosis patients have an increased number of these cells (Klimpel et al. 2003). γδ-T cells could act as a link between initial innate and later adaptive immune responses. Although they constitute only a small population (1–5%) of lymphocytes in blood, they are widespread in epithelial rich tissues and mucosal surfaces where they could play a major role in host defense as *Leptospira* spp. primarily infect through skin abrasions and mucosal routes (Bonneville et al. 2010). The role of CD4 T cells in host defense may be as helper cells that can differentiate into different phenotypes (Th1/Th2) secreting cytokines aiding in enhancement of the immune response. Although there is no evidence of direct role for CD8 T cells (CTLs) in protective immunity, specific studies have detected these cells in patients infected with different spirochetes. Mice depleted of CD8 T cells develop marked lung and kidney lesions and CD8 T cells obtained from

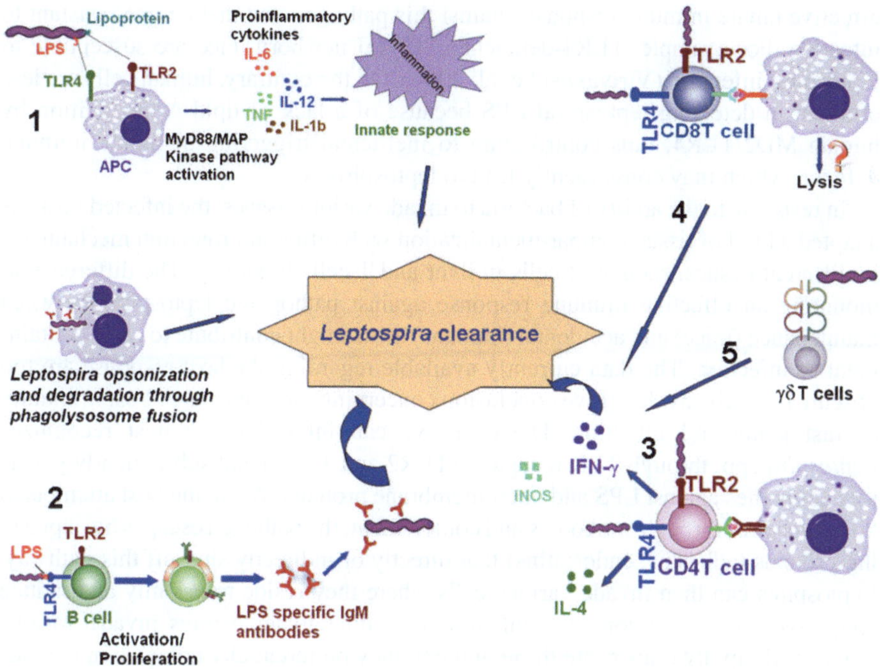

Fig. 8.5 Host immune response against *Leptospira* infection. (1) After gaining entry to the host, *Leptospira* LPS and lipoproteins interact with TLR4 and TLR2 expressed by Antigen-Presenting Cells (APCs), such as macrophages or dendritic cells, leading to induction of innate immune response and inflammation that contribute to bacterial clearance. (2) Leptospires may also interact with TLR4 on B cells through LPS, leading to B cell activation/proliferation and production of anti-LPS IgM antibodies. (3) APCs presenting processed *Leptospiral* antigens on MHC Class II molecules (MHCII) might interact with CD4 T cells leading to their activation and production of cytokines. (4) Processed *Leptospiral* antigens might be presented on MHC Class I (MHCI) molecules leading to interaction with CD8 T cells and the subsequent activation and perforin/granzyme-mediated lysis of infected APCs; however this mechanism is not yet confirmed. *Leptospira* may directly interact with CD4 and CD8 T cells via TLR4 and TLR2 through LPS and lipoproteins, leading to their activation and secretion of mediators like IFN-γ and iNOS. Leptospires may eventually be cleared from the host by these mediators. (5) Leptospires may also activate γδ-T cells present in the skin or mucous membranes at portals of entry through their lipid antigens/lipoproteins, causing them to secrete IFN-γ

leptospirosis patients proliferated in response to Lig protein peptides (Guo et al. 2010; van Voorhis et al. 1996; Busch et al. 1996).

A recent report highlights the importance of innate immune responses, particularly Toll-like receptors, in mounting an early defense against leptospiral infection (Chassin et al. 2009). Both TLR2 and TLR4 are required for effective innate defense against *Leptospira* spp. infection. Interestingly, differential recognition of leptospiral LPS, which is recognized by both TLR2 and TLR4 in mice and TLR2 and TLR1 in humans, makes the former resistant and latter more susceptible to leptospirosis (Nahori et al. 2005). Efficient recognition of LPS by murine cells may impart an effective innate immune response against this pathogen and make mice resistant to infection. For example, TLR4-deficient C3H/HeJ newborn mice are susceptible to leptospiral infection (Viriyakosol et al. 2006). On the contrary, human cells are less efficient in detecting leptospiral LPS because of a lack of lipid A recognition by human MD2/TLR4, thus contributing to inefficient triggering of innate immune defense, which may consequently lead to leptospirosis.

In response to the ability of bacteria to invade various tissues, the infected host has adapted a kind of tissue compartmentalization with different protection mechanisms in different tissues, such as B cells in liver and T cells in kidney. The difference in mounting an effective immune response against pathogenic leptospires between maintenance (mice) and accidental (humans) hosts might contribute to early containment of infection. The data currently available regarding the host response are too obscure to reach any definitive conclusion concerning mechanisms of host immunity against leptospiral infection. However, we can infer that the host recognizes *Leptospira* spp. through TLR receptors (TLR2 and TLR4) and subsequently generates antibodies against LPS and outer membrane proteins. While the host attempts to kill leptospires through antibodies and complement, the pathogen escapes by expressing proteins (adhesins, endostatins) that directly or indirectly shut off this pathway. Leptospires can then invade various cells where they reside transiently and induce apoptosis when conditions are unfavorable. Although leptospires invade various organs and rapidly translocate through them, they preferentially colonize in the kidney due the absence of complement, a possible immune evasion strategy. In spite of activation of T cells (CD4 and γδ-T cells) by the host in response to *Leptospira* spp., the response seems to be too weak to prevent infection.

8.6 Vaccines and Adjuvants

Current leptospiral vaccines used in humans and animals are mainly whole cell-based killed or attenuated vaccines. Unfortunately, these biologics are hobbled by several limitations including a high rate of adverse reaction due to the presence of LPS and culture media, short term and serovar specific immunity, and the requirement for annual booster doses to maintain immunity. Further, these vaccines are not effective in preventing carriage, which limits their use as an intervention to block transmission (therapeutic vaccine). Leptospiral vaccines have been used since the 1920s without much success in improving their efficacy or reducing their reactogenicity. Although current LPS-based vaccines are serovar specific, as the details of LPS structure and composition are analyzed, new avenues for development of vaccines and serodiagnosis are likely to emerge. Subunit vaccines based on outer membrane proteins (OMPs) are attractive alternatives to whole cell inactivated vaccines because of their antigenic conservation among serovars and high throughput technologies for vaccine production. Several OMPs of *Leptospira*, for example, OmpL1, LipL32, LipL41, LigA, LigB, Lp0607, Lp1118, and Lp1454 have shown promise in the development of subunit vaccines against leptospirosis (Chang et al. 2007; Haake et al. 1999; Palaniappan et al. 2006; Yan et al. 2009). However, initial attempts to exploit these OMPs as vaccine candidates have yielded disparate results. For example, LipL41 and OmpL1 provided protection synergistically but failed individually (Haake et al. 1999). LipL32, which is the most abundant outer membrane lipoprotein, failed to provide protection when used as a recombinant protein, but provided partial protection with low efficacy as a DNA vaccine delivered via adenovirus or BCG (Branger et al. 2001; Seixas et al. 2007). The recent report that a LipL32-deficient mutant retains virulence is surprising, but may partially explain the relative lack of efficacy of a LipL32-based vaccine (Liao et al. 2009; Murray et al. 2009a). As with other Gram-negative bacteria, several OMPs like Lp4337, Lp3685, and Lp0222 that have OmpA-like domains have been identified as potential immunogens. These, in particular, have been shown to confer partial protection against leptospirosis in a hamster model (Yan et al. 2010b). Strategies to use various combinations of key antigens in the so-called multicomponent vaccines in order make them more efficacious and cross-protective have shown some success in the past (Chang et al. 2007; Feng et al. 2009). Such a multicomponent vaccine would circumvent the problem of host genetic restriction and antigenic variability associated with single antigen-based vaccines. To date, Lig proteins (LigA and LigB) are the most promising subunit vaccine candidates and confer high-level protection in various animal models (Palaniappan et al. 2006; Yan et al. 2009; Silva et al. 2007). LigA has shown potential as both a protein and a DNA vaccine (Faisal et al. 2008). Further, high-level conservation and amino acid sequence identity (70–100%) for this protein among a wide range of pathogenic serovars suggests potential for imparting cross-protective immunity but this has yet to be determined (McBride et al. 2009).

The success of a subunit vaccine depends on the availability of potent yet non-toxic adjuvants (Burnette 1991). With the increased understanding of leptospiral pathogenesis and the immune response required for protection, it is tempting to evaluate new adjuvants and immunomodulators that could induce the desired repertoire of immune reactions leading to protective immunity. Recent studies suggest that alum could be successfully replaced by liposomes as an adjuvant for subunit vaccine antigens against leptospirosis (Faisal et al. 2009a). Apart from conventional liposomes, antigens associated with liposomes made up of lipids from non-pathogenic bacteria like *L. biflexa* serovar Potac (leptosomes) and *M. smegmatis* (smegmosomes) imparted significantly higher levels of protection than alum in a hamster model of leptospirosis (Faisal et al. 2009b, 2009c). The higher level of protection correlates with induction of significantly enhanced levels of both humoral and cell mediated immune responses. Fusing a protein with a stimulatory adjuvant (cholera toxin B subunit) to enhance the immune response is another promising strategy that could also be applied to other protective antigens (Habarta et al. 2010). Considering the fact that Lig proteins are the most promising antigens, strategies should be designed to develop a subunit vaccine based on this antigen. Coupling Lig protein with MPL (analog of LPS) in fusogenic liposomes could result in a potent, yet non-toxic, vaccine that could activate both innate and adaptive immune responses in a mechanism presented in Fig. 8.6. Although this strategy seems to activate the desired protective response against *Leptospira* spp, ease in vaccine preparation,

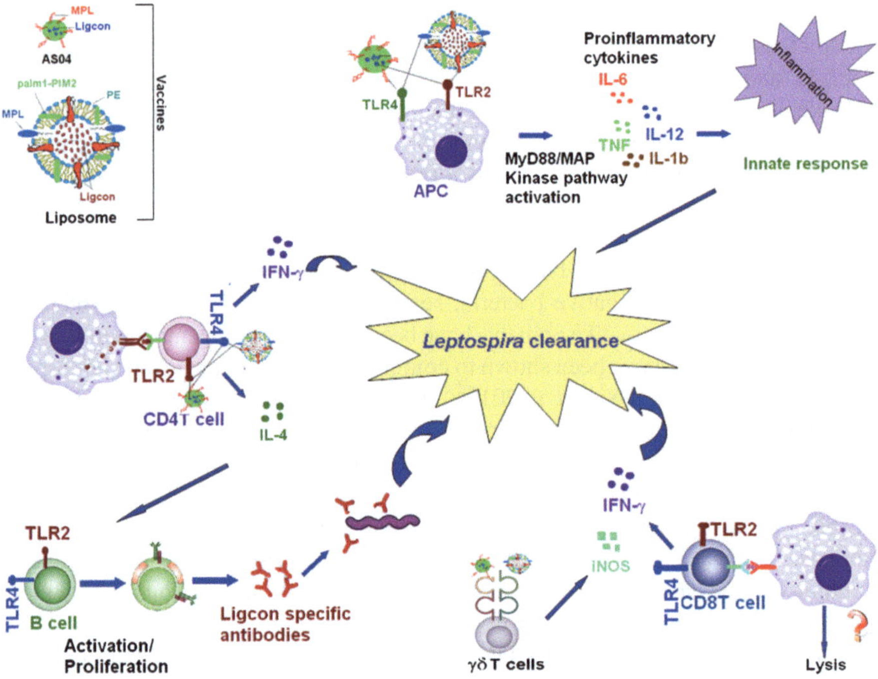

Fig. 8.6 Mechanism of protection mediated by Lig protein formulated with adjuvants. The conserved region of Lig protein (Ligcon) formulated with alum and MPL (Adjuvant System 04, AS04) or in Liposomes made up of various lipids (Phosphatidylethanolamine, PE and palmitoylated Phosphatidylinositol mannoside, palm1-PIM2) and MPL might interact with APCs via both TLR2 and TLR4 in the same way as whole *Leptospira*. This interaction may lead to activation of innate responses, phagocytosis of the vaccine, processing and subsequent presentation of Ligcon antigen peptides on both MHCI and MHCII which may facilitate interaction of APCs with both CD4 and CD8 T cells leading to their subsequent activation, proliferation and production of cytokines. B cells might produce Ligcon-specific antibodies that bind to bacteria to prevent their attachment to the host cells and may also cause opsonization, subsequent degradation and elimination

minimal batch to batch variation, dosage, toxicity, and manufacturing cost are some of the pivotal issues that need to be addressed and require extensive evaluation.

8.7 Conclusions and Future Research

Although significant progress has been made in understanding various aspects of *Leptospira* pathogenesis and the host immune response, several hurdles need to be overcome in order to make further progress. The major recent advancements include the availability of whole genome sequences of various leptospiral strains and the establishment of genetic tools like shuttle vectors and site-directed mutagenesis. A major obstacle to this is application of these tools in pathogenic leptospires with the same ease as in *L. biflexa*. Knocking out several related candidate virulence factor genes in pathogenic leptospires would circumvent the problem of redundancy as was observed with LipL32 and LigB single gene mutants.

Future research should be directed towards use of genetic tools in *L. biflexa* as a model organism to elucidate the function of 2000 core genes that are shared by pathogenic counterparts in order to gain better insight into the general biology of *Leptospira* spp. Employing powerful modern techniques like tiling array transcriptomics and RNA sequencing should pave way for better understanding of gene regulation and interaction among leptospiral proteins (Mader et al. 2010; Wang et al. 2009). This technique coupled with proteomics could be exploited to establish the expression of putative virulence factors and regulatory proteins when leptospires are subjected to conditions that mimic the host environment. It has been established that leptospires down regulate some proteins when they move to different environments, probably a strategy evolved to evade host immunity. Further studies on protein expression profiles would provide clues regarding factors involved in colonization, persistence in the host and immune evasion. With the availability of knockout models, biochemical and genetic studies can be employed to understand leptospiral pathogenesis and identify host determinants of disease susceptibility. Current research shows that differential recognition of LPS by TLR is key in determining host susceptibility or resistance to infection. Further research is required to identify factors other than LPS or innate receptors that likely play a major role in leptospiral pathogenesis.

With the elucidation of the complete structure of LPS, structural analogs could be prepared that would become active constituents in future vaccines. Although antibodies seem to play a major role, the invasive ability of leptospires and complement escape mechanisms indicate that future studies into the activation of CD4 and γδ-T cells should be directed towards understanding the mechanism of cell-mediated immune responses and designing strategies to activate and shape them. Understanding the pathogenic mechanisms and clinical symptoms of infection with different serovars will be an important step in development of an effective vaccine and a diagnostic kit to differentiate between vaccination and natural infection, which is still a major challenge. The availability of genome sequences of various serovars will speed up the process of identification of new vaccine candidates that protect against multiple *Leptospira* spp. With the availability of a large number of probable vaccine candidates identified using genomic and bioinformatic tools, a major hurdle is the lack of an in vitro system for evaluating protective immunity as high-throughput screening in experimental animals is impractical. Thus, the discovery of vaccine candidates will probably continue to rely on the search for new virulence factors and outer membrane proteins until epidemiologically validated immune correlates are identified. Virulence factors that are expressed during infection, like Lig proteins, are promising candidates but need to be coupled with effective adjuvants that can stimulate both innate and adaptive protective immune responses. Additionally, determining cross-protective immunity of conserved proteins like Lig proteins would be a significant step towards development of vaccines that could provide cross-protection against the large number of pathogenic serovars. Further, mechanisms of immunity need to be correlated to protection for evaluation in clinical vaccine trials. With the discovery of new virulence factors and the current pace of leptospira research on understanding the pathogenesis and host response, we anticipate that the next few years will bring major advances in improving diagnosis and state-of-the-art technologies for leptospiral vaccine development.

References

Adler B, de la Pena Moctezuma A (2010) Leptospira and leptospirosis. Vet Microbiol 140(3–4):287–296

Adler B, Faine S (1977) Host immunological mechanisms in the resistance of mice to leptospiral infections. Infect Immun 17(1):67–72

Adler B, Faine S (1978) The antibodies involved in the human immune response to leptospiral infection. J Med Microbiol 11(4):387–400

Ahmed N, Devi SM, Valverde Mde L, Vijayachari P, Machangu RS, Ellis WA, Hartskeerl RA (2006) Multilocus sequence typing method for identification and genotypic classification of pathogenic Leptospira species. Ann Clin Microbiol Antimicrob 5:28

Artiushin S, Timoney JF, Nally J, Verma A (2004) Host-inducible immunogenic sphingomyelinase-like protein, Lk73.5, of Leptospira interrogans. Infect Immun 72(2):742–749

Athanazio DA, Silva EF, Santos CS, Rocha GM, Vannier-Santos MA, McBride AJ, Ko AI, Reis MG (2008) Rattus norvegicus as a model for persistent renal colonization by pathogenic Leptospira interrogans. Acta Trop 105(2):176–180

Atzingen MV, Barbosa AS, De Brito T, Vasconcellos SA, de Morais ZM, Lima DM, Abreu PA, Nascimento AL (2008) Lsa21, a novel leptospiral protein binding adhesive matrix molecules and present during human infection. BMC Microbiol 8:70

Bajani MD, Ashford DA, Bragg SL, Woods CW, Aye T, Spiegel RA, Plikaytis BD, Perkins BA, Phelan M, Levett PN, Weyant RS (2003) Evaluation of four commercially available rapid serologic tests for diagnosis of leptospirosis. J Clin Microbiol 41(2):803–809

Barbosa AS, Abreu PA, Neves FO, Atzingen MV, Watanabe MM, Vieira ML, Morais ZM, Vasconcellos SA, Nascimento AL (2006) A newly identified leptospiral adhesin mediates attachment to laminin. Infect Immun 74(11):6356–6364

Barbosa AS, Abreu PA, Vasconcellos SA, Morais ZM, Goncales AP, Silva AS, Daha MR, Isaac L (2009) Immune evasion of leptospira species by acquisition of human complement regulator C4BP. Infect Immun 77(3):1137–1143

Barbosa AS, Monaris D, Silva LB, Morais ZM, Vasconcellos SA, Cianciarullo AM, Isaac L, Abreu PA (2010) Functional characterization of LcpA, a surface-exposed protein of Leptospira spp. that binds the human complement regulator C4BP. Infect Immun 78(7):3207–3216

Barocchi MA, Ko AI, Reis MG, McDonald KL, Riley LW (2002) Rapid translocation of polarized MDCK cell monolayers by Leptospira interrogans, an invasive but nonintracellular pathogen. Infect Immun 70(12):6926–6932

Bauby H, Saint Girons I, Picardeau M (2003) Construction and complementation of the first auxotrophic mutant in the spirochaete Leptospira meyeri. Microbiology 149(Pt 3):689–693

Bernheimer AW, Bey RF (1986) Copurification of Leptospira interrogans serovar pomona hemolysin and sphingomyelinase C. Infect Immun 54(1):262–264

Bharti AR, Nally JE, Ricaldi JN, Matthias MA, Diaz MM, Lovett MA, Levett PN, Gilman RH, Willig MR, Gotuzzo E, Vinetz JM (2003) Peru-United States Leptospirosis Consortium: Leptospirosis: a zoonotic disease of global importance. Lancet Infect Dis 3(12):757–771

Bolin CA, Alt DP (2001) Use of a monovalent leptospiral vaccine to prevent renal colonization and urinary shedding in cattle exposed to Leptospira borgpetersenii serovar hardjo. Am J Vet Res 62(7):995–1000

Bonneville M, O'Brien RL, Born WK (2010) Gammadelta T cell effector functions: a blend of innate programming and acquired plasticity. Nat Rev Immunol 10(7):467–478

Boonyod D, Poovorawan Y, Bhattarakosol P, Chirathaworn C (2005) LipL32, an outer membrane protein of Leptospira, as an antigen in a dipstick assay for diagnosis of leptospirosis. Asian Pac J Allergy Immunol 23(2–3):133–141

Bourhy P, Louvel H, Saint Girons I, Picardeau M (2005) Random insertional mutagenesis of Leptospira interrogans, the agent of leptospirosis, using a mariner transposon. J Bacteriol 187(9):3255–3258

Branger C, Sonrier C, Chatrenet B, Klonjkowski B, Ruvoen-Clouet N, Aubert A, Andre-Fontaine G, Eloit M (2001) Identification of the hemolysis-associated protein 1 as a cross-protective immunogen of Leptospira interrogans by adenovirus-mediated vaccination. Infect Immun 69(11):6831–6838

Brown RA, Blumerman S, Gay C, Bolin C, Duby R, Baldwin CL (2003) Comparison of three different leptospiral vaccines for induction of a type 1 immune response to Leptospira borgpetersenii serovar Hardjo. Vaccine 21(27–30):4448–4458

Bulach DM, Zuerner RL, Wilson P, Seemann T, McGrath A, Cullen PA, Davis J, Johnson M, Kuczek E, Alt DP, Peterson-Burch B, Coppel RL, Rood JI, Davies JK, Adler B (2006) Genome reduction in Leptospira borgpetersenii reflects limited transmission potential. Proc Natl Acad Sci USA 103(39):14560–14565

Burnette WN (1991) Recombinant subunit vaccines. Curr Opin Biotechnol 2(6):882–892

Burth P, Younes-Ibrahim M, Santos MC, Castro-Faria Neto HC, De Castro Faria MV (2005) Role of nonesterified unsaturated fatty acids in the pathophysiological processes of leptospiral infection. J Infect Dis 191(1):51–57

Busch DH, Jassoy C, Brinckmann U, Girschick H, Huppertz HI (1996) Detection of Borrelia burgdorferi-specific CD8+ cytotoxic T cells in patients with Lyme arthritis. J Immunol 157(8):3534–3541

Carvalho E, Barbosa AS, Gomez RM, Oliveira ML, Romero EC, Goncales AP, Morais ZM, Vasconcellos SA, Ho PL (2010) Evaluation of the expression and protective potential of Leptospiral sphingomyelinases. Curr Microbiol 60(2):134–142

Castiblanco-Valencia MM, Fraga TR, Silva LB, Monaris D, Abreu PA, Strobel S, Jozsi M, Isaac L, Barbosa AS (2012) Leptospiral immunoglobulin-like proteins interact with human complement regulators factor H, FHL-1, FHR-1, and C4BP. J Infect Dis 205(6):995–1004

Chang YF, Chen CS, Palaniappan RU, He H, McDonough SP, Barr SC, Yan W, Faisal SM, Pan MJ, Chang CF (2007) Immunogenicity of the recombinant leptospiral putative outer membrane proteins as vaccine candidates. Vaccine 25(48):8190–8197

Chassin C, Picardeau M, Goujon JM, Bourhy P, Quellard N, Darche S (2009) Badell E, d'Andon MF, Winter N, Lacroix-Lamande S, Buzoni-Gatel D, Vandewalle A, Werts C: TLR4- and TLR2-mediated B cell responses control the clearance of the bacterial pathogen, Leptospira interrogans. J Immunol 183(4):2669–2677

Choy HA, Kelley MM, Chen TL, Moller AK, Matsunaga J, Haake DA (2007) Physiological osmotic induction of Leptospira interrogans adhesion: LigA and LigB bind extracellular matrix proteins and fibrinogen. Infect Immun 75(5):2441–2450

Cinco M, Domenis R, Perticarari S, Presani G, Marangoni A, Blasi E (2006) Interaction of leptospires with murine microglial cells. New Microbiol 29(3):193–199

Croda J, Ramos JG, Matsunaga J, Queiroz A, Homma A, Riley LW, Haake DA, Reis MG, Ko AI (2007) Leptospira immunoglobulin-like proteins as a serodiagnostic marker for acute leptospirosis. J Clin Microbiol 45(5):1528–1534

Croda J, Figueira CP, Wunder EA Jr (2008) Santos CS, Reis MG, Ko AI, Picardeau M: Targeted mutagenesis in pathogenic Leptospira species: disruption of the LigB gene does not affect virulence in animal models of leptospirosis. Infect Immun 76(12):5826–5833

Croda J, Neto AN, Brasil RA, Pagliari C, Nicodemo AC, Duarte MI (2010) Leptospirosis pulmonary haemorrhage syndrome is associated with linear deposition of immunoglobulin and complement on the alveolar surface. Clin Microbiol Infect 16(6):593–599

Cullen PA, Cordwell SJ, Bulach DM, Haake DA, Adler B (2002) Global analysis of outer membrane proteins from Leptospira interrogans serovar Lai. Infect Immun 70(5):2311–2318

Davis JM, Haake DA, Ramakrishnan L (2009) Leptospira interrogans stably infects zebrafish embryos, altering phagocyte behavior and homing to specific tissues. PLoS Negl Trop Dis 3(6):e463

Ellis WA, Hovind-Hougen K, Moller S, Birch-Andresen A (1983) Morphological changes upon subculturing of freshly isolated strains of Leptospira interrogans serovar hardjo. Zentralbl Bakteriol Mikrobiol Hyg A 255(2–3):323–335

Faine S (1957) Virulence in leptospira. II. The growth in vivo of virulent Leptospira icterohaemorrhagiae. Br J Exp Pathol 38(1):8–14

Faisal SM, Yan W, Chen CS, Palaniappan RU, McDonough SP, Chang YF (2008) Evaluation of protective immunity of Leptospira immunoglobulin like protein A (LigA) DNA vaccine against challenge in hamsters. Vaccine 26(2):277–287

Faisal SM, Yan W, McDonough SP, Chang YF (2009a) Leptospira immunoglobulin-like protein A variable region (LigAvar) incorporated in liposomes and PLGA microspheres produces a robust immune response correlating to protective immunity. Vaccine 27(3):378–387

Faisal SM, Yan W, McDonough SP, Chang CF, Pan MJ, Chang YF (2009b) Leptosome-entrapped leptospiral antigens conferred significant higher levels of protection than those entrapped with PC-liposomes in a hamster model. Vaccine 27(47):6537–6545

Faisal SM, Yan W, McDonough SP, Mohammed HO, Divers TJ, Chang YF (2009c) Immune response and prophylactic efficacy of smegmosomes in a hamster model of leptospirosis. Vaccine 27(44):6129–6136

Feng CY, Li QT, Zhang XY, Dong K, Hu BY, Guo XK (2009) Immune strategies using single-component LipL32 and multi-component recombinant LipL32-41-OmpL1 vaccines against leptospira. Braz J Med Biol Res 42(9):796–803

Gamberini M, Gomez RM, Atzingen MV, Martins EA, Vasconcellos SA, Romero EC, Leite LC, Ho PL, Nascimento AL (2005) Whole-genome analysis of Leptospira interrogans to identify potential vaccine candidates against leptospirosis. FEMS Microbiol Lett 244(2):305–313

Girons IS, Bourhy P, Ottone C, Picardeau M, Yelton D, Hendrix RW, Glaser P, Charon N (2000) The LE1 bacteriophage replicates as a plasmid within Leptospira biflexa: construction of an L. biflexa-Escherichia coli shuttle vector. J Bacteriol 182(20):5700–5705

Gouveia EL, Metcalfe J, de Carvalho AL, Aires TS, Villasboas-Bisneto JC, Queirroz A, Santos AC, Salgado K, Reis MG, Ko AI (2008) Leptospirosis-associated severe pulmonary hemorrhagic syndrome, Salvador, Brazil. Emerg Infect Dis 14(3):505–508

Guegan R, Camadro JM, Saint Girons I, Picardeau M (2003) Leptospira spp. possess a complete haem biosynthetic pathway and are able to use exogenous haem sources. Mol Microbiol 49(3):745–754

Guo YJ, Wang KY, Sun SH (2010) Identification of an HLA-A*0201-restricted CD8(+) T-cell epitope encoded within Leptospiral immunoglobulin-like protein A. Microbes Infect 12(5):364–373

Haake DA (2006) Hamster model of leptospirosis. Curr Protoc Microbiol, Chapter 12:Unit 12E.2.

Haake DA, Mazel MK, McCoy AM, Milward F, Chao G, Matsunaga J, Wagar EA (1999) Leptospiral outer membrane proteins OmpL1 and LipL41 exhibit synergistic immunoprotection. Infect Immun 67(12):6572–6582

Haake DA, Chao G, Zuerner RL, Barnett JK, Barnett D, Mazel M, Matsunaga J, Levett PN, Bolin CA (2000) The leptospiral major outer membrane protein LipL32 is a lipoprotein expressed during mammalian infection. Infect Immun 68(4):2276–2285

Habarta A, Abreu PA, Olivera N, Hauk P, Cedola MT, Ferrer MF, Ho PL, Gomez RM (2010) Increased Immunogenicity to LipL32 of Leptospira interrogans when Expressed as a Fusion Protein with the Cholera Toxin B Subunit. Curr Microbiol 62(2):526–531

Hauk P, Macedo F, Romero EC, Vasconcellos SA, de Morais ZM, Barbosa AS, Ho PL (2008) In LipL32, the major leptospiral lipoprotein, the C terminus is the primary immunogenic domain and mediates interaction with collagen IV and plasma fibronectin. Infect Immun 76(6):2642–2650

He P, Sheng YY, Shi YZ, Jiang XG, Qin JH, Zhang ZM, Zhao GP, Guo XK (2007) Genetic diversity among major endemic strains of Leptospira interrogans in China. BMC Genomics 8:204

Hoke DE, Egan S, Cullen PA, Adler B (2008) LipL32 is an extracellular matrix-interacting protein of Leptospira spp. and Pseudoalteromonas tunicata. Infect Immun 76(5):2063–2069

Hospenthal DR, Murray CK (2003) In vitro susceptibilities of seven Leptospira species to traditional and newer antibiotics. Antimicrob Agents Chemother 47(8):2646–2648

Ito T, Yanagawa R (1987) Leptospiral attachment to extracellular matrix of mouse fibroblast (L929) cells. Vet Microbiol 15(1–2):89–96

Jost BH, Adler B, Vinh T, Faine S (1986) A monoclonal antibody reacting with a determinant on leptospiral lipopolysaccharide protects guinea pigs against leptospirosis. J Med Microbiol 22(3):269–275

Kariv R, Klempfner R, Barnea A, Sidi Y, Schwartz E (2001) The changing epidemiology of leptospirosis in Israel. Emerg Infect Dis 7(6):990–992

Katz AR, Ansdell VE, Effler PV, Middleton CR, Sasaki DM (2001) Assessment of the clinical presentation and treatment of 353 cases of laboratory-confirmed leptospirosis in Hawaii, 1974–1998. Clin Infect Dis 33(11):1834–1841

Kee SH, Kim IS, Choi MS, Chang WH (1994) Detection of leptospiral DNA by PCR. J Clin Microbiol 32(4):1035–1039

Klimpel GR, Matthias MA, Vinetz JM (2003) Leptospira interrogans activation of human peripheral blood mononuclear cells: preferential expansion of TCR gamma delta+T cells vs TCR alpha beta+T cells. J Immunol 171(3):1447–1455

Ko AI, Galvao Reis M, Ribeiro Dourado CM, Johnson WD Jr, Riley LW (1999) Urban epidemic of severe leptospirosis in Brazil. Salvador Leptospirosis Study Group. Lancet 354(9181):820–825

Ko AI, Goarant C, Picardeau M (2009) Leptospira: the dawn of the molecular genetics era for an emerging zoonotic pathogen. Nat Rev Microbiol 7(10):736–747

Koizumi N, Watanabe H (2003) Molecular cloning and characterization of a novel leptospiral lipoprotein with OmpA domain. FEMS Microbiol Lett 226(2):215–219

Koizumi N, Watanabe H (2004) Leptospiral immunoglobulin-like proteins elicit protective immunity. Vaccine 22(11–12):1545–1552

Kositanont U, Rugsasuk S, Leelaporn A, Phulsuksombati D, Tantitanawat S, Naigowit P (2007) Detection and differentiation between pathogenic and saprophytic Leptospira spp. by multiplex polymerase chain reaction. Diagn Microbiol Infect Dis 57(2):117–122

Leblebicioglu H, Sunbul M, Esen S, Eroglu C (2003) Jarisch-Herxheimer reaction in leptospirosis. Eur J Clin Microbiol Infect Dis 22(10):639, author reply 640

Lee SH, Kim KA, Park YG, Seong IW, Kim MJ, Lee YJ (2000) Identification and partial characterization of a novel hemolysin from Leptospira interrogans serovar lai. Gene 254(1–2):19–28

Lee SH, Kim S, Park SC, Kim MJ (2002) Cytotoxic activities of Leptospira interrogans hemolysin SphH as a pore-forming protein on mammalian cells. Infect Immun 70(1):315–322

Levett PN (2001) Leptospirosis. Clin Microbiol Rev 14(2):296–326

Li S, Ojcius DM, Liao S, Li L, Xue F, Dong H, Yan J (2010) Replication or death: distinct fates of pathogenic Leptospira strain Lai within macrophages of human or mouse origin. Innate Immun 16(2):80–92

Liao S, Sun A, Ojcius DM, Wu S, Zhao J, Yan J (2009) Inactivation of the fliY gene encoding a flagellar motor switch protein attenuates mobility and virulence of Leptospira interrogans strain Lai. BMC Microbiol 9:253

Lin YP, Chang YF (2008) The C-terminal variable domain of LigB from Leptospira mediates binding to fibronectin. J Vet Sci 9(2):133–144

Lin YP, Raman R, Sharma Y, Chang YF (2008) Calcium binds to leptospiral immunoglobulin-like protein, LigB, and modulates fibronectin binding. J Biol Chem 283(37):25140–25149

Lin YP, Greenwood A, Nicholson LK, Sharma Y, McDonough SP, Chang YF (2009a) Fibronectin binds to and induces conformational change in a disordered region of leptospiral immunoglobulin-like protein B. J Biol Chem 284(35):23547–23557

Lin YP, Greenwood A, Yan W, Nicholson LK, Sharma Y, McDonough SP, Chang YF (2009b) A novel fibronectin type III module binding motif identified on C-terminus of Leptospira immunoglobulin-like protein, LigB. Biochem Biophys Res Commun 389(1):57–62

Lin YP, Lee DW, McDonough SP, Nicholson LK, Sharma Y, Chang YF (2009c) Repeated domains of leptospira immunoglobulin-like proteins interact with elastin and tropoelastin. J Biol Chem 284(29):19380–19391
Lin YP, McDonough SP, Sharma Y, Chang YF (2010) The terminal immunoglobulin-like repeats of LigA and LigB of Leptospira enhance their binding to gelatin binding domain of fibronectin and host cells. PLoS One 5(6):e11301
Lin YP, McDonough SP, Sharma Y, Chang YF (2011) Leptospira immunoglobulin-like protein B (LigB) binding to the C-terminal fibrinogen alphaC domain inhibits fibrin clot formation, platelet adhesion and aggregation. Mol Microbiol 79(4):1063–1076
Lingappa J, Kuffner T, Tappero J, Whitworth W, Mize A, Kaiser R, McNicholl J (2004) HLA-DQ6 and ingestion of contaminated water: possible gene-environment interaction in an outbreak of Leptospirosis. Genes Immun 5(3):197–202
Liu Y, Zheng W, Li L, Mao Y, Yan J (2007) Pathogenesis of leptospirosis: interaction of Leptospira interrogans with in vitro cultured mammalian cells. Med Microbiol Immunol 196(4):233–239
Lo M, Bulach DM, Powell DR, Haake DA, Matsunaga J, Paustian ML, Zuerner RL, Adler B (2006) Effects of temperature on gene expression patterns in Leptospira interrogans serovar Lai as assessed by whole-genome microarrays. Infect Immun 74(10):5848–5859
Longhi MT, Oliveira TR, Romero EC, Goncales AP, de Morais ZM, Vasconcellos SA, Nascimento AL (2009) A newly identified protein of Leptospira interrogans mediates binding to laminin. J Med Microbiol 58(Pt 10):1275–1282
Louvel H, Picardeau M (2007) Genetic manipulation of Leptospira biflexa. Curr Protoc Microbiol, Chapter 12:Unit 12E.4.
Louvel H, Saint Girons I, Picardeau M (2005) Isolation and characterization of FecA- and FeoB-mediated iron acquisition systems of the spirochete Leptospira biflexa by random insertional mutagenesis. J Bacteriol 187(9):3249–3254
Louvel H, Bommezzadri S, Zidane N, Boursaux-Eude C, Creno S, Magnier A, Rouy Z, Medigue C, Saint Girons C, Bouchier I, Picardeau M (2006) Comparative and functional genomic analyses of iron transport and regulation in Leptospira spp. J Bacteriol 188(22):7893–7904
Mader U, Nicolas P, Richard H, Bessieres P, Aymerich S (2010) Comprehensive identification and quantification of microbial transcriptomes by genome-wide unbiased methods. Curr Opin Biotechnol 22(1):32–41
Mariya R, Chaudhary P, Kumar AA, Thangapandian E, Amutha R, Srivastava SK (2006) Evaluation of a recombinant LipL41 antigen of Leptospira interrogans serovar canicola in ELISA for serodiagnosis of bovine leptospirosis. Comp Immunol Microbiol Infect Dis 29(5–6):269–277
Matsunaga J, Barocchi MA, Croda J, Young TA, Sanchez Y, Siqueira I, Bolin CA, Reis MG, Riley LW, Haake DA, Ko AI (2003) Pathogenic Leptospira species express surface-exposed proteins belonging to the bacterial immunoglobulin superfamily. Mol Microbiol 49(4):929–945
Matsunaga J, Lo M, Bulach DM, Zuerner RL, Adler B, Haake DA (2007) Response of Leptospira interrogans to physiologic osmolarity: relevance in signaling the environment-to-host transition. Infect Immun 75(6):2864–2874
McBride AJ, Athanazio DA, Reis MG, Ko AI (2005) Leptospirosis. Curr Opin Infect Dis 18(5):376–386
McBride AJ, Cerqueira GM, Suchard MA, Moreira AN, Zuerner RL, Reis MG, Haake DA, Ko AI, Dellagostin OA (2009) Genetic diversity of the Leptospiral immunoglobulin-like (Lig) genes in pathogenic Leptospira spp. Infect Genet Evol 9(2):196–205
McCurry J (2009) Philippines struggles to recover from typhoons. Lancet 374(9700):1489
Meri T, Murgia R, Stefanel P, Meri S, Cinco M (2005) Regulation of complement activation at the C3-level by serum resistant leptospires. Microb Pathog 39(4):139–147
Merien F, Perolat P, Mancel E, Persan D, Baranton G (1993) Detection of Leptospira DNA by polymerase chain reaction in aqueous humor of a patient with unilateral uveitis. J Infect Dis 168(5):1335–1336
Merien F, Baranton G, Perolat P (1997) Invasion of Vero cells and induction of apoptosis in macrophages by pathogenic Leptospira interrogans are correlated with virulence. Infect Immun 65(2):729–738

Merien F, Truccolo J, Baranton G, Perolat P (2000) Identification of a 36-kDa fibronectin-binding protein expressed by a virulent variant of Leptospira interrogans serovar icterohaemorrhagiae. FEMS Microbiol Lett 185(1):17–22

Minette HP, Shaffer MF (1968) Experimental leptospirosis in monkeys. Am J Trop Med Hyg 17(2):202–212

Monahan AM, Callanan JJ, Nally JE (2008) Proteomic analysis of Leptospira interrogans shed in urine of chronically infected hosts. Infect Immun 76(11):4952–4958

Murray GL, Srikram A, Hoke DE, Wunder EA Jr, Henry R, Lo M, Zhang K, Sermswan RW, Ko AI, Adler B (2009a) Major surface protein LipL32 is not required for either acute or chronic infection with Leptospira interrogans. Infect Immun 77(3):952–958

Murray GL, Morel V, Cerqueira GM, Croda J, Srikram A, Henry R, Ko AI, Dellagostin OA, Bulach DM, Sermswan RW, Adler B, Picardeau M (2009b) Genome-wide transposon mutagenesis in pathogenic Leptospira species. Infect Immun 77(2):810–816

Murray GL, Srikram A, Henry R, Puapairoj A, Sermswan RW, Adler B (2009c) Leptospira interrogans requires heme oxygenase for disease pathogenesis. Microbes Infect 11(2):311–314

Murray GL, Srikram A, Henry R, Hartskeerl RA, Sermswan RW, Adler B (2010) Mutations affecting Leptospira interrogans lipopolysaccharide attenuate virulence. Mol Microbiol 78(3):701–709

Nahori MA, Fournie-Amazouz E, Que-Gewirth NS, Balloy V, Chignard M, Raetz CR, Saint Girons I, Werts C (2005) Differential TLR recognition of leptospiral lipid A and lipopolysaccharide in murine and human cells. J Immunol 175(9):6022–6031

Naiman BM, Alt D, Bolin CA, Zuerner R, Baldwin CL (2001) Protective killed Leptospira borgpetersenii vaccine induces potent Th1 immunity comprising responses by CD4 and gammadelta T lymphocytes. Infect Immun 69(12):7550–7558

Naiman BM, Blumerman S, Alt D, Bolin CA, Brown R, Zuerner R, Baldwin CL (2002) Evaluation of type 1 immune response in naive and vaccinated animals following challenge with Leptospira borgpetersenii serovar Hardjo: involvement of WC1(+) gammadelta and CD4 T cells. Infect Immun 70(11):6147–6157

Nally JE, Chantranuwat C, Wu XY, Fishbein MC, Pereira MM, Da Silva JJ, Blanco DR, Lovett MA (2004) Alveolar septal deposition of immunoglobulin and complement parallels pulmonary hemorrhage in a guinea pig model of severe pulmonary leptospirosis. Am J Pathol 164(3):1115–1127

Nally JE, Chow E, Fishbein MC, Blanco DR, Lovett MA (2005) Changes in lipopolysaccharide O antigen distinguish acute versus chronic Leptospira interrogans infections. Infect Immun 73(6):3251–3260

Nally JE, Whitelegge JP, Bassilian S, Blanco DR, Lovett MA (2007) Characterization of the outer membrane proteome of Leptospira interrogans expressed during acute lethal infection. Infect Immun 75(2):766–773

Nascimento AL, Ko AI, Martins EA, Monteiro-Vitorello CB, Ho PL, Haake DA, Verjovski-Almeida S, Hartskeerl RA, Marques MV, Oliveira MC, Menck CF, Leite LC, Carrer H, Coutinho LL, Degrave WM, Dellagostin OA, El-Dorry H, Ferro ES, Ferro MI, Furlan LR, Gamberini M, Giglioti EA, Goes-Neto A, Goldman GH, Goldman MH, Harakava R, Jeronimo SM, Junqueira-de-Azevedo IL, Kimura ET, Kuramae EE, Lemos EG, Lemos MV, Marino CL, Nunes LR, de Oliveira RC, Pereira GG, Reis MS, Schriefer A, Siqueira WJ, Sommer P, Tsai SM, Simpson AJ, Ferro JA, Camargo LE, Kitajima JP, Setubal JC, Van Sluys MA (2004) Comparative genomics of two Leptospira interrogans serovars reveals novel insights into physiology and pathogenesis. J Bacteriol 186(7):2164–2172

Okuda M, Sakai Y, Matsuuchi M, Oikawa T, Watanabe M, Itamoto K, Iwata H, Kano R, Hasegawa A, Onishi T, Inokuma H (2005) Enzyme-linked immunosorbent assay for the detection of canine Leptospira antibodies using recombinant OmpL1 protein. J Vet Med Sci 67(3):249–254

Palaniappan RU, Chang YF, Jusuf SS, Artiushin S, Timoney JF, McDonough SP, Barr SC, Divers TJ, Simpson KW, McDonough PL, Mohammed HO (2002) Cloning and molecular characterization of an immunogenic LigA protein of Leptospira interrogans. Infect Immun 70(11):5924–5930

Palaniappan RU, Chang YF, Hassan F, McDonough SP, Pough M, Barr SC, Simpson KW, Mohammed HO, Shin S, McDonough P, Zuerner RL, Qu J, Roe B (2004) Expression of leptospiral immunoglobulin-like protein by Leptospira interrogans and evaluation of its diagnostic potential in a kinetic ELISA. J Med Microbiol 53(Pt 10):975–984

Palaniappan RU, Chang YF, Chang CF, Pan MJ, Yang CW, Harpending P, McDonough SP, Dubovi E, Divers T, Qu J, Roe B (2005) Evaluation of lig-based conventional and real time PCR for the detection of pathogenic leptospires. Mol Cell Probes 19(2):111–117

Palaniappan RU, McDonough SP, Divers TJ, Chen CS, Pan MJ, Matsumoto M, Chang YF (2006) Immunoprotection of recombinant leptospiral immunoglobulin-like protein A against Leptospira interrogans serovar Pomona infection. Infect Immun 74(3):1745–1750

Palaniappan RU, Ramanujam S, Chang YF (2007) Leptospirosis: pathogenesis, immunity, and diagnosis. Curr Opin Infect Dis 20(3):284–292

Pappas G, Cascio A (2006) Optimal treatment of leptospirosis: queries and projections. Int J Antimicrob Agents 28(6):491–496

Pappas G, Papadimitriou P, Siozopoulou V, Christou L, Akritidis N (2008) The globalization of leptospirosis: worldwide incidence trends. Int J Infect Dis 12(4):351–357

Pereira MM, Andrade J, Marchevsky RS, Ribeiro dos Santos R (1998) Morphological characterization of lung and kidney lesions in C3H/HeJ mice infected with Leptospira interrogans serovar icterohaemorrhagiae: defect of CD4+ and CD8+ T-cells are prognosticators of the disease progression. Exp Toxicol Pathol 50(3):191–198

Pereira MM, Da Silva JJ, Pinto MA, Da Silva MF, Machado MP, Lenzi HL, Marchevsky RS (2005) Experimental leptospirosis in marmoset monkeys (Callithrix jacchus): a new model for studies of severe pulmonary leptospirosis. Am J Trop Med Hyg 72(1):13–20

Picardeau M (2008) Conjugative transfer between Escherichia coli and Leptospira spp. as a new genetic tool. Appl Environ Microbiol 74(1):319–322

Picardeau M, Brenot A, Saint Girons I (2001) First evidence for gene replacement in Leptospira spp. Inactivation of L. biflexa flaB results in non-motile mutants deficient in endoflagella. Mol Microbiol 40(1):189–199

Picardeau M, Bulach DM, Bouchier C, Zuerner RL, Zidane N, Wilson PJ, Creno S, Kuczek ES, Bommezzadri S, Davis JC, McGrath A, Johnson MJ, Boursaux-Eude C, Seemann T, Rouy Z, Coppel RL, Rood JI, Lajus A, Davies JK, Medigue C, Adler B (2008) Genome sequence of the saprophyte Leptospira biflexa provides insights into the evolution of Leptospira and the pathogenesis of leptospirosis. PLoS One 3(2):e1607

Pinne M, Choy HA, Haake DA (2010) The OmpL37 surface-exposed protein is expressed by pathogenic Leptospira during infection and binds skin and vascular elastin. PLoS Negl Trop Dis 4(9):e815

Reis RB, Ribeiro GS, Felzemburgh RD, Santana FS, Mohr S, Melendez AX, Queiroz A, Santos AC, Ravines RR, Tassinari WS, Carvalho MS, Reis MG, Ko AI (2008) Impact of environment and social gradient on Leptospira infection in urban slums. PLoS Negl Trop Dis 2(4):e228

Ren SX, Fu G, Jiang XG, Zeng R, Miao YG, Xu H, Zhang YX, Xiong H, Lu G, Lu LF, Jiang HQ, Jia J, Tu YF, Jiang JX, Gu WY, Zhang YQ, Cai Z, Sheng HH, Yin HF, Zhang Y, Zhu GF, Wan M, Huang HL, Qian Z, Wang SY, Ma W, Yao ZJ, Shen Y, Qiang BQ, Xia QC, Guo XK, Danchin A, Saint Girons I, Somerville RL, Wen YM, Shi MH, Chen Z, Xu JG, Zhao GP (2003) Unique physiological and pathogenic features of Leptospira interrogans revealed by whole-genome sequencing. Nature 422(6934):888–893

Riley LW, Ko AI, Unger A, Reis MG (2007) Slum health: diseases of neglected populations. BMC Int Health Hum Rights 7:2

Ristow P, Bourhy P, Kerneis S, Schmitt C, Prevost MC, Lilenbaum W, Picardeau M (2008) Biofilm formation by saprophytic and pathogenic leptospires. Microbiology 154(Pt 5):1309–1317

Segers RP, van Gestel JA, van Eys GJ, van der Zeijst BA, Gaastra W (1992) Presence of putative sphingomyelinase genes among members of the family Leptospiraceae. Infect Immun 60(4):1707–1710

Seguro AC, Lomar AV, Rocha AS (1990) Acute renal failure of leptospirosis: nonoliguric and hypokalemic forms. Nephron 55(2):146–151

Sehgal SC, Sugunan AP, Murhekar MV, Sharma S, Vijayachari P (2000) Randomized controlled trial of doxycycline prophylaxis against leptospirosis in an endemic area. Int J Antimicrob Agents 13(4):249–255

Seixas FK, da Silva EF, Hartwig DD, Cerqueira GM, Amaral M, Fagundes MQ, Dossa RG, Dellagostin OA (2007) Recombinant Mycobacterium bovis BCG expressing the LipL32 antigen of Leptospira interrogans protects hamsters from challenge. Vaccine 26(1):88–95

Shimizu T, Matsusaka E, Nagakura N, Takayanagi K, Masuzawa T, Iwamoto Y, Morita T, Mifuchi I, Yanagihara Y (1987) Chemical properties of lipopolysaccharide-like substance (LLS) extracted from Leptospira interrogans serovar canicola strain Moulton. Microbiol Immunol 31(8):717–725

Silva EF, Medeiros MA, McBride AJ, Matsunaga J, Esteves GS, Ramos JG, Santos CS, Croda J, Homma A, Dellagostin OA, Haake DA, Reis MG, Ko AI (2007) The terminal portion of leptospiral immunoglobulin-like protein LigA confers protective immunity against lethal infection in the hamster model of leptospirosis. Vaccine 25(33):6277–6286

Silva EF, Santos CS, Athanazio DA, Seyffert N, Seixas FK, Cerqueira GM, Fagundes MQ, Brod CS, Reis MG, Dellagostin OA, Ko AI (2008) Characterization of virulence of Leptospira isolates in a hamster model. Vaccine 26(31):3892–3896

Sonrier C, Branger C, Michel V, Ruvoen-Clouet N, Ganiere JP, Andre-Fontaine G (2000) Evidence of cross-protection within Leptospira interrogans in an experimental model. Vaccine 19(1):86–94

Srikram A, Zhang K, Bartpho T, Lo M, Hoke DE, Sermswan RW, Adler B, Murray GL (2011) Cross-protective Immunity Against Leptospirosis Elicited by a Live, Attenuated Lipopolysaccharide Mutant. J Infect Dis 203(6):870–879

Sterling CR, Thiermann AB (1981) Urban rats as chronic carriers of leptospirosis: an ultrastructural investigation. Vet Pathol 18(5):628–637

Stevenson B, Choy HA, Pinne M, Rotondi ML, Miller MC, Demoll E, Kraiczy P, Cooley AE, Creamer TP, Suchard MA, Brissette CA, Verma A, Haake DA (2007) Leptospira interrogans endostatin-like outer membrane proteins bind host fibronectin, laminin and regulators of complement. PLoS One 2(11):e1188

Tchamedeu Kameni AP, Couture-Tosi E, Saint-Girons I, Picardeau M (2002) Inactivation of the spirochete recA gene results in a mutant with low viability and irregular nucleoid morphology. J Bacteriol 184(2):452–458

Tian YC, Hung CC, Li YJ, Chen YC, Chang MY, Yen TH, Hsu HH, Wu MS, Phillips A, Yang CW (2010) Leptospira santorosai serovar Shermani detergent extract induced an increases in fibronectin production through a toll-like receptor2-mediated pathway. Infect Immun 79(3):1134–1142

Trueba G, Zapata S, Madrid K, Cullen P, Haake D (2004) Cell aggregation: a mechanism of pathogenic Leptospira to survive in fresh water. Int Microbiol 7(1):35–40

Tu V, Adler B, Faine S (1982) The role of macrophages in the protection of mice against leptospirosis: in vitro and in vivo studies. Pathology 14(4):463–468

van Voorhis WC, Barrett LK, Nasio JM, Plummer FA, Lukehart SA (1996) Lesions of primary and secondary syphilis contain activated cytolytic T cells. Infect Immun 64(3):1048–1050

Verma A, Hellwage J, Artiushin S, Zipfel PF, Kraiczy P, Timoney JF, Stevenson B (2006) LfhA, a novel factor H-binding protein of Leptospira interrogans. Infect Immun 74(5):2659–2666

Vernel-Pauillac F, Merien F (2006) Proinflammatory and immunomodulatory cytokine mRNA time course profiles in hamsters infected with a virulent variant of Leptospira interrogans. Infect Immun 74(7):4172–4179

Victoriano AF, Smythe LD, Gloriani-Barzaga N, Cavinta LL, Kasai T, Limpakarnjanarat K, Ong BL, Gongal G, Hall J, Coulombe CA, Yanagihara Y, Yoshida S, Adler B (2009) Leptospirosis in the Asia Pacific region. BMC Infect Dis 9:147

Vieira ML, Vasconcellos SA, Goncales AP, de Morais ZM, Nascimento AL (2009) Plasminogen acquisition and activation at the surface of leptospira species lead to fibronectin degradation. Infect Immun 77(9):4092–4101

Vieira ML, Atzingen MV, Oliveira TR, Oliveira R, Andrade DM, Vasconcellos SA, Nascimento AL (2010a) In vitro identification of novel plasminogen-binding receptors of the pathogen Leptospira interrogans. PLoS One 5(6):e11259

Vieira ML, de Morais ZM, Goncales AP, Romero EC, Vasconcellos SA, Nascimento AL (2010b) Lsa63, a newly identified surface protein of Leptospira interrogans binds laminin and collagen IV. J Infect 60(1):52–64

Vinh T, Adler B, Faine S (1986) Ultrastructure and chemical composition of lipopolysaccharide extracted from Leptospira interrogans serovar copenhageni. J Gen Microbiol 132(1):103–109

Viriyakosol S, Matthias MA, Swancutt MA, Kirkland TN, Vinetz JM (2006) Toll-like receptor 4 protects against lethal Leptospira interrogans serovar icterohaemorrhagiae infection and contributes to in vivo control of leptospiral burden. Infect Immun 74(2):887–895

Vivian JP, Beddoe T, McAlister AD, Wilce MC, Zaker-Tabrizi L, Troy S, Byres E, Hoke DE, Cullen PA, Lo M, Murray GL, Adler B, Rossjohn J (2009) Crystal structure of LipL32, the most abundant surface protein of pathogenic Leptospira spp. J Mol Biol 387(5):1229–1238

Wang Z, Gerstein M, Snyder M (2009) RNA-Seq: a revolutionary tool for transcriptomics. Nat Rev Genet 10(1):57–63

Wangroongsarb P, Chanket T, Gunlabun K, Long do H, Satheanmethakul P, Jetanadee S, Thaipadungpanit J, Wuthiekanun V, Peacock SJ, Blacksell SD, Smythe LD, Bulach DM, Kalambaheti T (2007) Molecular typing of Leptospira spp based on putative O-antigen polymerase gene (wzy), the benefit over 16S rRNA gene sequence. FEMS Microbiol Lett 271(2):170–179

Watt G, Padre LP, Tuazon ML, Calubaquib C, Santiago E, Ranoa CP, Laughlin LW (1988) Placebo-controlled trial of intravenous penicillin for severe and late leptospirosis. Lancet 1(8583):433–435

Werts C, Tapping RI, Mathison JC, Chuang TH, Kravchenko V, Saint Girons I, Haake DA, Godowski PJ, Hayashi F, Ozinsky A, Underhill DM, Kirschning CJ, Wagner H, Aderem A, Tobias PS, Ulevitch RJ (2001) Leptospiral lipopolysaccharide activates cells through a TLR2-dependent mechanism. Nat Immunol 2(4):346–352

Xue F, Yan J, Picardeau M (2009) Evolution and pathogenesis of Leptospira spp.: lessons learned from the genomes. Microbes Infect 11(3):328–333

Tucunduva de Faria M, Athanazio DA, Goncalves Ramos EA, Silva EF, Reis MG, Ko AI (2007) Morphological alterations in the kidney of rats with natural and experimental Leptospira infection. J Comp Pathol 137(4):231–238

Xue F, Dong H, Wu J, Wu Z, Hu W, Sun A, Troxell B, Yang XF, Yan J (2010) Transcriptional responses of Leptospira interrogans to host innate immunity: significant changes in metabolism, oxygen tolerance, and outer membrane. PLoS Negl Trop Dis 4(10):e857

Yan W, Faisal SM, McDonough SP, Divers TJ, Barr SC, Chang CF, Pan MJ, Chang YF (2009) Immunogenicity and protective efficacy of recombinant Leptospira immunoglobulin-like protein B (rLigB) in a hamster challenge model. Microbes Infect 11(2):230–237

Yan W, Faisal SM, Divers T, McDonough SP, Akey B, Chang YF (2010a) Experimental Leptospira interrogans serovar Kennewicki infection of horses. J Vet Intern Med 24(4):912–917

Ristow P, Bourhy P, da Cruz McBride FW, Figueira CP, Huerre M, Ave P, Girons IS, Ko AI, Picardeau M (2007) The OmpA-like protein Loa22 is essential for leptospiral virulence. PLoS Pathog 3(7):e97

Yan W, Faisal SM, McDonough SP, Chang CF, Pan MJ, Akey B, Chang YF (2010b) Identification and characterization of OmpA-like proteins as novel vaccine candidates for Leptospirosis. Vaccine 28(11):2277–2283

Yang CW (2007) Leptospirosis renal disease: understanding the initiation by Toll-like receptors. Kidney Int 72(8):918–925

Yuri K, Takamoto Y, Okada M, Hiramune T, Kikuchi N, Yanagawa R (1993) Chemotaxis of leptospires to hemoglobin in relation to virulence. Infect Immun 61(5):2270–2272

Zhang YX, Geng Y, Yang JW, Guo XK, Zhao GP (2008) Cytotoxic activity and probable apoptotic effect of Sph2, a sphigomyelinase hemolysin from Leptospira interrogans strain Lai. BMB Rep 41(2):119–125

Chapter 9
Immune Resistance by Relapsing Fever Spirochetes

Monica E. Embers and Job E. Lopez

9.1 Introduction

9.1.1 Disease Associated with Relapsing Fever Spirochetes

Relapsing fever (RF) spirochetosis is a neglected global disease primarily afflicting those in resource-poor countries (Fig. 9.1). Infection with the causative *Borrelia* spp. spirochetes can be either louse-borne (epidemic) or tick-borne (endemic). Signs and symptoms largely overlap, but vary between the two in important ways that affect their respective mortality. Where mortality with endemic RF is rare (<5%), epidemic RF can reach up to 40% mortality in an outbreak (Raoult and Roux 1999). The true global impact of this disease is masked by the non-specific symptoms caused by the spirochetal infection, such that it may be attributed to other pathogens that are prevalent in a population. For instance, in Sudan an outbreak of louse-borne RF affected 20,000 villagers with a 10% mortality rate, yet was originally diagnosed as being caused by a hemorrhagic fever virus (Piesman and Schwan 2010). Further complicating our understanding of the potential significance of this disease is the geographical overlap of RF spirochetes with additional febrile episode-inducing illnesses (Fig. 9.1), with one study reporting that 18.2% of RF borreliosis cases were misdiagnosed as malaria (Nordstrand et al. 2007).

M.E. Embers, Ph.D.
Division of Bacteriology and Parasitology, Tulane National Primate Research Center,
Tulane University Health Sciences, 18703 Three Rivers Rd. Covington, LA 70433, USA
email: members@tulane.edu

J.E. Lopez, Ph.D. (✉)
Department of Biological Sciences, 219 Harned Hall,
295 E. Lee Boulevard, Mississippi State, MS 39762, USA

M.E. Embers (ed.), *The Pathogenic Spirochetes: Strategies for Evasion of Host Immunity and Persistence*, DOI 10.1007/978-1-4614-5404-5_9,

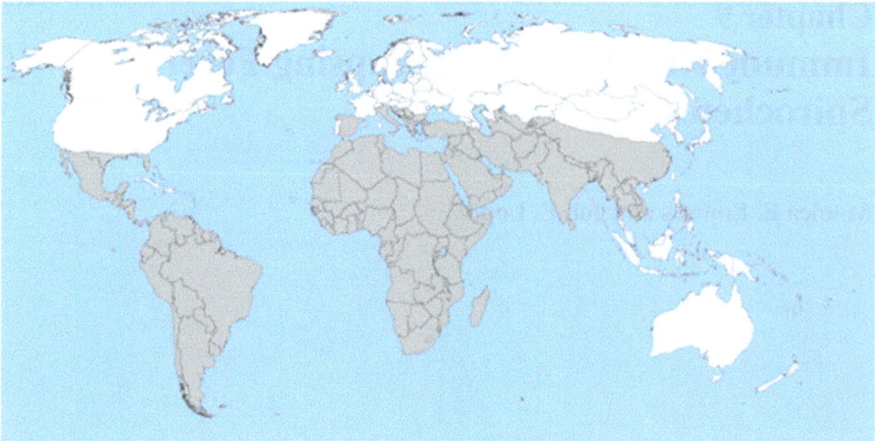

Fig. 9.1 The estimated geographic overlap of RF spirochetes, trypanosomes, and malaria parasites is *shaded in grey*

9.1.2 Historical Perspective and Epidemiological Notes

RF spirochetes were first reported to cause human disease in 1905 by J. Everett Dutton and John L. Todd. However, from clinical and epidemiological descriptions, the spirochetes may have caused disease as far back as 550 A.D. (Bryceson et al. 1970). In 1907, the Office International d'Hygiene Publique was established, which recorded the epidemiological data on RF spirochetes, bringing to light the severity of this disease. Louse-borne RF caused major epidemics in the first and second World Wars as well as the Korean War (Bryceson et al. 1970). It is estimated that between 1910 and 1950, there were over 15 million cases of louse-borne RF occurring throughout Europe and Africa causing severe morbidity and mortality (Bryceson et al. 1970). Today, the primary foci of louse-borne RF continue to be Ethiopia and Sudan.

RF spirochetosis is truly a global disease with human cases ranging from Africa, the Middle East, India, and the Americas. Old world species from Africa include *Borrelia hispanica, Borrelia crocidurae, Borrelia duttonii,* and *Borrelia recurrentis. B. hispanica, B. crocidurae,* and *B. duttonii* are transmitted by the soft tick *Ornithodoros erraticus, Ornithodoros sonrai,* and *Ornithodoros moubata*, respectively, while *B. recurrentis* is transmitted by the human body louse, *Pediculus humanus humanus. B. hispanica* and *B. crocidurae* are distributed throughout Northern and West Africa, respectively (Sarih et al. 2009; Trape et al. 1991), while *B. recurrentis and B. duttoni* occur in East Africa.

An epidemiological study in northwestern Morocco surveyed six locations within the Kenitra District (Sarih et al. 2009). Of the patients with unexplained fever, 20.5% had been infected with RF spirochetes, and not surprisingly, the highest incidence of infections was reported in locations where the patients lived in traditional mud huts (Sarih et al. 2009), where *Ornithodoros* ticks frequently reside within cracks (Talbert et al. 1998). Extensive epidemiological surveys conducted in Senegal and Togo (Trape et al. 1991; Trape et al. 1996a, 1996b) produced evidence that tick-borne RF is a leading cause of hospital visits in Senegal, with high dwelling infestations by *O. sonrai* and a 31% vector infection rate (Vial et al. 2006). Further compounding our understanding of disease severity is the misdiagnosis of RF borreliosis, as reported by Nordstrand et al. (2007). Given the similar distribution pattern and clinical manifestation, RF borreliosis is often misdiagnosed as malaria (Nordstrand et al. 2007). In Togo, there was an 8.8% prevalence of RF borreliosis, and 33% of patients were co-infected with malaria parasites. However, only 1 of 18 patients was accurately treated for RF borreliosis, while the others received antimalarial drugs (Nordstrand et al. 2007).

In East Africa, tick- (endemic) and louse- (epidemic) borne RF continue to burden the livelihood of the inhabitants. Interestingly, Sudan and Ethiopia remain two regions still endemic to louse-borne RF spirochetes, and an animal reservoir for tick- and louse-borne RF spirochetes has yet to be identified. Situations promoting outbreaks of louse-borne RF borreliosis include diminished hygiene and areas of overcrowding, as has been observed in war camps and during famine. For example, in 1991, transit camps were established for 300,000–500,000 prisoners of war being relocated from the Eritrea region of Ethiopia to Addis Ababa (Sundnes and Haimanot 1993). An outbreak followed between July and September in two of the relocation camps located in Mekele and Bahr Dar. Human population densities in the camps were approximately 2 soldiers per m^2 with no running water for bathing or washing clothes (Sundnes and Haimanot 1993). Measures taken to contain the outbreak included delousing, vector control, and antibiotics. At the resolution of the epidemic, it was concluded that antibiotic treatment alone was insufficient to control the outbreak (Sundnes and Haimanot 1993). Recently, promising results from a 2008 survey in Ethiopia reported a decrease of louse-borne RF borreliosis in regions south of Addis Ababa, which could likely be attributed to modernizing living conditions, personal hygiene, and insecticide use (Ramos et al. 2008).

B. duttonii, a causative agent of tick-borne (endemic) RF spirochetosis, continues to be distributed throughout most of East Africa (Larsson et al. 2009) and infection significantly affects both maternal and perinatal health (Jongen et al. 1997; Melkert 1988, 1991). In pregnant women, the disease causes preterm labor, miscarriage, and maternal and child mortality (Melkert 1988; Melkert 1991). Diminished living conditions contribute to the continued occurrence of this disease. Investigators reported that in Tanzania, 88–91% of dwellings in villages studied were infested with ticks (Talbert et al. 1998). While insecticidal treatment was effective in controlling tick infestation, the retail price of each treatment per household was $12.00 and two treatments were needed (Talbert et al. 1998). To put the cost of vector control into perspective, according the estimates from the Central Intelligence Agency, in 2010 the daily gross domestic income in Tanzania was approximately $4.00 per day per household. Furthermore, the efficacy of insecticidal treatment is unknown and reinfestation is likely to occur (Talbert et al. 1998).

Throughout Middle Eastern countries, Egypt, the southern countries of the former Soviet Union, India, and Kashmir, *Borrelia persica* is a primary causative agent of tick-borne RF borreliosis (Sidi et al. 2005; Parola and Raoult 2001; Assous and Wilamowski 2009; Assous et al. 2006; Aher et al. 2008). The spirochetes are transmitted by *Ornithodoros tholozani,* and these extremely hardy ticks can survive 5–10 years between blood meals (Pospelova-Shtrom 1962), with the most recent epidemiological studies having been conducted in Israel and Iran (Sidi et al. 2005; Masoumi Asl et al. 2009; Moemenbellah-Fard et al. 2009). Throughout Israel, *O. tholozani* are frequently found in caves and arid regions. Though the incidence of tick-borne RF borreliosis in the civilian population has decreased over the last 50 years, military personnel are 36 times more likely to contract the disease than civilians (Sidi et al. 2005). In Iran, the primary demographic affected by RF spirochetes continues to be young people living in rural areas in mud and thatch houses (Masoumi Asl et al. 2009; Moemenbellah-Fard et al. 2009). While *B. persica* is the primary causative agent of tick-borne RF borreliosis, two additional species that may cause human disease are *Borrelia caucasica* and *Borrelia latyschewii* (Assous and Wilamowski 2009). As molecular data continue to be acquired from tick and human infections (Assous et al. 2006), a better understanding of the species' causing human disease will be attained.

In the Americas, *B. hermsii, B. turicatae,* and possibly *B. parkeri* are the three species of RF spirochetes that cause human disease. Little is known about the potential of *B. parkeri* to cause human disease, but the spirochetes have been linked to a case of equine abortion in San Diego county (Walker et al. 2002). While the tick vector, *O. parkeri,* is distributed throughout the western United States (Cooley et al. 1944), the range of *B. parkeri* is unclear. *B. hermsii* is the most well characterized species of RF spirochete. *B. hermsii* is distributed throughout the western United States with well-defined endemic foci in eastern Washington and western Montana (Schwan et al. 2003; Schwan et al. 2007; Schwan et al. 2009). With over 30 isolates of *B. hermsii* collected throughout the western United States (Schwan et al. 2007), significant progress has been made toward developing diagnostic tools to determine the distribution of *B. hermsii* (Schwan et al. 2007, 2009; Lopez et al. 2009, 2010; Porcella et al. 2000; Schwan et al. 1995, 1996). Moreover, given the ecological overlap between RF spirochetes and Lyme disease-causing spirochetes (Oliver et al. 2003; Zingg et al. 1993; Lane and Brown 1991; Lin et al. 2001; Brown and Lane 1994; Brown et al. 2006), and the serological cross-reactivity in mammals infected with a given species (Lopez et al. 2010; Magnarelli et al. 1987), the identification of species-specific diagnostic antigens will improve epidemiological surveys.

Two diagnostic antigens that can discriminate between Lyme and RF borrelioses are glycerophosphodiester phosphodiesterase (GlpQ) and the *Borrelia* immunogenic protein A (BipA) (Lopez et al. 2010; Schwan et al. 1996). Recombinant GlpQ and BipA are currently being used in serological surveys to determine the range of *B. turicatae*. The distribution of *B. turicatae* is less characterized than that of *B. hermsii,* but human cases were reported throughout the 1990s (Rawlings 1995; Davis et al. 2002), and isolates have been obtained from sick dogs in Texas and Florida (Breitschwerdt et al. 1994; Schwan et al. 2005; Whitney et al. 2007).

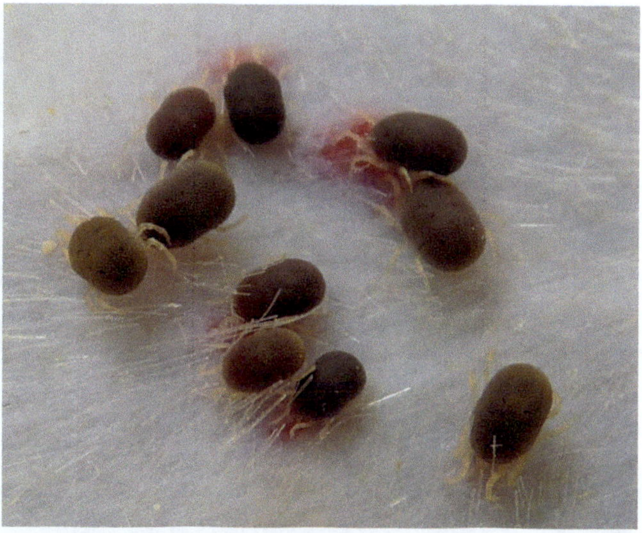

Fig. 9.2 Depending on the developmental stage, *O. turicata* will complete feeding shortly after attaching. Nymphal ticks are shown feeding to repletion within 20 min after attaching onto the host

However, a human isolate has yet to be obtained, and serological studies have not been conducted to identify the vertebrate host for *B. turicatae*. Previous reports described the distribution of *Ornithodoros talaje, Ornithodoros rudis,* and *O. turicata* (Fig. 9.2) in Central and South America (Davis 1956; Felsenfeld 1973), but at the time the molecular tools had not been developed to determine the species' of RF spirochetes causing disease.

9.2 Pathogenesis of RF Spirochetes

9.2.1 Spirochete-Vector Interactions

In general, transmission of tick-borne RF spirochetes can occur through the saliva or coxal fluid, whereas transmission of louse-borne RF spirochetes occurs through crushed lice or contamination of louse feces at the bite site (Varma 1962; Houhamdi and Raoult 2005). While not completely defined, the rate of transovarial transmission of RF spirochetes varies between species of ticks, and the longevity of *Ornithodoros* ticks can be upwards of 20 years when fed regularly (Balashov 1972). Dr. Edward Francis first observed the longevity of *O. turicata* and the transmissibility of *B. turicatae* after the ticks had been starved for over 5 years (Francis 1938). Some pathogens, such as *Bacillus anthracis,* respond to extended periods of starvation and extreme environments by spore formation (Fritze 2004; Nicholson 2002). Genomic analysis of *B. turicatae* has not, however, identified homologues

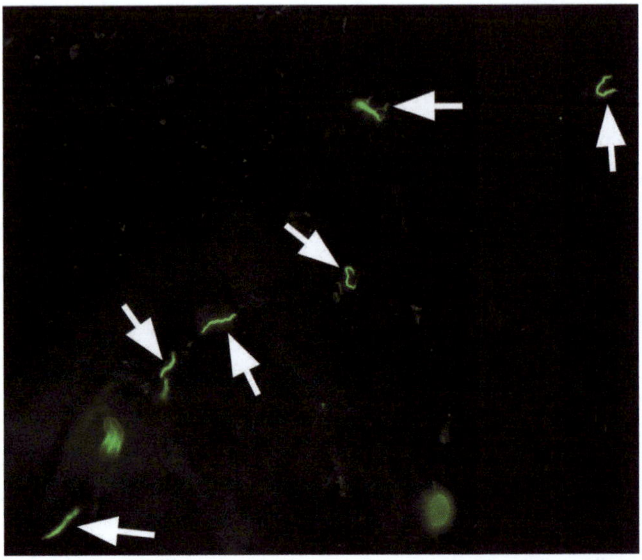

Fig. 9.3 Salivary gland colonization by *B. turicatae*. RF spirochetes have evolved mechanisms to populate the salivary glands of *Ornithodoros* ticks. The immunoflourescent assay shown demonstrates salivary gland colonization by *B. turicatae* (white arrows). Upon feeding, spirochetes expelled through the saliva and begin to replicate in mammalian blood, reaching upwards of 10^4 spirochetes per μl of blood

to spore forming genes (unpublished observations), and the molecular constituents necessary for spirochete adaptation within the tick remain unknown.

Even during periods of starvation, we hypothesize that RF spirochetes remain motile, in contrast to recent evidence indicating that Lyme disease-causing spirochetes enter a dormant non-motile stage in the midgut of *Ixodes scapularis* (Dunham-Ems et al. 2009). Support of this hypothesis includes the rapidity of engorgement (15–45 min) by species of *Ornithodoros* (Varma 1962) (Fig. 9.2) and the ability of RF spirochetes to quickly colonize and reach upwards of 10^7 organisms per ml in mammalian blood within 3 days after tick bite (Lopez et al. 2011). For these events to occur, RF spirochetes have adapted to long-term salivary gland colonization (Fig. 9.3) and need to remain motile in order to quickly colonize the mammalian host. Unlike RF spirochetes, *Borrelia burgdorferi* becomes motile and migrates from the midgut to the salivary glands of *I. scapularis,* an event that takes 48 h once the tick attaches (Dunham-Ems et al. 2009; Piesman et al. 1987).

An intriguing aspect of RF spirochete adaptation within the tick is the vector specificity of the bacteria (Davis 1956; Schwan 1996). In 1956, studies by Dr. Gordon E. Davis demonstrated that *Borrelia mazzottii,* a novel species of RF spirochete recovered in Mexico, could be acquired and subsequently transmitted by *O. talaje* that originated from Mexico and Guatemala. However, different species of *Ornithodoros* originating from Panama, Puerto Rico, Colombia, and Mexico failed to transmit *B. mazzottii* (Davis 1956). Moreover, mice did not become infected when injected with triturated *Ornithodoros* ticks that originally failed to transmit *B. mazzottii* by tick bite, suggesting that *B. mazzottii* was unable to colonize certain species' of *Ornithodoros* (Davis 1956).

More recently, studies have been conducted where *O. hermsi, O. turicata,* and *O. parkeri* were fed on mice infected with *B. hermsii* (Schwan 1996). While *O. turicata* and *O. parkeri* were unable to transmit *B. hermsii,* the spirochetes colonized the central ganglion (an anatomic region anterior to the salivary glands) of the non-transmitting ticks. This observation demonstrated the importance of salivary gland colonization in the transmissibility of RF spirochetes, and indicated that unique molecular mechanisms between a given species of tick and RF spirochete are critical to ensure the continued life cycle of the bacteria (Schwan 1996).

Additional work on spirochete-vector interactions has addressed the events required for tick colonization and subsequent transmission. For example, the volume of blood ingested during feeding was determined for *O. hermsi* nymphs, males, and females, demonstrating that female ticks and the first three nymphal stages of *O. hermsi* take in the largest volumes, increasing their body weight by over 500% (McCoy et al. 2010). By understanding the blood meal volumes ingested, calculations have indicated that *B. hermsii* may fail to replicate within *O. hermsi,* and within 5 days after the blood meal, spirochete densities drop tenfold within the tick (McCoy et al. 2010). Interestingly, while as few as 15 spirochetes per μl of blood is sufficient to colonize *O. hermsi* (Lopez et al. 2011), assessment of heavily colonized salivary glands suggested that spirochetes may replicate specifically within these tissues. Alternatively, during tick feeding, spirochetes may concentrate at the bite site, increasing the amount of bacteria ingested during low-level infections. Regardless, there are several aspects of spirochete-vector interactions needing further investigation.

9.2.2 Transmission and Clinical Manifestation of RF Borreliosis

Species of *Ornithodoros* are nocturnal feeders and patients are often unaware of the tick bite. Upon transmission into the mammal, RF spirochetes up-regulate a family of genes coding for the variable major proteins (Vmps) (Restrepo et al. 1992; Stoenner et al. 1982; Dai et al. 2006; Barbour et al. 2006). The Vmps are surface proteins that undergo antigenic variation, allowing for immune evasion and spirochete persistence in the blood. Additionally, RF spirochetes replicate within the blood reaching upwards of 10^4 spirochetes or more per μl (Coffey and Eveland 1967). Recurrent febrile episodes are observed during periods of high spirochete densities in the blood (Francis 1938) with additional symptoms including chills, nausea, vomiting, and diarrhea. Transplacental transmission occurs during pregnancy resulting in fetal death (Melkert 1988, 1991; Melkert and Stel 1991). A unique physical sign of disease observed during louse-borne RF borreliosis includes conjunctival hemorrhaging (Bryceson et al. 1970), and invasion of the central nervous system (CNS) has been reported with *B. duttonii, B. recurrentis,* and *B. turicatae* infection (Bryceson et al. 1970; Larsson et al. 2006; Cadavid and Barbour 1998a).

9.2.3 Neurotropism and Persistence by RF Spirochetes

A comprehensive review of the literature by Cadavid and Barbour (Cadavid and Barbour 1998b) indicated that neurologic disease can be a relatively frequent occurrence in infections by RF spirochetes, especially with the tick-borne species. The first direct evidence for central nervous system (CNS) involvement in RF was reported by Leboeuf and Gambier in 1918 when they detected spirochetes in the cerebrospinal fluid of sub-Saharan African patients with relapsing fever (Leboeuf and Gambier 1918). Subsequent reports commonly described meningismus and facial palsy as frequent complications in RF. Encephalitis, myelitis, radiculitis, and neuropsychiatric dysfunction were also reported (Cadavid and Barbour 1998b). Neurotropism is known to vary between species, and even among different strains within a species. As such, the frequency of all types of neurological manifestations varies from none to >50% (Cadavid and Barbour 1998b).

Both animal models and wild reservoir hosts have been used to document the residual brain infection with RF spirochetes. One study indicated that infection of experimental animals persisted in the brain for up to 3 years (Cadavid and Barbour 1998b). Also, residual spirochetes were susceptible to immune serum from the infected animal, suggesting that the brain provided a protective niche from the immune response (Cadavid and Barbour 1998b). More recently, persistent brain infection in mice was established with *B. duttonii*. Spirochete quantification within murine brains indicated that *B. duttonii* infection remained 80 days after infection. Injecting uninfected mice with homogenized brains demonstrated the viability of the spirochetes (Larsson et al. 2006). Silent infection was reactivated by treating the mice with the immunosuppressant methylprednisolone and remarkably, *B. duttonii* failed to cause an inflammatory response or tissue damage in murine brains (Larsson et al. 2006). Moreover, *B. duttonii* isolated from the brain displayed unique Vmp profiles compared to the original isolate used to infect mice, suggesting a role for Vmps in persistent brain infection in *B. duttonii*. In the absence of detection elsewhere, and with little to no brain pathology, this silent infection shown to be reactivated with immunosuppression, thus indicated that the brain could serve as a cache in reservoir hosts (Larsson et al. 2006).

The molecules involved in CNS invasion have been best characterized with *B. turicatae* (Cadavid and Barbour 1998a; Cadavid et al. 1993, 1997, 1994) infection. There are two serotypes (A and B) of the Ozona strain of *B. turicatae* (Cadavid et al. 1994) that were originally characterized by the production of the variable major protein (Vmp) A or B (Cadavid et al. 1994). Vmp A and B have been further characterized and renamed as variable small proteins (Vsps) A and B (Pennington et al. 1999a), and subsequently Vsp1 and 2, respectively (Lawson et al. 2006). Spirochete populations that express *vsp1* are neurotropic in mice, while those that express *vsp2* cause significantly enlarged joints and severe arthritis (Cadavid et al. 1994). The crystal structure of Vsp1 has indicated that the protein forms a dimer, and further characterization of the electrostatic potential of Vsp1 and Vsp2 suggests that differences in the proteins' charge may facilitate tissue tropism (Cadavid et al. 1997; Lawson et al. 2006).

Spirochetal infection was one of the first infections treated with the arsenic compounds often called "magic bullet" therapy of Ehrlich and Hata (Ehrlich 1910). Salvarsan and other antimicrobials were found to be effective against residual brain infection if administered early, before the first episode of spirochetemia (Cadavid and Barbour 1998b). Studies in mice have shown, in addition, that prevention of residual brain infection with antibiotics is highly dose-dependent (Lapierre et al. 1971) as well as time-dependent. As a general observation, reports on RF in human and animal studies indicate that: (1) neurological involvement can be frequent, depending on the species and strain; (2) tick-borne RF more commonly leads to neurological manifestations; and (3) antibiotics that are not administered early and do not penetrate the blood–brain barrier may fail to eliminate infection (Cadavid and Barbour 1998b).

9.2.4 Variable Major Proteins and Immune Evasion

As stated above, RF spirochetes undergo antigenic variation as a mechanism to evade the host immune response, prolonging infection and increasing the chances of bacterial acquisition by the arthropod vector. The mechanism of antigenic variation has most extensively been characterized in *B. hermsii* and to a lesser degree in *B. turicatae.* Several decades ago, Stoenner and colleagues showed that multiple variants (serotypes) could arise from a single organism (Stoenner et al. 1982). It is now known that the bacteria contain a family of genes that code for the immunodominant surface variable small proteins (Vsps) and variable large proteins (Vlps) (Dai et al. 2006; Barbour et al. 2006; Pennington et al. 1999a; Meier et al. 1985; Plasterk et al. 1985). The molecular mechanism of antigenic variation occurs through gene conversion, where a transcriptionally silent *vsps* or *vlps* will replace the *vsp* or *vlp* in the expression locus, generating a new variant (Barbour et al. 1991; Kitten and Barbour 1990). The recombination event has been mapped to the up- and down-stream homology sequences (UHS and DHS) flanking each *vsp or vlp*, and through gene conversion, *B. hermsii* can generate 60–70 antigenically distinct variants during host infection (Dai et al. 2006; Barbour et al. 2006). Interestingly, the recombination rates have also been defined to the sequence identity of the UHS between the expressed and silent gene, and to the proximity of the DHS to the silent loci, with DHS elements that are closer to the silent loci recombining into the expression site at a greater frequency (Dai et al. 2006). There is also evidence of partial gene conversion, where portions of silent *vsps* or *vlps* will recombine into the expression locus (Restrepo et al. 1994), and hypermutation of the 5′ end of the expression site occurs (Restrepo and Barbour 1994), further diversifying the genetic repertoire of *B. hermsii.* In addition to the genetic variation of the major surface proteins, research in *B. hermsii* indicates that expression site switching can add new serotypes and lend an additional strategy for immune evasion (Barbour et al. 2000). Vsps and Vlps have also been identified in *B. duttonii* (Tabuchi et al. 2002), *B. recurrentis* (Vidal et al. 1998), *B. crocidurae* (Shamaei-Tousi et al. 1999), and

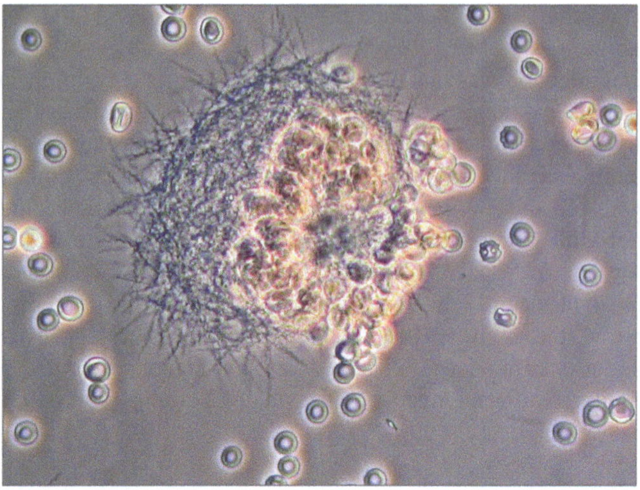

Fig. 9.4 Rosetting of erythrocytes by RF spirochetes. Photo by Marie Andersson, provided by Sven Bergstrom, Department of Molecular Biology, Umea University, Umea Sweden

B. turicatae (Pennington et al. 1999b). While the protein's primary role is immune evasion, studies in *B. turicatae* indicate that they may also be responsible for tissue tropism in the mammal (Cadavid et al. 1993; Cadavid et al. 1997; Cadavid et al. 1994; Pennington et al. 1999b, 1997).

9.2.5 Additional Mechanisms of Vertebrate Host Adaptation by RF Spirochetes

As RF spirochetes have adapted to the vertebrate host, the bacteria have developed additional mechanisms to escape the host immune response. Erythrocyte rosetting is the ability of RF spirochetes to adhere to, and cause aggregation of erythrocytes, so the bacteria are covered with red blood cells (Shamaei-Tousi et al. 1999, 2001; Burman et al. 1998; Mooser 1958). Rosetting (Fig. 9.4) has been observed with *B. duttonii, B. hispanica,* and *B. crocidurae* and has been proposed as an additional mechanism of immune evasion. For *B. crocidurae,* the delayed humoral response against the spirochetes has been associated with rosetting, and disassociating erythrocytes from the spirochetes increased the susceptibility of the bacteria to antibodies (Burman et al. 1998). Investigators have oxidized erythrocytes with sodium periodate, destroying most glycan epitopes and diminishing erythrocyte rosetting; further studies demonstrated that rosette-inducing RF spirochetes specifically bind neolacto glycan on the surface of human erythrocytes (Guo et al. 2009). While the spirochete receptor has yet to be identified, as the surface proteome of the bacteria continues to be defined, the pathogenic mechanisms of RF spirochetes will be better understood.

Additional proteins that play a role in RF spirochete adaptation within the vertebrate host are the Factor H-binding protein A (FhbA) (Hovis et al. 2008, 2006), the human complement plasminogen A (HcpA) protein (Grosskinsky et al. 2009), and the *B. hermsii* serine protease designated BhpA (Guyard et al. 2006). FhbA has been identified as a lipidated surface protein that binds to the complement regulatory protein, Factor H (Hovis et al. 2004; McDowell et al. 2003) (see also Chap. 4). Factor H is a serum protein that contributes to the regulation of alternative complement pathway (Ruddy and Austen 1969, 1971), and the ability of RF spirochetes to bind Factor H has been hypothesized to be important for bacterial dissemination within the host (Hovis et al. 2006). However, the tools to genetically manipulate genes hypothesized to be necessary for spirochete adaptation within the host have only recently been developed (Battisti et al. 2008). As the genetic toolbox for RF spirochetes is expanded (Fine et al. 2011), a better understanding of host adaptation will be developed.

An alternative method to gene deletion has been the ectopic expression of RF spirochete genes in an organism lacking the genetic background of interest, and one such organism has been *Borrelia burgdorferi* (Grosskinsky et al. 2009; Guyard et al. 2006). While RF spirochetes and Lyme disease (LD) spirochetes are related, the mammalian-spirochete-tick interactions significantly differ. For example, RF spirochetes attain high spirochete densities in mammalian blood (Coffey and Eveland 1967), whereas *Borrelia burdoferi sensu lato* are exist transiently in the blood and have been identified in the heart, joints, liver, ear, spleen, kidney, bladder, and uterus (Yang et al. 1994; Wang et al. 2002). Moreover, RF and LD spirochetes are transmitted by *Ornithodoros* and *Ixodes* ticks, respectively. Tick-borne RF spirochetes reside in the salivary glands of the tick while LD spirochetes are restricted to the midgut and have to migrate through the salivary glands for transmission to occur once the tick feeds (Schwan and Hinnebusch 1998; Piesman et al. 2001). Genomic analyses between the two species of *Borrelia* have identified subsets of RF spirochete genes that are absent from *B. burgdorferi,* including the *bhpA* (Guyard et al. 2006) gene. *BhpA* contains sequence identity to a family of serine proteases involved with defense against oxidative stress (Pallen and Wren 1997), and to test the role of *bhpA* during spirochete pathogenesis, *B. burgdorferi* was transformed with *B. hermsii bhpA* (Guyard et al. 2006). *B. burgdorferi* transformants were more resistant to neutrophil killing and oxidative stress compared to control spirochetes (Guyard et al. 2006). Furthermore, this study provided insight into the differences between the pathogenesis of RF and LD spirochetes in that LD spirochetes may not be able to colonize host blood given that the bacteria lack essential genes necessary for survival (Guyard et al. 2006).

9.3 Immune Response Elicited Against RF Spirochetes

9.3.1 The Primary Immune Response

Rodents are major natural reservoir hosts for RF spirochetes, and, in general, recapitulate the relapses seen in human infection. Not surprisingly, the vast majority of studies on the immune response to infection have been conducted in the mouse. The infectious dose for RF spirochetes is thought to be very low (Stoenner et al. 1982) and the onset of symptoms (the acute phase) usually begins within a week of infection (Larsson et al. 2009). As expected, the abundant variable proteins are major targets of the host's antibody response. The periods of fever and bacteremia are accompanied by antigenic variation. With each subsequent relapse, the bacterial loads during peak bacteremia decline, due to the serum antibody response. Recombination at the variable protein gene loci occurs at a somewhat predictable rate, and with each resolution of bacteremia, a small population of spirochetes escapes immune recognition by the expression of Vsps and Vlps that have not been encountered by host immune cells (Barbour 1990)

9.3.2 Role of B Cells and the Memory Response

The reduction in spirochete levels and eventual clearance has been shown in animal models to be a direct result of the antibody response. The T cell independence of immune-mediated clearance was established by Newman and Johnson, using nude mice, thymectomized mice, and reconstitution of irradiated mice with T-cell depleted splenocytes (Newman and Johnson 1984). Here, the clearance was not impaired by the lack of functional T cells, suggesting not only the importance of the B cell response, but that of a T-independent B cell response. Subsequent studies have demonstrated neutralization (Barbour and Bundoc 2001) and a directly bactericidal (Connolly and Benach 2001) role for serum IgM in combating RF spirochetemia.

A central mechanism of adaptive immunity is the memory response. The principles of immunity to reinfection and vaccination rely on generating populations of lymphocytes that persist for long periods and when antigen is re-introduced after a length of time, expand clonally. In the case of B cells, the memory is generated by lymphocytes that expand as antibody-secreting plasma cells at antigen encounter. The conventional generation of B cell memory typically occurs in the germinal centers of the lymphoid organs, where isotype switch and affinity maturation occur. More recently, B cell subsets with different roles in immunity have been delineated (Vaughan et al. 2011). Mature B cells can actually be divided into follicular (FO) B cells, present in the lymphoid follicles, Marginal zone (MZ) B cells, located in the marginal sinus of the spleen, and B1 cells predominantly found in the mouse peritoneum. These are subdivided into B1a and B1b cells. B1 and MZ B cells are known to engage in the T-cell independent antibody response. In contrast to the LD spirochete, which induces a massive expansion of MZ B cells upon infection, RF spirochetes induce a loss of MZ B cells (Malkiel et al. 2009). This may reflect the presence of LD spirochetes in the spleen versus the blood-borne RF spirochetes and/or the differential responses to particular antigens.

The B1a cells have been thought of as the innate, polyspecific subset of B cells (Vaughan et al. 2011) that produce low affinity antibodies, whereas B1b cells may confer specific, long-lasting immunity. This was demonstrated with murine infection using *Borrelia hermsii* (Alugupalli et al. 2004). Here, adoptive transfer of B1b lymphocytes from convalescent mice to Rag 1−/− mice (lacking mature B and T lymphocytes) conferred protection; this protection consisted of a specific IgM response that occurred when challenged 60 days after the reconstitution, indicating that this population alone could confer memory and afford protection (Alugupalli et al. 2004). Given the ability of these spirochetes to engage in antigenic variation, the presumably invariant target(s), of the specific memory response became the subject of further inquiry. As such, the complement Factor H-binding protein of *B. hermsii* was shown to be a primary target for the IgM-secreting B1b response. These data indicate that, unlike any other known response to pathogen, this B cell subset drives the immune effort to control infection with RF spirochetes. Importantly, this particular B cell subset has not been identified in humans, so it remains to be seen if the same mechanism to control infection occurs in RF patients.

9.4 Future Directions

The study of RF spirochetes and the disease they cause has many open questions. While the variable membrane proteins are a hallmark of their immune evasion capabilities, these spirochetes likely utilize many tactics to avoid recognition by the host's immune cells. We have touched on the complement-binding proteins and the physical mechanism of erythrocyte rosetting. In many cases, RF spirochete infection is self-limiting, yet the illness lasts for weeks to months. The time it takes the immune response to "catch up" and how exactly the infection is cleared remain enigmatic.

In addition, the tools that have led to the identification and diagnosis of other spirochetal infections are not available or in common use for RF. Major strides toward the accurate diagnosis of infections need to be made in the near future, as the re-emergence and misdiagnosis of RF can prevail in endemic regions (Cutler 2006). The development of reliable diagnostics and low-cost vaccines for RF spirochete infections could significantly reduce the impact of this treatable, yet shrouded disease.

Expanding on current efforts to identify targets for recognition by antibodies in the host should be a priority. The potential for vaccination again both the bacteria and the vector is real, once good candidates are identified. The mouse model has provided significant insight into interactions between host and pathogen. The non-human primate was shown to be highly susceptible to RF spirochete infection decades ago (Francis 1938). Re-development of this animal model, with the ability to characterize immune responses and examine phenotypic diversity of the bacteria *in vivo* will no doubt lead to better understanding of the requirements for detection and control of the RF spirochetoses.

References

Aher AR, Shah H, Rastogi V, Tukaram PK, Choudhury RC (2008) A case report of relapsing fever. Indian J Pathol Microbiol 51(2):292–293

Alugupalli KR, Leong JM, Woodland RT, Muramatsu M, Honjo T, Gerstein RM (2004) B1b lymphocytes confer T cell-independent long-lasting immunity. Immunity 21(3):379–390

Assous MV, Wilamowski A (2009) Relapsing fever borreliosis in Eurasia–forgotten, but certainly not gone! Clin Microbiol Infect 15(5):407–414

Assous MV, Wilamowski A, Bercovier H, Marva E (2006) Molecular characterization of tickborne relapsing fever Borrelia, Israel. Emerg Infect Dis 12(11):1740–1743

Balashov YS (1972) Bloodsucking ticks (Ixodoidea)- vectors of diseases of man and animals. Misc Publ Entomol Soc Am 8:161–376

Barbour AG (1990) Antigenic variation of a relapsing fever Borrelia species. Annu Rev Microbiol 44:155–171

Barbour AG, Burman N, Carter CJ, Kitten T, Bergström S (1991) Variable antigen genes of the relapsing fever agent *Borrelia hermsii* are activated by promoter addition. Mol Microbiol 5:489–493

Barbour AG, Bundoc V (2001) In vitro and in vivo neutralization of the relapsing fever agent Borrelia hermsii with serotype-specific immunoglobulin M antibodies. Infect Immun 69(2): 1009–1015

Barbour AG, Carter CJ, Sohaskey CD (2000) Surface protein variation by expression site switching in the relapsing fever agent Borrelia hermsii. Infect Immun 68(12):7114–7121

Barbour AG, Dai Q, Restrepo BI, Stoenner HG, Frank SA (2006) Pathogen escape from host immunity by a genome program for antigenic variation. Proc Natl Acad Sci USA 103: 18290–18295

Battisti JM, Raffel SJ, Schwan TG (2008) A system for site-specific genetic manipulation of the relapsing fever spirochete *Borrelia hermsii*. Methods Mol Biol 431:69–84

Breitschwerdt EB, Nicholson WL, Kiehl AR, Steers C, Meuten DJ, Levine JF (1994) Natural infections with *Borrelia* spirochetes in two dogs from Florida. J Clin Microbiol 32:352–357

Brown RN, Lane RS (1994) Natural and experimental *Borrelia burgdorferi* infections in woodrats and deer mice from California. J Wildl Dis 30(3):389–398

Brown RN, Peot MA, Lane RS (2006) Sylvatic maintenance of *Borrelia burgdorferi* (Spirochaetales) in Northern California: untangling the web of transmission. J Med Entomol 43(4):743–751

Bryceson AD, Parry EH, Perine PL, Warrell DA, Vukotich D, Leithead CS (1970) Louse-borne relapsing fever. Q J Med 39(153):129–170

Burman N, Shamaei-Tousi A, Bergström S (1998) The spirochete *Borrelia crocidurae* causes erythrocyte rosetting during relapsing fever. Infect Immun 66:815–819

Cadavid D, Bundoc V, Barbour AG (1993) Experimental infection of the mouse brain by a relapsing fever *Borrelia* species: a molecular analysis. J Infect Dis 168:143–151

Cadavid D, Pennington PM, Kerentseva TA, Bergström S, Barbour AG (1997) Immunologic and genetic analyses of VmpA of a neurotropic strain of *Borrelia turicatae*. Infect Immun 65: 3352–3360

Cadavid D, Thomas DD, Crawley R, Barbour AG (1994) Variability of a bacterial surface protein and disease expression in a possible mouse model of systemic Lyme borreliosis. J Exp Med 179:631–642

Cadavid D, Barbour AG (1998) Neuroborreliosis during relapsing fever: review of the clinical manifestations, pathology, and treatment of infections in humans and experimental animals. Clin Infect Dis 26(1):151–164

Coffey EM, Eveland WC (1967) Experimental relapsing fever initiated by *Borrelia hermsi*. II. Sequential appearance of major serotypes in the rat. J Infect Dis 117:29–34

Connolly SE, Benach JL (2001) Cutting edge: the spirochetemia of murine relapsing fever is cleared by complement-independent bactericidal antibodies. J Immunol 167(6):3029–3032

Cooley RA, Kohls GM: The Agarasidae of North America, Central America, and Cuba, Monography No. 1 edn; 1944.

Cutler SJ (2006) Possibilities for relapsing fever reemergence. Emerg Infect Dis 12(3):369–374

Dai Q, Restrepo BI, Porcella SF, Raffel SJ, Schwan TG, Barbour AG (2006) Antigenic variation by *Borrelia hermsii* occurs through recombination between extragenic repetitive elements on linear plasmids. Mol Microbiol 60:1329–1343

Davis H, Vincent JM, Lynch J (2002) Tick-borne relapsing fever caused by *Borrelia turicatae*. Pediatr Infect Dis J 21(7):703–705

Davis GE (1956) A relapsing fever spirochete, *Borrelia mazzottii* (sp. nov.) from Ornithodoros talaje from Mexico. Am J Hyg 63(1):13–17

Dunham-Ems SM, Caimano MJ, Pal U, Wolgemuth CW, Eggers CH, Balic A, Radolf JD (2009) Live imaging reveals a biphasic mode of dissemination of *Borrelia burgdorferi* within ticks. J Clin Invest 119(12):3652–3665

Ehrlich PH (1910) The experimental chemotherapy of Spirilloses. Julius Springer, Berlin

Felsenfeld O (1973) The problem of relapsing fever in the Americas. IMS Ind Med Surg 42(3):7–10

Fine LM, Earnhart CG, Marconi RT (2011) Genetic transformation of the relapsing fever spirochete *Borrelia hermsii*: stable integration and expression of green fluorescent protein from linear plasmid 200. J Bacteriol 193(13):3241–3245

Francis E (1938) Longevity of the tick *Ornithodoros turicata* and of *Spirochaeta recurrentis* with this tick. Publ Hlth Rep 53:2220–2241

Francischetti IM, Mather TN, Ribeiro JM (2003) Cloning of a salivary gland metalloprotease and characterization of gelatinase and fibrin(ogen)lytic activities in the saliva of the Lyme disease tick vector *Ixodes scapularis*. Biochem Biophys Res Commun 305(4):869–875

Fritze D (2004) Taxonomy of the genus bacillus and related genera: the aerobic endospore-forming bacteria. Phytopathology 94(11):1245–1248

Grosskinsky S, Schott M, Brenner C, Cutler SJ, Kraiczy P, Zipfel PF, Simon MM, Wallich R (2009) *Borrelia recurrentis* employs a novel multifunctional surface protein with anti-complement, anti-opsonic and invasive potential to escape innate immunity. PLoS One 4(3):e4858

Guo BP, Teneberg S, Munch R, Terunuma D, Hatano K, Matsuoka K, Angstrom J, Boren T, Bergstrom S (2009) Relapsing fever Borrelia binds to neolacto glycans and mediates rosetting of human erythrocytes. Proc Natl Acad Sci USA 106(46):19280–19285

Guyard C, Battisti JM, Raffel SJ, Schrumpf ME, Whitney AR, Krum JG, Porcella SF, Rosa PA, DeLeo FR, Schwan TG (2006) Relapsing fever spirochetes produce a serine protease that provides resistance to oxidative stress and killing by neutrophils. Mol Microbiol 60(3): 710–722

Houhamdi L, Raoult D (2005) Excretion of living *Borrelia recurrentis* in feces of infected human body lice. J Infect Dis 191(11):1898–1906

Hovis KM, Freedman JC, Zhang H, Forbes JL, Marconi RT (2008) Identification of an anti-parallel coiled-coil/loop domain required for ligand binding by the *Borrelia hermsii* FhbA protein: additional evidence for the role of FhbA in the host-pathogen interaction. Infect Immun 76(5):2113–2122

Hovis KM, Jones JP, Sadlon T, Raval G, Gordon DL, Marconi RT (2006) Molecular analyses of the interaction of *Borrelia hermsii* FhbA with the complement regulatory proteins Factor H and Factor H-like protein 1. Infect Immun 74:2007–2014

Hovis KM, McDowell JV, Griffin L, Marconi RT (2004) Identification and characterization of a linear-plasmid-encoded Factor H-binding protein (FhbA) of the relapsing fever spirochete *Borrelia hermsii*. J Bacteriol 186:2612–2618

Jongen VH, van Roosmalen J, Tiems J, Van Holten J, Wetsteyn JC (1997) Tick-borne relapsing fever and pregnancy outcome in rural Tanzania. Acta Obstet Gynecol Scand 76(9):834–838

Kitten T, Barbour AG (1990) Juxtaposition of expressed variable antigen genes with a conserved telomere in the bacterium *Borrelia hermsii*. Proc Natl Acad Sci USA 87:6077–6081

Lawson CL, Yung BH, Barbour AG, Zuckert WR (2006) Crystal structure of neurotropism-associated variable surface protein 1 (Vsp1) of *Borrelia turicatae*. J Bacteriol 188(12):4522–4530

Lapierre J, Hien TV, Roose A (1971) HT, Roose A: Action de la cephaloridine dans les fievres recurrentes experimentales a *Borrelia duttoni* chez la souris. C R Seances Soc Biol Fil 165:282–284

Lopez JE, McCoy BN, Krajacich BJ, Schwan TG (2011) Acquisition and subsequent transmission of Borrelia hermsii by the soft tick Ornithodoros hermsi. J Med Entomol 48(4):891–895
Lane RS, Brown RN (1991) Wood rats and kangaroo rats: potential reservoirs of the Lyme disease spirochete in California. J Med Entomol 28(3):299–302
Lin T, Oliver JH Jr, Gao L, Kollars TM Jr (2001) Clark KL: Genetic heterogeneity of *Borrelia burgdorferi* sensu lato in the southern United States based on restriction fragment length polymorphism and sequence analysis. J Clin Microbiol 39(7):2500–2507
Larsson C, Andersson M, Pelkonen J, Guo BP, Nordstrand A, Bergstrom S (2006) Persistent brain infection and disease reactivation in relapsing fever borreliosis. Microbes Infect 8(8): 2213–2219
Leboeuf A, Gambier A (1918) Sur deux cas de spirochetose humaine a Brazzaville (Moyen Congo). Bull Soc Pathol Exot 11:359–364
Larsson C, Andersson M, Bergstrom S (2009) Current issues in relapsing fever. Curr Opin Infect Dis 22(5):443–449
Lopez JE, Porcella SF, Schrumpf ME, Raffel SJ, Hammer CH, Zhao M, Robinson MA, Schwan TG (2009) Identification of conserved antigens for early serodiagnosis of relapsing fever Borrelia. Microbiology 155(Pt 8):2641–2651
Lopez JE, Schrumpf ME, Nagarajan V, Raffel SJ, McCoy BN, Schwan TG (2010) A novel surface antigen of relapsing fever spirochetes can discriminate between relapsing fever and Lyme borreliosis. Clinical Vaccine Immunol 17(4):564–571
Magnarelli LA, Anderson JF, Johnson RC (1987) Cross-reactivity in serological tests for Lyme disease and other spirochetal infections. J Infect Dis 156:183–187
Malkiel S, Kuhlow CJ, Mena P, Benach JL (2009) The loss and gain of marginal zone and peritoneal B cells is different in response to relapsing fever and Lyme disease Borrelia. J Immunol 182(1):498–506
Masoumi Asl H, Goya MM, Vatandoost H, Zahraei SM, Mafi M, Asmar M, Piazak N, Aghighi Z (2009) The epidemiology of tick-borne relapsing fever in Iran during 1997–2006. Travel Med Infect Dis 7(3):160–164
McCoy BN, Raffel SJ, Lopez JE, Schwan TG (2010) Bloodmeal size and spirochete acquisition of Ornithodoros hermsi (Acari: Argasidae) during feeding. J Med Entomol 47(6):1164–1172
McDowell JV, Tran E, Hamilton D, Wolfgang J, Miller K, Marconi RT (2003) Analysis of the ability of spirochete species associated with relapsing fever, avian borreliosis, and epizootic bovine abortion to bind factor H and cleave C3b. J Clin Microbiol 41:3905–3910
Meier JT, Simon MI, Barbour AG (1985) Antigenic variation is associated with DNA rearrangements in a relapsing fever Borrelia. Cell 41:403–409
Melkert PW (1988) Relapsing fever in pregnancy: analysis of high-risk factors. Br J Obstet Gynaecol 95(10):1070–1072
Melkert PW, Stel HV (1991) Neonatal Borrelia infections (relapsing fever): report of 5 cases and review of the literature. East Afr Med J 68(12):999–1005
Melkert PW (1991) Mortality in high risk patients with tick-borne relapsing fever analysed by the Borrelia-index. East Afr Med J 68(11):875–879
Moemenbellah-Fard MD, Benafshi O, Rafinejad J, Ashraf H (2009) Tick-borne relapsing fever in a new highland endemic focus of western Iran. Ann Trop Med Parasitol 103(6):529–537
Mooser H (1958) Erythrocyte adhesion and hemagglomeration by relapsing fever spirochetes. Z Tropenmed Parasitol 9(2):93–111
Newman K, Johnson RC (1984) T-cell-independent elimination of Borrelia turicatae. Infect Immun 45(3):572–576
Nicholson WL (2002) Roles of Bacillus endospores in the environment. Cell Mol Life Sci 59(3):410–416
Nordstrand A, Bunikis I, Larsson C, Tsogbe K, Schwan TG, Nilsson M, Bergström S (2007) Tickborne relapsing fever diagnosis obscured by malaria, Togo. Emerg Infect Dis 13(1):117–123
Oliver JH Jr, Lin T, Gao L, Clark KL, Banks CW, Durden LA, James AM, Chandler FW Jr (2003) An enzootic transmission cycle of Lyme borreliosis spirochetes in the southeastern United States. Proc Natl Acad Sci USA 100(20):11642–11645
Pallen MJ, Wren BW (1997) The HtrA family of serine proteases. Mol Microbiol 26(2):209–221

Parola P, Raoult D (2001) Ticks and tickborne bacterial diseases in humans: an emerging infectious threat. Clin Infect Dis 32(6):897–928

Pennington PM, Allred CD, West CS, Alvarez R, Barbour AG (1997) Arthritis severity and spirochete burden are determined by serotype in the *Borrelia turicatae*-mouse model of Lyme disease. Infect Immun 65(1):285–292

Pennington PM, Cadavid D, Bunikis J, Norris SJ, Barbour AG (1999a) Extensive interplasmidic duplications change the virulence phenotype of the relapsing fever agent *Borrelia turicatae*. Mol Microbiol 34(5):1120–1132

Pennington PM, Cadavid D, Barbour AG (1999b) Characterization of VspB of Borrelia turicatae, a major outer membrane protein expressed in blood and tissues of mice. Infect Immun 67(9):4637–4645

Plasterk RHA, Simon MI, Barbour AG (1985) Transposition of structural genes to an expression sequence on a linear plasmid causes antigenic variation in the bacterium *Borrelia hermsii*. Nature 318:257–263

Piesman J, Mather TN, Sinsky RJ, Spielman A (1987) Duration of tick attachment and *Borrelia burgdorferi* transmission. J Clin Microbiol 25:557–558

Piesman J, Schneider BS, Zeidner NS (2001) Use of quantitative PCR to measure density of *Borrelia burgdorferi* in the midgut and salivary glands of feeding tick vectors. J Clin Microbiol 39(11):4145–4148

Piesman J, Schwan TG (2010) Ecology of Borreliae and Their Arthropod Vectors. Caister Academic Press, Norfolk, UK

Pospelova-Shtrom MV (1962) Ornithodoros ticks and their epidemiological significance. Joint Publications Research Service, DC, Washington

Porcella SF, Raffel SJ, Schrumpf ME, Schriefer ME, Dennis DT, Schwan TG (2000) Serodiagnosis of louse-borne relapsing fever with glycerophosphodiester phosphodiesterase (GlpQ) from *Borrelia recurrentis*. J Clin Microbiol 38:3561–3571

Ramos JM, Malmierca E, Reyes F, Tesfamariam A (2008) Results of a 10-year survey of louse-borne relapsing fever in southern Ethiopia: a decline in endemicity. Ann Trop Med Parasitol 102(5):467–469

Raoult D, Roux V (1999) The Body Louse as a Vector of Reemerging Human Diseases. Clin Infect Dis 29(4):888–911

Rawlings JA (1995) An overview of tick-borne relapsing fever with emphasis on outbreaks in Texas. Tex Med 91:56–59

Restrepo BI, Kitten T, Carter CJ, Infante D, Barbour AG (1992) Subtelomeric expression regions of *Borrelia hermsii* linear plasmids are highly polymorphic. Mol Microbiol 6:3299–3311

Restrepo BI, Carter CJ, Barbour AG (1994) Activation of a *vmp* pseudogene in *Borrelia hermsii*: an alternate mechanism of antigenic variation during relapsing fever. Mol Microbiol 13:287–299

Restrepo BI, Barbour AG (1994) Antigen diversity in the bacterium *B. hermsii* through "somatic" mutations in rearranged *vmp* genes. Cell 78:867–876

Ruddy S, Austen KF (1969) C3 inactivator of man. I. Hemolytic measurement by the inactivation of cell-bound C3. J Immunol 102(3):533–543

Ruddy S, Austen KF (1971) C3b inactivator of man. II. Fragments produced by C3b inactivator cleavage of cell-bound or fluid phase C3b. J Immunol 107(3):742–750

Sarih M, Garnier M, Boudebouch N, Bouattour A, Rihani A, Hassar M, Gern L, Postic D, Cornet M (2009) *Borrelia hispanica* relapsing fever, Morocco. Emerg Infect Dis 15(10):1626–1629

Schwan TG, Policastro PF, Miller Z, Thompson RL, Damrow T, Keirans JE (2003) Tick-borne relapsing fever caused by *Borrelia hermsii*, Montana. Emerg Infect Dis 9:1151–1154

Schwan TG, Raffel SJ, Schrumpf ME, Porcella SF (2007) Diversity and distribution of *Borrelia hermsii*. Emerg Infect Dis 13:436–442

Schwan TG, Gage KL, Hinnebusch BJ (1995) Analysis of relapsing fever spirochetes from the western United States. J Spirochetal Tick-Borne Dis 2:3–8

Schwan TG, Schrumpf ME, Hinnebusch BJ, Anderson DE, Konkel ME (1996) GlpQ: an antigen for serological discrimination between relapsing fever and Lyme borreliosis. J Clin Microbiol 34:2483–2492

Schwan TG (1996) Ticks and *Borrelia*: model systems for investigating pathogen-arthropod interactions. Infect Agents Dis 5(3):167–181

Schwan TG, Hinnebusch BJ (1998) Bloodstream- versus tick-associated variants of a relapsing fever bacterium. Science 280:1938–1940

Schwan TG, Raffel SJ, Schrumpf ME, Policastro PF, Rawlings JA, Lane RS, Breitschwerdt EB, Porcella SF (2005) Phylogenetic analysis of the spirochetes *Borrelia parkeri* and *Borrelia turicatae* and the potential for tick-borne relapsing fever in Florida. J Clin Microbiol 43: 3851–3859

Schwan TG, Raffel SJ, Schrumpf ME, Webster LS, Marques AR, Spano R, Rood M, Burns J, Hu R (2009) Tick-borne relapsing fever and *Borrelia hermsii*, Los Angeles County, California, USA. Emerg Infect Dis 15(7):1026–1031

Shamaei-Tousi A, Martin P, Bergh A, Burman N, Brannstrom T, Bergstrom S (1999) Erythrocyte-aggregating relapsing fever spirochete *Borrelia crocidurae* induces formation of microemboli. J Infect Dis 180(6):1929–1938

Shamaei-Tousi A, Collin O, Bergh A, Bergstrom S (2001) Testicular damage by microcirculatory disruption and colonization of an immune-privileged site during *Borrelia crocidurae* infection. J Exp Med 193(9):995–1004

Stoenner HG, Dodd T, Larsen C (1982) Antigenic variation of *Borrelia hermsii.* J Exp Med 156:1297–1311

Sidi G, Davidovitch N, Balicer RD, Anis E, Grotto I, Schwartz E (2005) Tickborne relapsing fever in Israel. Emerg Infect Dis 11(11):1784–1786

Sundnes KO, Haimanot AT (1993) Epidemic of louse-borne relapsing fever in Ethiopia. Lancet 342(8881):1213–1215

Tabuchi N, Mitani H, Seino S, Fukunaga M (2002) The 44-kb linear plasmid molecule in the relapsing fever agent *Borrelia duttonii* strain Ly serve as a preservation of *vmp* genes. Microbiol Immunol 46(3):159–165

Talbert A, Nyange A, Molteni F (1998) Spraying tick-infested houses with lambda-cyhalothrin reduces the incidence of tick-borne relapsing fever in children under five years old. Trans R Soc Trop Med Hyg 92(3):251–253

Trape JF, Duplantier JM, Bouganali H, Godeluck B, Legros F, Cornet JP, Camicas JL (1991) Tick-borne borreliosis in west Africa. Lancet 337(8739):473–475

Trape JF, Godeluck B, Diatta G, Rogier C, Legros F, Albergel J, Pepin Y, Duplantier JM (1996a) The spread of tick-borne borreliosis in West Africa and its relationship to sub-Saharan drought. Am J Trop Med Hyg 54(3):289–293

Trape JF, Godeluck B, Diatta G, Rogier C, Legros F, Albergel J, Pepin Y, Duplantier JM (1996b) Tick-borne borreliosis in west Africa: recent epidemiological studies. Rocz Akad Med Bialymst 41(1):136–141

van Dam AP, van Gool T, Wetsteyn JCFM, Dankert J (1999) Tick-borne relapsing fever imported from West Africa: diagnosis by quantitative buffy coat analysis and in vitro culture of *Borrelia crocidurae*. J Clin Microbiol 37:2027–2030

Varma MGR (1962) Transmission of relapsing fever spirochetes by ticks. In: Symposia of the Zoological Society of London: 1962; Regent's Park, London: The Zoological Society of London, 61–82

Vaughan AT, Roghanian A, Cragg MS (2011) B cells—Masters of the immunoverse. Int J Biochem 43(3):280–285

Vial L, Diatta G, Tall A, Ba el H, Bouganali H, Durand P, Sokhna C, Rogier C, Renaud F, Trape JF (2006) Incidence of tick-borne relapsing fever in west Africa: longitudinal study. Lancet 368(9529):37–43

Vidal V, Scragg IG, Cutler SJ, Rockett KA, Fekade D, Warrell DA, Wright DJ, Kwiatkowski D (1998) Variable major lipoprotein is a principal TNF-inducing factor of louse-borne relapsing fever. Nat Med 4(12):1416–1420

Walker RL, Read DH, Hayes DC, Nordhausen RW (2002) Equine abortion associated with the *Borrelia parkeri-B. turicatae* tick-borne relapsing fever spirochete group. J Clin Microbiol 40:1558–1562

Wang G, Ojaimi C, Wu H, Saksenberg V, Iyer R, Liveris D, McClain SA, Wormser GP, Schwartz I (2002) Disease severity in a murine model of lyme borreliosis is associated with the genotype of the infecting *Borrelia burgdorferi sensu stricto* strain. J Infect Dis 186(6):782–791

Whitney MS, Schwan TG, Sultemeier KB, McDonald PS, Brillhart MN (2007) Spirochetemia caused by *Borrelia turicatae* infection in 3 dogs in Texas. Vet Clin Pathol 36(2):212–216

Yang L, Weis JH, Eichwald E, Kolbert CP, Persing DH, Weis JJ (1994) Heritable susceptibility to severe *Borrelia burgdorferi*-induced arthritis is dominant and is associated with persistence of large numbers of spirochetes in tissues. Infect Immun 62(2):492–500

Zingg BC, Brown RN, Lane RS, LeFebvre RB (1993) Genetic diversity among *Borrelia burgdorferi* isolates from wood rats and kangaroo rats in California. J Clin Microbiol 31(12):3109–3114

Part III
Persistence

Chapter 10
Neuroborreliosis Caused by *Borrelia burgdorferi*: Possible Role of Glial Cells

Mario T. Philipp, Geeta Ramesh, and Geetha Parthasarathy

10.1 Foreword

The central nervous system has traditionally been considered as an immune-privileged site, largely due to a lack of lymphatic draining and low levels of immune surveillance (Carson et al. 2006). It is therefore tempting to speculate that spirochetes have evolved to home to, among other organs, the central nervous system (CNS), as a strategy for immune evasion. While it is possible that the adaptive immune response to *Borrelia burgdorferi* may be slower and less effective in the CNS than it is in the periphery, an area of research that remains largely to be studied, this spirochete elicits a powerful innate, pro-inflammatory response in nervous system glial cells. How this response may contribute to Lyme neuroborreliosis pathogenesis is the subject of this chapter.

10.2 Introduction

Lyme borreliosis, the most frequently reported arthropod-borne disease in the USA, is prevalent worldwide (Jaenson 1991). Its etiologic agent, the spirochete *B. burgdorferi* (*sensu lato*), infects humans via vector ticks of the genus *Ixodes*. Following infection, spirochetes remain in the periphery of the site of the tick bite for a limited time, to disperse then to multiple organs, including the skin, joints, heart, as well as the nervous system. This ubiquitous organ tropism by *B. burgdorferi* is likely the

M.T. Philipp, Ph.D. (✉) • G. Ramesh, Ph.D. • G. Parthasarathy, Ph.D.
Bacteriology and Parasitology, Tulane National Primate Research Center,
Covington Louisiana, USA
e-mail: philipp@tulane.edu; gramesh@tulane.edu; gparthas@tulane.edu

M.E. Embers (ed.), *The Pathogenic Spirochetes: Strategies for Evasion of Host Immunity and Persistence*, DOI 10.1007/978-1-4614-5404-5_10,

cause of the diverse manifestations of Lyme borreliosis. These include the skin rash known as erythema migrans, arthritis, carditis, and neuroborreliosis (Steere et al. 2004).

The pathogenesis of Lyme borreliosis has a strong inflammatory component. In patients, perivascular cellular infiltrates have been found in peripheral nerves after tick bites (Vallat et al. 1987), and in the brain (Benach and Garcia-Monco 1992), as well as in numerous tissues in animal models of Lyme disease (Philipp et al. 1993; Roberts et al. 1995; Barthold 1996). Our studies in the rhesus macaque, the only animal model to exhibit signs of Lyme borreliosis in all of the organ systems that are affected by the disease in humans (Philipp et al. 1993; Roberts et al. 1995; England et al. 1997; Pachner et al. 1995, 2001, 2002; Cadavid et al. 2000, 2004; Bai et al. 2004), provide numerous examples of typical inflammatory lesions. Erythema migrans (EM), the skin-rash that in humans and nonhuman primates signals the point of entry of *B. burgdorferi*, is shown in Fig. 10.1a, b; the figure portrays EM lesions in rhesus macaques. The inflammatory nature of this lesion is made evident by the perivascular lymphocytic and monocytic infiltrate (Fig. 10.1c). Spirochetes or spirochetal antigens are sometimes found within these infiltrates (Fig. 10.1d).

10.3 Lyme Neuroborreliosis

As with other manifestations of Lyme disease, neuroborreliosis, which is perhaps the most morbid form of the disease, also presents with inflammatory signs and symptoms. In the USA it is manifest in about 15% of untreated patients (Steere 2001). Lyme neuroborreliosis may affect not only the CNS but also the peripheral nervous system (PNS).

10.3.1 PNS Neuroborreliosis

PNS neuroborreliosis may manifest very early in the infection process, as a cranial neuropathy or acute radiculoneuropathy (Logigian 1997). Within the first 4 weeks of the appearance of EM, 50–75% of early neuroborreliosis patients will show facial palsy. More rarely, other cranial nerves, alone or in combination, also may be affected (Logigian 1997). Early Lyme radiculoneuropathy often presents with pain in the distribution of the affected nerves and nerve roots. Pain may radiate from the spine into an extremity or the trunk. Signs and symptoms such as sensory loss, weakness, or hyporeflexia may be focal but are more often distributed multifocally and asymmetrically, and can affect any or all of the limbs and the trunk (Logigian 1997). Up to 90% of patients with Lyme radiculoneuropathy also show a lymphocytic pleocytosis in the cerebrospinal fluid (CSF). This particular combination of PNS and CNS signs and symptoms is the Garin-Bujadoux-Bannwarth syndrome

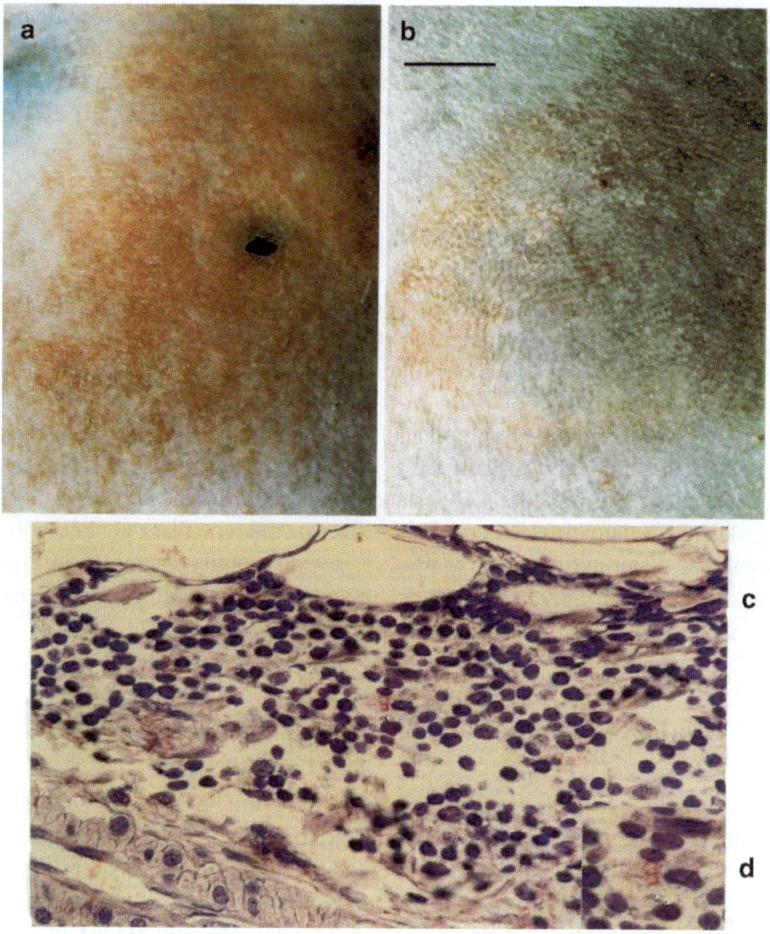

Fig. 10.1 Erythema migrans in the skin of a rhesus macaque infected with *B. burgdorferi* by needle inoculation (**a**), and via ticks (**b**). Perivascular lymphocytic and monocytic infiltrate (**c**), in the dermis subjacent to erythema shown in (**b**). Inset (**d**) shows *B. burgdorferi* spirochetes stained with a specific MAb. From Philipp et al. (1993)

(Logigian 1997). Chronic PNS involvement appears on average during the second year of infection and manifests most commonly as a distal paresthesia with symptoms of "numbness" and "pins and needles." Signs of sensory loss to all modalities can be present as well (Logigian 1997). Less frequently, patients with chronic PNS disease present with pain in the distribution of cervical, thoracic, or lumbosacral dermatomes. Electrophysiological studies in patients with this radicular form of late Lyme neuropathy show mild sensorimotor axonal loss, whereas the electrophysiological abnormalities in patients with distal paresthesia are more severe but equally consistent with neurodegeneration, in particular, axonal loss (Logigian and Steere 1992). In an electrophysiological study in rhesus macaques (England et al. 1997), our group was able to demonstrate mononeuropathy multiplex in animals with long-term *B. burgdorferi* infections. Altogether, 5 of 8 infected monkeys demonstrated primarily axonal-loss-variety multifocal neuropathies. Pathologically, peripheral nerve specimens showed multifocal axonal degeneration and regeneration and perivascular inflammatory cellular infiltrates (England et al. 1997).

10.3.2 CNS Neuroborreliosis

CNS neuroborreliosis may manifest early on, within the first few weeks to months of the appearance of EM, as a meningitis (inflammation of the extraparenchymal CNS) often as part of the Garin-Bujadoux-Bannwarth syndrome or, more seriously, as a parenchymal inflammation of brain and spinal cord (encephalomyelitis) (Halperin 1997, 2010). The latter is less common than the meningitis but shares with it the presence of lymphocytic pleocytosis of the CSF and intrathecal anti-*B. burgdorferi* antibody production. This disorder's importance is derived from the fact that it can lead to irreversible CNS damage (Halperin 1997, 2010) of the type that may be attributed to neuronal loss. Later in the course of disease, several months postinfection, encephalopathies may appear. Patients with this late symptom complain of specific memory and/or intellectual impairment, often associated with incapacitating fatigue.

10.4 Glial Cells in the Pathogenesis of Lyme Neuroborreliosis

We have argued that the inflammatory context generated by a *B. burgdorferi* infection of the CNS or the PNS could lead to programmed cell death (apoptosis) of glial cells and neurons and that neuroglial apoptosis is the key event leading to nervous system Lyme disease. Here we review the evidence for glial cell involvement in the inflammatory process that accompanies experimental infection of the CNS and PNS of rhesus macaques by *B. burgdorferi*.

Neurons and glia are the major cell classes that compose the PNS and CNS. Neurons differ from glia in that the former can be excited electrically. The main PNS glia are the Schwann cells, of which there are four types with different functions: myelinating and nonmyelinating Schwann cells, terminal Schwann-like cells of the sensory neurites, and presynaptic Schwann cells of the neuromuscular junction. The main function of the myelinating Schwann cells is to produce the myelin sheath that insulates the neuronal axons in the PNS. The other Schwann cell types perform a variety of PNS maintenance functions, such as signaling, and regulation of neuronal homeostasis. Glial cells of the CNS are the astrocytes, oligodendrocytes, ependymal cells, and microglia. About 80% of the CNS cells are astrocytes. There are numerous types of astrocytes, with different morphology and function, depending on their localization. Most astrocytes interact physically with neurons and with the blood–brain barrier, and are thought to play a major role in neuronal and blood–brain barrier structural and functional integrity. Oligodendrocytes are the myelinating cells of the CNS. There are four main types of oligodendrocytes, which differ both in their morphology and CNS localization (Types I–IV). Together they represent about 5% of the CNS cells. Ependymal cells are glia that localize both to the brain ventricle walls and the central canal of the spinal cord. Their main function is the production and dynamics of the CSF, and the separation and interaction between the CSF and the rest of the CNS. About 5% of CNS cells are ependymal cells. Unlike astrocytes, oligodendrocytes, and ependymal cells, which are all of neural or ectodermal ontogeny, microglia are myelomonocytic in origin. They are the monocytes/macrophages of the CNS and their abundance is between 10 and 15% (Verkhratsky and Butt 2007). In addition to the functions described above, glial cells are known to participate in inflammatory responses and thought to contribute as such to the pathogenesis of neurodegenerative diseases, e.g., Alzheimer's disease, and the AIDS–dementia complex (Amor et al. 2010), and, as we posit, to the pathogenesis of Lyme neuroborreliosis.

Astroglial and neuronal proteins, anti-myelin antibodies and cells secreting antibodies to myelin basic protein have been detected in the CSF of patients with Lyme neuroborreliosis, indicating possible glial and neuronal damage in the CNS parenchyma (Suchanek et al. 1986; Baig et al. 1991; Dotevall et al. 1999). There is evidence that *B. burgdorferi* spirochetes can adhere to neurons, CNS glia, and Schwann cells from studies in neuronal and glial cell lines and primary rat brain cultures (Garcia-Monco and Benach 1997), and that *B. burgdorferi* can adhere to and perhaps invade human neuroglial and cortical neuronal cells (Livengood and Gilmore 2006).

We have established, not only via experiments in vivo and in vitro but also ex vivo, that live *B. burgdorferi* spirochetes elicit inflammatory mediators from astrocytes and microglia. In this context, neurons and oligodendrocytes undergo apoptosis. Initial experiments in vivo made use of a modification of the rhesus model of Lyme disease that involved stereotaxic inoculations of live *B. burgdorferi* spirochetes into the animals' gray and white matter (Ramesh et al. 2008). An example of inflammatory mediators produced in this context by glial cells is shown in Fig. 10.2, which depicts astrocytes producing IL-6, an inflammatory cytokine. Importantly, apoptosis of oligodendrocytes was also observed (Fig. 10.3, Ramesh et al. 2008).

These results were further substantiated with experiments performed ex vivo, whereby live *B. burgdorferi* were allowed to migrate into freshly collected brain tissue explants from rhesus macaques that were not infected with *B. burgdorferi*. Tissue explants from the frontal cortex were cultured either for 4 or 8 h with or without *B. burgdorferi*, and production of inflammatory mediators and apoptosis were assessed immunohistochemically in glial cells and neurons (Ramesh et al. 2008). When exposed to *B. burgdorferi*, astrocytes and microglia produced the potent pro-inflammatory mediators IL-8 and COX-2, and microglia made the monocyte chemotactic protein CCL2 (Fig. 10.4, and Ramesh et al. 2008). In this inflammatory environment generated ex vivo, both oligodendrocytes and neurons underwent apoptosis (Fig. 10.5, Ramesh et al. 2008).

Our most all-encompassing in vivo approach to proving that live spirochetes are able to elicit in glial cells an inflammatory milieu not only in the CNS but also in

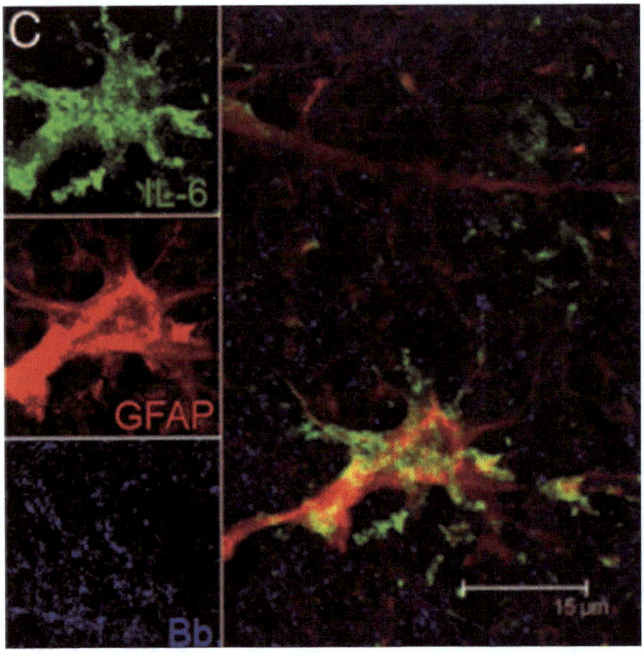

Fig. 10.2 Frontal cortex astrocytes produce IL-6 in the presence of *B. burgdorferi*. Tissue section taken at the site of inoculation from an animal given stereotaxic inoculations with live *B. burgdorferi*. IL-6-producing astrocytes appear yellow because of co-localization of antibody to IL-6 (*green*) and antibody to the astrocyte marker GFAP (*red*) in the vicinity of *B. burgdorferi* antigen stained *blue*. Reprinted from Ramesh et al. (2008) with permission from Elsevier

the PNS, involved intrathecal inoculations of live *B. burgdorferi* into the *cisterna magna* of rhesus macaques. This study included seven animals: three were studied for 6 weeks postinoculation (PI) and the other four for 12 weeks. In each of these two groups, one of the animals served as control and was inoculated only with culture medium. The other animals were given live *B. burgdorferi* intrathecally. Inflammatory changes induced in the CNS, and dorsal root nerves and ganglia (DRG) were examined temporally and postmortem. Cytokine and chemokine ELISAs of the CSF showed elevated concentrations of IL-6, IL-8, CCL2, and CXCL13, as early as 1 week postinoculation, accompanied by primarily lymphocytic and monocytic pleocytosis (Ramesh et al. 2009). In contrast, onset of the acquired immune response, evidenced by the appearance of anti-C6 serum antibodies, was first detectable after 3 weeks postinoculation. This indicates that innate immunity involving glial cells is a key inflammatory component in the onset of Lyme neuroborreliosis. CSF cell pellets and CNS tissues were culture-positive for *B. burgdorferi* in all but one of the five animals that were inoculated with *B. burgdorferi*, thus indicating that this in vivo approach resulted in a high yield of infection of the CNS. Histopathology revealed signs of acute Lyme neuroborreliosis, namely, severe multifocal leptomeningitis, radiculitis, and DRG inflammatory

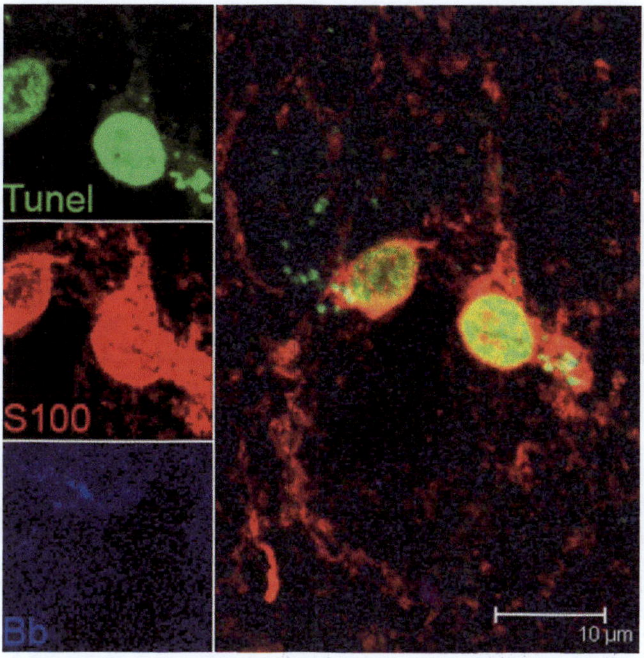

Fig. 10.3 Oligodendrocyte apoptosis in a rhesus macaque given stereotaxic inoculations with live *B. burgdorferi*. Apoptotic oligodendrocytes show TUNEL-positive nuclei stained green. Oligodendrocytes appear red because of staining with antibody to S-100. *B. burgdorferi* antigen appears *blue* in color. Reprinted from Ramesh et al. (2008) with permission from Elsevier

lesions (Figs. 10.6 and 10.7). Immunofluorescence staining and confocal microscopy evinced *B. burgdorferi* antigen in the CNS and DRG. IL-6 was detected in astrocytes and neurons in the spinal cord and neurons in the DRG of infected animals (Ramesh et al. 2009). CCL2 and the B-cell chemokine CXCL13 were found in microglia as well as in endothelial cells, macrophages, and T cells. The latter were the two most prominent mediators found in the CSF, and their CSF concentrations were at most time points higher than their corresponding serum concentrations. This is evidence to suggest that CCL2 and CXCL13 are both primarily produced in the CNS and may be major contributors to the observed pathology. High levels of CXCL13 are found in patients with acute, untreated Lyme neuroborreliosis (Schmidt et al. 2011). Importantly, the DRG of infected animals showed significant levels of satellite cell and neuronal apoptosis (Fig. 10.8). These in vivo results support the notion that innate responses of glia against *B. burgdorferi* initiate/mediate the inflammation that is seen in acute Lyme neuroborreliosis and show that neuronal apoptosis occurs in this context. Taken together with the results of the ex vivo experiments they entail evidence that *B. burgdorferi* spirochetes can generate in the CNS/PNS an inflammatory microenvironment, and that the latter is concomitant with limited glial/neuronal loss by apoptosis.

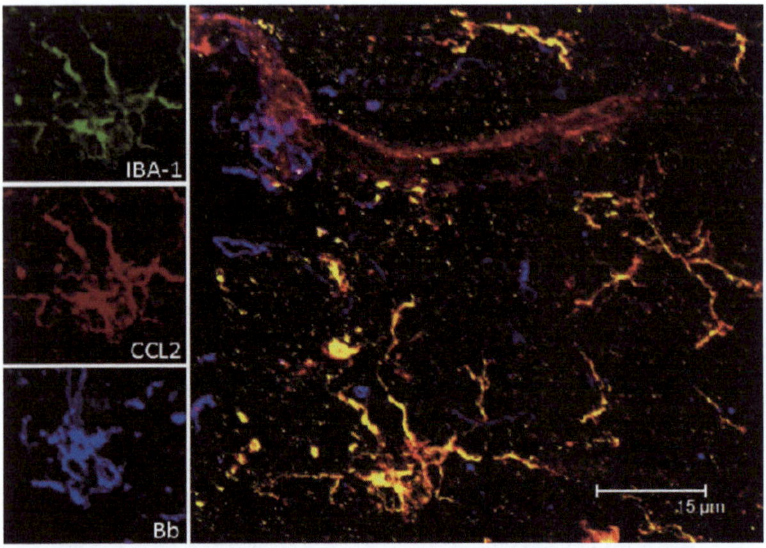

Fig. 10.4 Visualization of the production of CCL2 by glial cells in *B. burgdorferi*-exposed frontal cortex tissue explants. CCL2-producing microglia appear yellow because of co-localization of antibody to the microglial marker IBA-1 (*green*) and antibody to CCL2 (*red*). Spirochetes appear *blue*

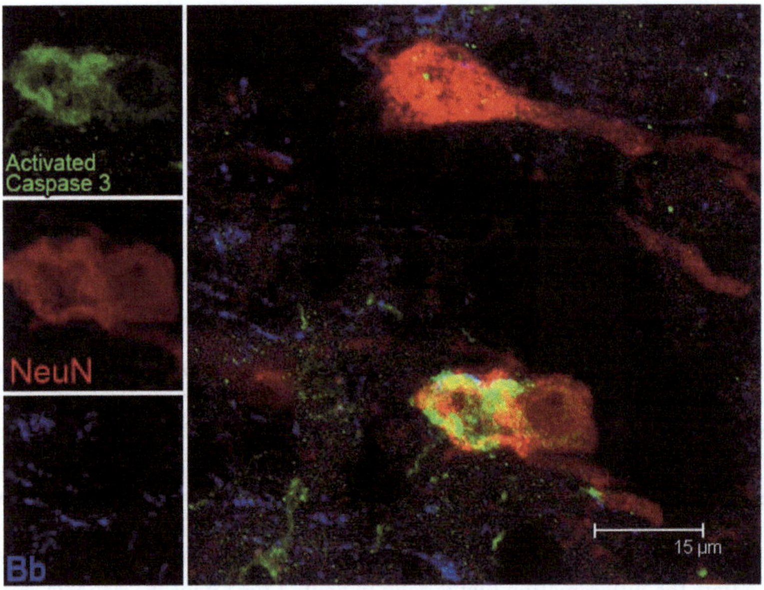

Fig. 10.5 Neuronal apoptosis in frontal cortex brain tissue exposed to live *B. burgdorferi*. Confocal micrograph showing apoptotic neurons as yellow because of co-localization of antibody to active caspase-3 (*green*) and antibody to the neuronal marker NeuN (*red*). Spirochetes appear *blue* in color. Reprinted from Ramesh et al. (2008) with permission from Elsevier

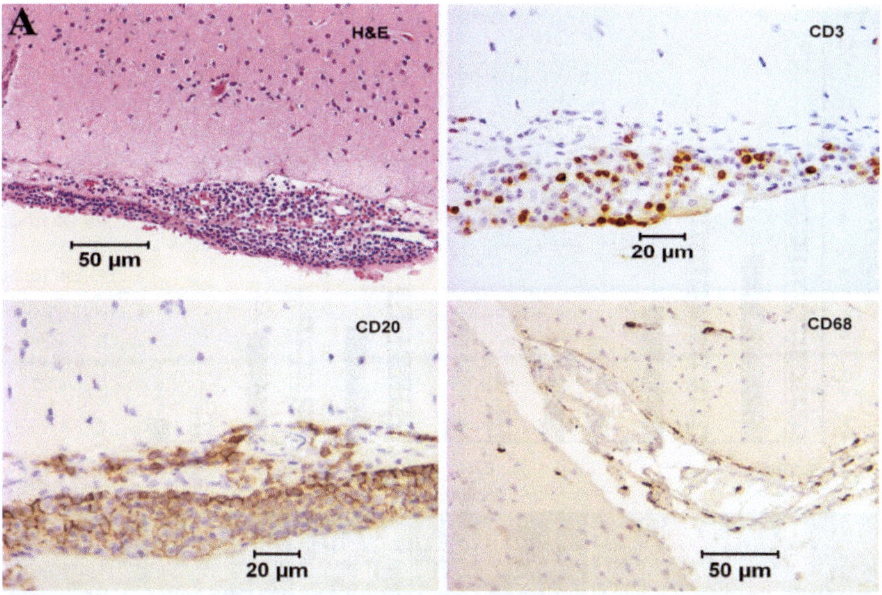

Fig. 10.6 Leptomeningitis in the CNS of a *B. burgdorferi*-inoculated rhesus macaque showing lymphocytic, plasmacytic and monocytic infiltration by hematoxylin and eosin staining, as well as immunohistochemical staining showing CD3-labeled T cells, CD-20-labeled B cells and CD68-labeled macrophages. Lymphocytes, particularly B cells, predominate in the infiltrates. From Ramesh et al. (2009)

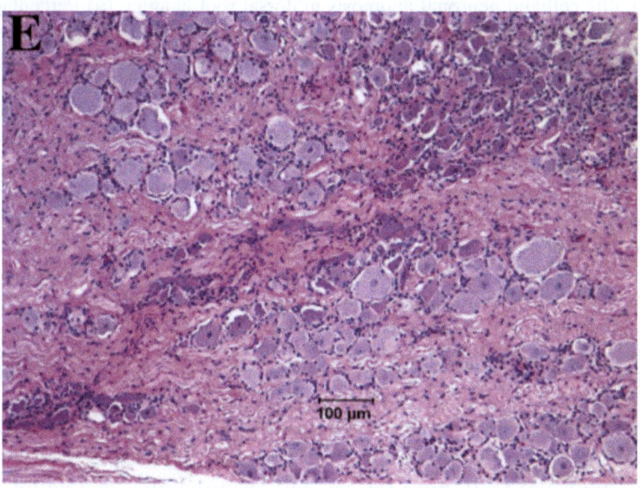

Fig. 10.7 Inflammation of dorsal root ganglia from a rhesus macaque inoculated intrathecally with *B. burgdorferi* showing infiltrates composed of lymphocytes, plasma cells, and monocytes. From Ramesh et al. (2009)

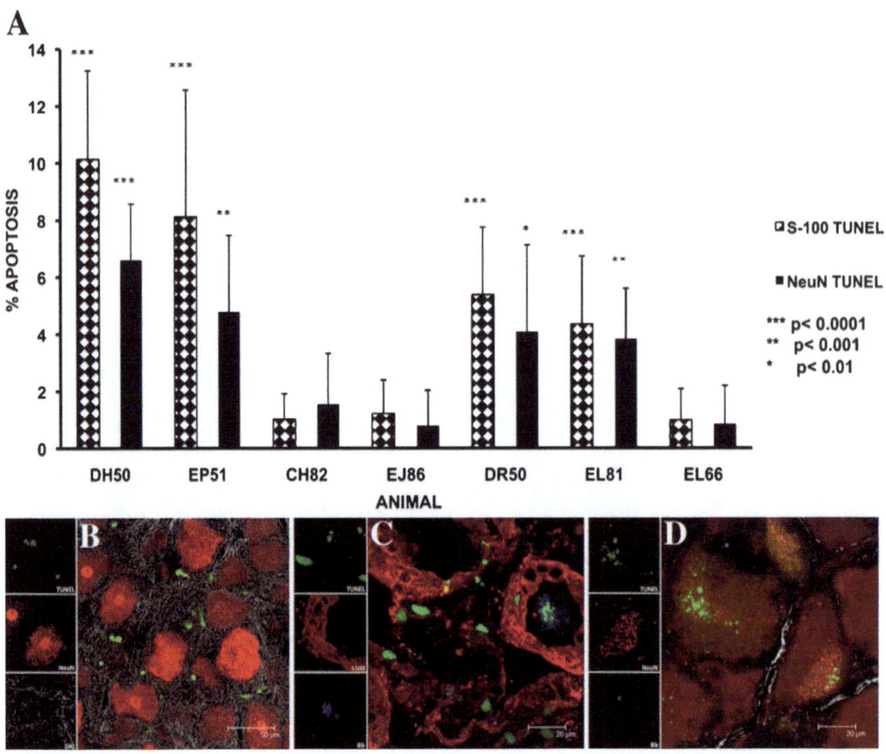

Fig. 10.8 Satellite cell and neuronal apoptosis in dorsal root ganglia: (**a**) Percentage of satellite cells and neurons dying by apoptosis as revealed by the in situ TUNEL assay performed on tissues from the DRG of all of the animals in the intrathecal inoculation study. Data represent the mean values ± SD of the counts from ten fields. The *p* values represent the statistical significance with respect to the control animal for each group (EL66 for the 6-week group and EJ86 for the 12-week one). (**b**) A qualitative view of the extent of the apoptosis (*green*) around neurons stained with antibody to NeuN (*red*) observed in the DRG of animal DH50 following the in situ TUNEL assay and immunofluorescence staining and confocal microscopy. (**c**) Apoptosis of DRG satellite cells. DRG sections from animal DH50 were labeled with antibody to S-100 (*red*). Cells undergoing apoptosis, stained using the TUNEL assay, are shown in *green*. Apoptotic satellite cells surround a neuron that appears unstained, but is also showing the TUNEL signal in the center. (**d**) Neurons stained with anti-NeuN antibody (*red*), and undergoing apoptosis (*green*) are shown in the vicinity of residual *B. burgdorferi* antigen (*blue*). The confocal micrograph is from a DRG section of animal DH50, following application of the in situ TUNEL assay and immunofluorescence staining. From Ramesh et al. (2009)

To evaluate the relative contribution of microglia and astrocytes to inflammation as caused by exposure to *B. burgdorferi*, and examine glial cell apoptosis, we isolated both types of glial cells from the brain of uninfected rhesus macaques. In experiments where *B. burgdorferi* was cocultured in vitro with primary microglia for 5 days we observed robust expression and release of IL-6 (22.8 ± 0.66 ng/ml), IL-8 (51.3 ± 0.53 ng/ml), CCL2 (272 ± 30.0 ng/ml), CCL4 (18.6 ± 0.00 ng/ml), and CCL5 (1.73 ± 0.02 ng/ml) (Myers et al. 2009). Astrocytes expressed these pro-inflammatory mediators as well but at much lower levels; for example, the concentration of CCL2 produced in 5 days by microglia stimulated with *B. burgdorferi* (MOI 10:1) was 27 times higher than that produced by astrocytes under the same conditions; the concentration of CCL4 was over 300 times higher (Myers et al. 2009). Quantitatively, CCL2 emerges as a prominent chemokine produced in response to *B. burgdorferi*, both in the CSF in vivo and by microglia in vitro. When neurons of the SHSY-5Y neuroblastoma cell line were cultured in the presence of *B. burgdorferi*, the cells produced lower levels of these cytokines and chemokines (Myers et al. 2009). Neither neurons nor glial cells (astrocytes or microglia) showed significant levels of apoptosis when each cell type was cultured alone with *B. burgdorferi*. Neurons did not die when cocultured with astrocytes and *B. burgdorferi*. However, when neurons were cocultured with microglia and *B. burgdorferi*, significant neuronal apoptosis occurred (Myers et al. 2009). Confocal microscopy imaging of neuron/microglia/*B. burgdorferi* cell cultures stained for apoptosis and with cell type-specific markers confirmed that it was predominantly the neurons that were dying (Myers et al. 2009). Microarray analysis further demonstrated an intense microglia-mediated inflammatory response to *B. burgdorferi* including upregulation in gene transcripts for pro-inflammatory cytokines, Toll-like receptor 2 (TLR-2), and NF kappa B1 and B2. Significant transcript alterations in essential p53 pathway genes also occurred in neuronal cells cultured in the presence of microglia and *B. burgdorferi*, which indicated a shift from cell survival to preparation for apoptosis when compared to neurons cultured in the presence of *B. burgdorferi* alone (Myers et al. 2009).

Taken together, these findings indicate that *B. burgdorferi* is not directly toxic to neurons; rather, neurons become distressed and die in the inflammatory surroundings generated by microglia through a bystander effect. If, as we hypothesized, neuronal and glial apoptosis is the key pathogenic event in Lyme neuroborreliosis, then, targeting microglial responses may be a significant therapeutic approach for the treatment of this form of Lyme disease.

Toll-like receptors (TLR) on cells that engage in immune responses recognize pathogen-associated molecular patterns and signal the cell to produce inflammatory mediators. To begin to evaluate the role of TLR in the inflammatory response of glial cells to live *B. burgdorferi*, we examined whether rhesus microglia and astrocytes expressed TLR 1, 2, 5, and 9 constitutively and/or in response to *B. burgdorferi*, and analyzed the functional response of these cells to flagellin (which binds to TLR5), to CpG DNA (TLR9), and the *B. burgdorferi* lipoprotein OspA (TLR1 and TLR2). This was done under the hypothesis that at the very least *B. burgdorferi*, a Gram-negative bacterium that does not have LPS, would have these molecular patterns available for stimulation of the above TLRs (Bernardino et al. 2008). We first assessed TLR gene expression by quantitative real-time RT-PCR. TLR1, 2, 5, and 9 genes were constitutively expressed in both microglia and astrocytes; however the transcription of TLR9 was found to be low in comparison to the housekeeping gene (Bernardino et al. 2008). We then investigated the expression of individual TLRs after stimulation with live and sonicated *B. burgdorferi*, L-OspA (purified recombinant lipidated OspA), CpG ODN M362 (unmethylated CpG oligodeoxynucleotides),

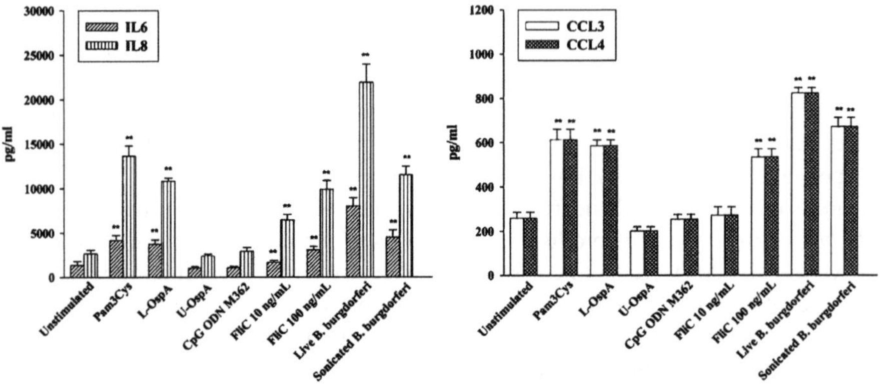

Fig. 10.9 Secretion of pro-inflammatory mediators by microglia. Concentrations of IL6, IL8, CCL3, and CCL4 were determined after 24 h incubations of microglia with no stimulant, 0.25 μg/ml L-OspA, 0.25 μg/ml U-OspA, 12.5 ng/ml Pam3Cys, 1.0 μM CpG ODN M362, 10 ng/ml or 100 ng/ml FliC, and live and sonicated *B. burgdorferi* (10 MOI). Shown are the mean values ± SD obtained from triplicate specimens. *Double asterisks* ($P<0.005$) indicates significant differences between unstimulated and stimulated cells. Similar results were obtained with supernatants obtained from cells of three additional rhesus macaques. From Bernardino et al. (2008)

and FliC (flagellin isolated from *Salmonella typhimurium*). We observed a significant upregulation of TLR2 transcript in microglia stimulated for 8 h with L-OspA and live or sonicated *B. burgdorferi* (Bernardino et al. 2008). In contrast, no significant differences were detected in astrocytes. The expression of TLR5 and TLR9 genes in both cell types remained unchanged after stimulation, and the upregulation in TLR1 transcript was restricted to microglia after stimulation with live and sonicated *B. burgdorferi*. With regard to TLR protein expression, both microglia and astrocytes showed constitutive production of TLR1, 2, and 5 as assessed by Western blotting. However, expression of TLR1 was more marked than that of TLR2 and 5. No increase in protein expression levels was detectable by Western blot upon stimulation with L-OspA, FliC, or *B. burgdorferi* (Bernardino et al. 2008). When assessed by fluorescent confocal microscopy 24 h poststimulation with live *B. burgdorferi*, activated/phagocytic microglia did simultaneously express TLR1, 2, and 5, and these receptors were localized on the cell surface and/or intracellularly. Interestingly, these TLRs were upregulated only in microglia that also contained spirochetes or spirochetal fragments. Microglia that had not internalized spirochetal components expressed much lower levels of TLR1 and 5 and no detectable levels of TLR2. Microglia that were stimulated with *B. burgdorferi* also showed upregulation of TLR4 in phagocytic, but not in unstimulated, cells. Functionally, both microglia (Fig. 10.9, Bernardino et al. 2008) and astrocytes (Bernardino et al. 2008) upregulated the production of IL-6, IL-8, and the macrophage inflammatory chemokines CCL3 and CCL4 when these cells were stimulated with the lipohexapeptide Pam3Cys (a synthetic TLR2 agonist), L-OspA, FliC, and both sonicated and live *B. burgdorferi*, but not with U-OspA (unlipidated OspA) or CpG ODN M362. Taken together, these results provide proof of the concept that astrocyte and microglial TLR1, 2, and 5 are involved in the in vivo response of primate glial cells to *B. burgdorferi*. The pro-inflammatory molecules elicited by these TLR-mediated responses could be a significant factor in the pathogenesis of Lyme neuroborreliosis. Microglia, moreover, are the key mediators of neuronal apoptosis caused by *B. burgdorferi.*

To more fully evaluate the role of TLRs in the response to *B. burgdorferi* and as a preliminary approach to the silencing of primary phagocytic cells such as microglia, we studied the effect of silencing TLR pathway members in cells of the human monocytic cell line THP-1 on the inflammatory response of these cells to live *B. burgdorferi* (Dennis et al. 2009). We first focused on myeloid differentiation primary response protein 88 (MyD88), an adapter molecule that is essential in most of the TLR pathways. Real-time PCR, flow cytometry, and confocal microscopy experiments revealed that MyD88 was maximally expressed in THP-1 cells after 24-h stimulation of these cells with live *B. burgdorferi* (Dennis et al. 2009). Transfection of THP-1 cells with MyD88-specific short-interfering double-stranded RNA molecules that inhibit gene expression (siRNA) significantly reduced the expression of the *MYD88* gene that was induced in these cells by live *B. burgdorferi* ($P<0.001$). Expression of the MyD88 protein was also diminished. The mean fluorescence intensity due to MyD88 expression in control siRNA-transfected cells stimulated with live *B. burgdorferi* increased by 2.23-fold with respect to that of unstimulated cells but increased only by 1.10-fold in cells transfected with MyD88 siRNA, further validating the efficiency of the *MYD88* knockdown in cells. After transfection for 24 h with either control siRNA or MyD88 siRNA and stimulation of cells with live *B. burgdorferi*, cell culture supernatants were collected at 24 h poststimulation to determine the levels of production of TNF, IL-6, and IL-8. Unstimulated cells did not produce significant levels of any of these mediators. Silencing of *MYD88* led to significant inhibition of TNF, IL-8, and IL-6 secretion ($P<0.001$) induced by live *B. burgdorferi* in THP-1 cells, by 24%, 35%, and 84%, respectively, compared to that induced in control siRNA-transfected cells stimulated with live *B. burgdorferi* (Dennis et al. 2009). *MYD88*-silenced cells that were stimulated with L-OspA showed a similar pattern of inhibition of these inflammatory mediators. Our results indicate that recognition of live *B. burgdorferi* by human monocytes and the downstream secretion of TNF, IL-8, and IL-6 inflammatory mediators require, in part, the MyD88 adapter molecule. These are data to further suggest that the TLR pathway is involved in the induction of inflammatory mediators by live *B. burgdorferi*. Specific silencing of the *TLR1*, *TLR2*, or *TLR5* genes by RNA interference revealed that silencing of the *TLR1* and *TLR2* genes alone or combined, but not the *TLR5* gene, caused a downregulation of IL-6, IL-8, and TNF expression in live *B. burgdorferi*-stimulated THP-1 cells (Dennis et al. 2009). Overall, similar results were obtained for THP-1 cells when these cells were stimulated with purified lipoproteins instead of *B. burgdorferi*. Our results indicate that the TLR pathway mediates, at least in part, the release of inflammatory mediators in human monocytes stimulated with live *B. burgdorferi* spirochetes and suggest, moreover, that the TLR-dependent interaction between these cells and live spirochetes is mediated by spirochetal lipoproteins but not by flagellin.

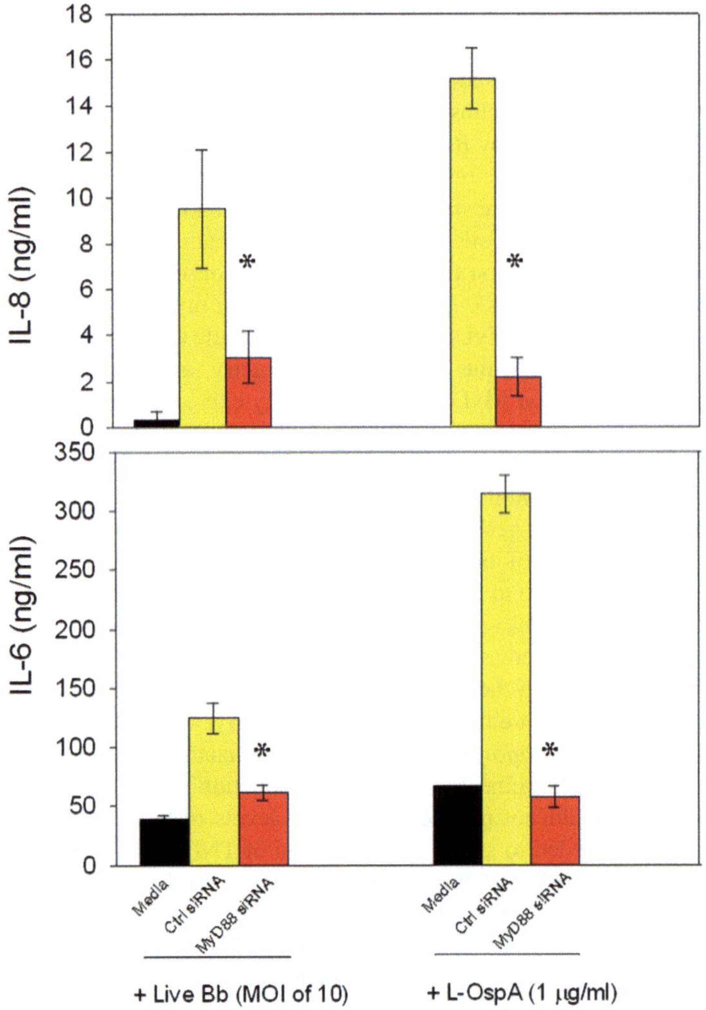

Fig. 10.10 MyD88 siRNA transfection of rhesus microglia. Purified rhesus microglia (0.5×10^5) were transfected with 100 nM nontargeting control siRNA or with 100 nM of MyD88 siRNA and cultured in medium alone for 24 h. They were then stimulated with live *B. burgdorferi* (MOI of 10) for an additional 24 h. Supernatants were analyzed by antibody capture ELISA for IL-8 and IL-6 production. The lower limit of detection of the ELISA was 15 pg/ml. Each bar represents the mean ± SDs of two independent experiments. *Asterisks* indicate significant differences ($P < 0.05$) in the production of inflammatory mediators in cells transfected with MyD88 siRNA compared to that of the control siRNA-transfected cells

In an initial attempt to evaluate the role of TLR in the pro-inflammatory response of microglia to *B. burgdorferi*, we used *MYD88*-specific siRNA (and control siRNA) to transfect rhesus microglia. Stimulation of specifically silenced cells with either live *B. burgdorferi* or L-OspA resulted in significantly diminished production of both IL-8 and IL-6 (Fig. 10.10). This experiment indicates that the TLR pathway(s) is involved in the activation of microglia by *B. burgdorferi*.

The parallelism between microglia-mediated inflammation and neuroglial apoptosis that we have demonstrated suggests itself as a causal relationship, namely, that the inflammatory response elicited from microglia by *B. burgdorferi* is the cause, at least in part, of neuroglial apoptosis and thus of Lyme neuroborreliosis of the CNS. In addition, and because we have shown that DRG neurons also die in the immediate proximity of elevated concentrations of CSF inflammatory mediators caused by *B. burgdorferi* in vivo (Ramesh et al. 2009), we argue that CNS inflammation also affects the PNS.

In addition to microglia, it is possible that oligodendrocytes also contribute to generating a pro-inflammatory microenvironment in the CNS in the presence of *B. burgdorferi*. We evaluated the ability of live *B. burgdorferi* to elicit inflammation in vitro in differentiated human MO3.13 oligodendrocytes (Buntinx et al. 2003) and in differentiated primary human oligodendrocytes, by measuring the concentration of immune mediators in culture supernatants. Concomitant apoptosis also was quantified in these cultures. The above responses were evaluated after 48 h of stimulation with *B. burgdorferi*. This spirochete induced enhanced levels of the cytokine IL-6 and the chemokines IL-8 and CCL2 in MO3.13 cells as compared to basal levels, and IL-8 and CCL2 in primary human oligodendrocytes, in a dose-dependent manner. Oligodendrocyte cultures also showed significantly elevated levels of apoptosis when compared with medium controls (Ramesh et al. 2012). As for the involvement of TLRs in this response, in preliminary experiments a pattern of immune modulation similar to that induced by *B. burgdorferi* was elicited by FliC (flagellin), a TLR5 agonist, rather than by the synthetic triacylated lipopeptide Pam3Csk4, a molecule that binds TLR2. In addition, both *B. burgdorferi* (in a dose-dependent manner) and FliC (10 ng/ml) induced significant activation of nuclear factor (NF) kB (RelA) in MO3.13 oligodendrocytes, unlike Pam3Csk4, which induced only minimal NF-kB (RelA) activation by 30 min at 200 ng/ml. A role for mitogen-activated protein kinase (MAPK) pathways, particularly extracellular signal-regulated kinase (ERK) and c-Jun N-terminal kinase (JNK), for the pro-inflammatory response of oligodendrocytes to *B. burgdorferi* also was identified.

10.5 Concluding Remarks and Future Studies

Our findings indicate that the inflammatory response elicited by the Lyme disease spirochete in glial cells may contribute to neural cell damage. In addition, the abundant release of chemokines by glial cells exposed to live *B. burgdorferi* could cause the leukocyte pleocytosis that signals the onset of Lyme neuroborreliosis. This incoming leukocyte population may further contribute to accentuate the inflammatory environment in the CNS. As oligodendrocytes are vital to the functioning and survival of neurons, the inflammation and subsequent apoptosis of oligodendrocytes induced by *B. burgdorferi* could contribute, together with the effect of microglia on neurons, to the pathogenesis of Lyme neuroborreliosis.

The fact that microglia are contributors to *B. burgdorferi*-induced inflammation is very relevant. This is so because these glial cells, once stimulated, may remain activated for extended periods of time (Qin et al. 2007). These authors showed that a single systemic (intraperitoneal) administration of lipopolysaccharide (LPS) to adult mice resulted in increases in brain concentration of TNF that lasted for 10 months. In contrast, systemic concentrations of this cytokine subsided by 9 h in serum and by 1 week in liver (Qin et al. 2007). Systemic LPS administration activated microglia and increased expression of brain pro-inflammatory factors such as TNF, CCL2, IL-1, and NF-kB p65. Moreover, the number of tyrosine hydroxylase-immunoreactive neurons in the substantia nigra was reduced by 23% at 7-month posttreatment and by 47% at 10 months (Qin et al. 2007). It follows that microglia initially stimulated by a *B. burgdorferi* CNS infection could remain activated, and continue to mediate neuronal apoptosis long after the infection has been removed by antibiotic treatment. This phenomenon could contribute to the symptoms of Lyme disease that sometimes linger after antibiotic treatment for Lyme neuroborreliosis and are known as posttreatment Lyme disease syndrome of the nervous system.

Future studies will focus on demonstrating a cause–effect relationship between inflammation and neuroglial apoptosis and on furthering our understanding of the inflammatory and pro-apoptotic pathways elicited in glial cells by *B. burgdorferi.*

References

Amor S, Puentes F, Baker D, van der Valk P (2010) Inflammation in neurodegenerative diseases. Immunology 129(2):154–169

Bai Y, Narayan K, Dail D, Sondey M, Hodzic E, Barthold SW, Pachner AR, Cadavid D (2004) Spinal cord involvement in the nonhuman primate model of Lyme disease. Lab invest 84(2):160–172

Baig S, Olsson T, Hojeberg B, Link H (1991) Cells secreting antibodies to myelin basic protein in cerebrospinal fluid of patients with Lyme neuroborreliosis. Neurology 41(4):581–587

Barthold S (1996) Lyme borreliosis in the laboratory mouse. J Spirochetal Tick-Borne Dis 3:22–44

Benach JL, Garcia-Monco JC (1992) Aspects of the pathogenesis of neuroborreliosis. In: Schutzer S (ed) Lyme disease; molecular and immunological approaches. Cold Spring Habor Laboratory Press, Cold Spring Harbor, New York

Bernardino AL, Myers TA, Alvarez X, Hasegawa A, Philipp MT (2008) Toll-like receptors: insights into their possible role in the pathogenesis of lyme neuroborreliosis. Infect Immun 76(10):4385–4395

Buntinx M, Vanderlocht J, Hellings N, Vandenabeele F, Lambrichts I, Raus J, Ameloot M, Stinissen P, Steels P (2003) Characterization of three human oligodendroglial cell lines as a model to study oligodendrocyte injury: morphology and oligodendrocyte-specific gene expression. J Neurocytol 32(1):25–38

Cadavid D, O'Neill T, Schaefer H, Pachner AR (2000) Localization of *Borrelia burgdorferi* in the nervous system and other organs in a nonhuman primate model of lyme disease. Lab invest 80(7):1043–1054

Cadavid D, Bai Y, Hodzic E, Narayan K, Barthold SW, Pachner AR (2004) Cardiac involvement in non-human primates infected with the Lyme disease spirochete Borrelia burgdorferi. Lab invest 84(11):1439–1450

Carson MJ, Doose JM, Melchior B, Schmid CD, Ploix CC (2006) CNS immune privilege: hiding in plain sight. Immunol Rev 213:48–65

Dennis VA, Dixit S, O'Brien SM, Alvarez X, Pahar B, Philipp MT (2009) Live *Borrelia burgdorferi* spirochetes elicit inflammatory mediators from human monocytes via the Toll-like receptor signaling pathway. Infect Immun 77(3):1238–1245

Dotevall L, Hagberg L, Karlsson JE, Rosengren LE (1999) Astroglial and neuronal proteins in cerebrospinal fluid as markers of CNS involvement in Lyme neuroborreliosis. Eur J Neurol 6(2):169–178

England JD, Bohm RP Jr, Roberts ED, Philipp MT (1997) Mononeuropathy multiplex in rhesus monkeys with chronic Lyme disease. Ann Neurol 41(3):375–384

Garcia-Monco JC, Benach JL (1997) Mechanisms of injury in Lyme neuroborreliosis. Semin Neurol 17(1):57–62

Halperin JJ (1997) Neuroborreliosis: central nervous system involvement. Semin Neurol 17(1):19–24

Halperin JJ (2010) Nervous system Lyme disease. J R Coll Physicians Edinb 40(3):248–255

Jaenson TG (1991) The epidemiology of lyme borreliosis. Parasitol Today 7(2):39–45

Livengood JA, Gilmore RD Jr (2006) Invasion of human neuronal and glial cells by an infectious strain of *Borrelia burgdorferi*. Microbes Infect 8(14–15):2832–2840

Logigian EL (1997) Peripheral nervous system Lyme borreliosis. Semin Neurol 17(1):25–30

Logigian EL, Steere AC (1992) Clinical and electrophysiologic findings in chronic neuropathy of Lyme disease. Neurology 42(2):303–311

Myers TA, Kaushal D, Philipp MT (2009) Microglia are mediators of *Borrelia burgdorferi*-induced apoptosis in SH-SY5Y neuronal cells. PLoS Pathog 5(11):e1000659

Pachner AR, Delaney E, O'Neill T (1995) Neuroborreliosis in the nonhuman primate: *Borrelia burgdorferi* persists in the central nervous system. Ann Neurol 38(4):667–669

Pachner AR, Cadavid D, Shu G, Dail D, Pachner S, Hodzic E, Barthold SW (2001) Central and peripheral nervous system infection, immunity, and inflammation in the NHP model of Lyme borreliosis. Ann Neurol 50(3):330–338

Pachner AR, Dail D, Narayan K, Dutta K, Cadavid D (2002) Increased expression of B-lymphocyte chemoattractant, but not pro-inflammatory cytokines, in muscle tissue in rhesus chronic Lyme borreliosis. Cytokine 19(6):297–307

Philipp MT, Aydintug MK, Bohm RP Jr, Cogswell FB, Dennis VA, Lanners HN, Lowrie RC Jr, Roberts ED, Conway MD, Karacorlu M et al (1993) Early and early disseminated phases of Lyme disease in the rhesus monkey: a model for infection in humans. Infect Immun 61(7):3047–3059

Qin L, Wu X, Block ML, Liu Y, Breese GR, Hong JS, Knapp DJ, Crews FT (2007) Systemic LPS causes chronic neuroinflammation and progressive neurodegeneration. Glia 55(5):453–462

Ramesh G, Borda JT, Dufour J, Kaushal D, Ramamoorthy R, Lackner AA, Philipp MT (2008) Interaction of the Lyme disease spirochete *Borrelia burgdorferi* with brain parenchyma elicits inflammatory mediators from glial cells as well as glial and neuronal apoptosis. Am J Pathol 173(5):1415–1427

Ramesh G, Borda JT, Gill A, Ribka EP, Morici LA, Mottram P, Martin DS, Jacobs MB, Didier PJ, Philipp MT (2009) Possible role of glial cells in the onset and progression of Lyme neuroborreliosis. J Neuroinflammation 6:23

Ramesh G, Benge S, Pahar B, Philipp MT (2012) A possible role for inflammation in mediating apoptosis of oligodendrocytes as induced by the Lyme disease spirochete *Borrelia burgdorferi*. J Neuroinflammation 9(1):72

Roberts ED, Bohm RP Jr, Cogswell FB, Lanners HN, Lowrie RC Jr, Povinelli L, Piesman J, Philipp MT (1995) Chronic lyme disease in the rhesus monkey. Lab Invest 72(2):146–160

Schmidt C, Plate A, Angele B, Pfister HW, Wick M, Koedel U, Rupprecht TA (2011) A prospective study on the role of CXCL13 in Lyme neuroborreliosis. Neurology 76(12):1051–1058

Steere AC (2001) Lyme disease. N Eng J Med 345(2):115–125

Steere AC, Coburn J, Glickstein L (2004) The emergence of Lyme disease. J Clin Invest 113(8):1093–1101

Suchanek G, Kristoferitsch W, Stanek G, Bernheimer H (1986) Anti-myelin antibodies in cerebrospinal fluid and serum of patients with meningopolyneuritis Garin-Bujadoux-Bannwarth and other neurological diseases. Zentralbl Bakteriol Mikrobiol Hyg A 263(1–2):160–168

Vallat JM, Hugon J, Lubeau M, Leboutet MJ, Dumas M, Desproges-Gotteron R (1987) Tick-bite meningoradiculoneuritis: clinical, electrophysiologic, and histologic findings in 10 cases. Neurology 37(5):749–753

Verkhratsky A, Butt A (2007) Glial neurobiology: a textbook. Wiley, Chichester, UK

Chapter 11
Antibiotic Resistance in *Treponema pallidum* subsp. *pallidum*, the Syphilis Agent

Lola V. Stamm

11.1 Introduction: Syphilis, a "New" Disease

Emerging infectious diseases (EIDs) are an important challenge to public health (Morens et al. 2008). This phenomenon is not new since EIDs that cause significant morbidity and mortality have occurred throughout recorded history. Several factors are involved in disease emergence including travel and commerce, conflict (war), human demographics and behavior, host susceptibility, and microbial adaptation and change. In the late fifteenth century, an EID appeared in Spain after the return of Columbus' expedition to the New World (Morens et al. 2008; Knell 2004). Although controversy exists concerning the origin of this disease, members of Columbus' crew, as well as enslaved Hispaniola natives who were carried to Spain, were reportedly afflicted with the "serpentine" disease (Morens et al. 2008; Baker and Armelagos 1988; de Melo et al. 2010; Rothschild 2005). In 1494, King Charles VIII of France invaded Italy with an army of over 30,000 mercenaries who had been recruited from throughout Europe. His forces reached the city of Naples, which was primarily defended by Spanish soldiers, in February 1495. After occupying the city for a few months, the French army disbanded. Mercenaries infected with the new disease, presumably by Napalese prostitutes, returned home or traveled elsewhere to wage war, thus spreading the EID throughout Europe. During this period of exploration, conflict, and commerce, the disease spread eastward reaching India (1498) and China (1505). Scholars used a variety of names to refer to the new disease (e.g., the French disease, the Disease of Naples, the Great Pox, etc.). It was not until Fracastoro's poem (*Syphilis sive morbus gallicus*), published in 1530, that the disease acquired its modern appellation (i.e., syphilis) (Tognotii 2009; Waugh 1982).

L.V. Stamm, M.S., Ph.D. (✉)
Department of Epidemiology, Gillings School of Global Public Health,
University of North Carolina, Chapel Hill, NC 27599–7435, USA
e-mail: lstamm@email.unc.edu

M.E. Embers (ed.), *The Pathogenic Spirochetes: Strategies for Evasion of Host Immunity and Persistence*, DOI 10.1007/978-1-4614-5404-5_11,

Numerous descriptions by those who were infected and detailed accounts of physicians differentiate syphilis from other diseases present at the time and provide an accurate picture of the clinical manifestations of the primary and secondary stages of syphilis, which predominated during the early part of the epidemic (Tognotii 2009; Waugh 1982). Details of the tertiary stage of syphilis appeared in later writings. Transmission of syphilis was acknowledged as venereal, though non-venereal transmission (e.g., congenital, etc.) was also recognized (Waugh 1982). As observed with some other EIDs, the severity of syphilis changed over time from an acute, severe disease to a more chronic, somewhat less severe one (Knell 2004; Tognotii 2009). This may have been the result of better transmission of less virulent strains of the syphilis agent since infection with the more virulent strains would have caused severe, potentially disabling symptoms, thus reducing the likelihood of their transmission. Nonetheless, syphilis was hardly innocuous and it continued to plague mankind causing morbidity and mortality and sparing neither kings nor commoners (Wright 1971; Frazen 2008).

11.2 Discovery of the Syphilis Agent and Introduction of Modern Chemotherapy

The beginning of the twentieth century marked several important advances for syphilis research. In 1905, the causative agent of syphilis was identified by Fritz Schaudinn and Erich Hoffman, who used a modification of the Giemsa azur-eosin stain to visualize delicate, tightly coiled organisms (i.e., spirochetes), designated as *Spirochaeta pallida*, in material from an eroded labial lesion of a patient with secondary syphilis (Waugh 2005; Kohler 2007). Later, Schaudinn proposed creation of a new genus, *Treponema*, and changed *S. pallida* to *Treponema pallidum*. Multiple attempts to culture *T. pallidum* failed and this bacterium has remained, to the present, refractory to in vitro cultivation (Stamm 2001). In 1906, Wasserman and colleagues developed the first serologic test for diagnosis of syphilis (Kohler 2007; Bialynicki-Birula 2008). That same year, Bertarelli reported successful transmission of syphilis to the rabbit, an animal model that would prove invaluable for testing potential chemotherapeutic agents. Additionally, Landsteiner and Mucha used dark-field microscopy to visualize live, motile *T. pallidum* from syphilis lesions (Kohler 2007).

Despite major advances in syphilis research, the treatment of syphilis had changed little since the end of fifteenth century when physicians advocated the use of mercury in the form of ointment or rubs (Parascandola 2009; O'Shea 1990). At the end of the eighteenth century, internal administration of mercury (e.g., orally or as enemas or vapors) supplanted external administration. Although mercury was toxic and the "cures" it produced were likely symptomatic rather than biologic, it remained the preferred treatment for syphilis into the beginning of the twentieth century and was used sporadically until the 1950s. Fortunately, the search for an efficacious treatment for syphilis continued. In April 1910, Paul Ehrlich and Sahachiro Hata reported at the

Congress for Internal Medicine (Wiesbaden, Germany) the discovery of compound No. 606 (arsphenamine) and described their encouraging results with experimentally infected rabbits and patients with acquired primary syphilis (Bosch and Rosich 2008). This announcement led to an avalanche of requests for arsphenamine that resulted in Ehrlich dispensing 65,000 free samples. Due to the high demand, Hoechst AG marketed the drug as Salvarsan ('the arsenic that saves'). This was the first truly effective drug against syphilis. However, treatment with Salvarsan was prolonged, requiring weekly injections for over a year, and combination therapy (i.e. Salvarsan alternated with mercury or bismuth) was necessary to affect a biologic cure and thus prevent infectious relapse. Because Ehrlich's "magic bullet" was not without toxicity, he introduced compound No. 914 (Neosalvarsan) in 1914, the year before his death. This water-soluble formulation shortened treatment time and was less toxic than Salvarsan (Bosch and Rosich 2008). Although these drugs became established as the standard treatment for syphilis, research on arsenicals continued resulting in reevaluation of oxophenarine (compound No. 599; Mapharsen), the active metabolite of Salvarsan. When Mapharsen was given twice weekly in combination with bismuth, curative treatment could be completed in 2 months rather than 2 years. Subsequently, Chargin and Leifer (1956) found that Mapharsen, delivered for 5 days by the slow intravenous drip method to hospitalized patients, was curative for early syphilis and less toxic than Neosalvarsan. Mapharsen became the drug of choice for treatment of syphilis until the introduction of penicillin.

11.3 Penicillin: the "Miracle Drug"

Curative treatment of syphilis with arsenicals was difficult to achieve due to the requirement for prolonged therapy, which often led to patient noncompliance; the complexity and lack of standardized treatment regimens; the substantial toxicity, which could be fatal; and the lack of efficacy for neurosyphilis. In 1943, Mahoney et al. (1943) reported the curative effect of penicillin on early syphilis. Experiments were initially conducted with the rabbit model and then with a group of four patients who had dark-field positive primary syphilis. Parameters for the optimal penicillin regimen (i.e., dose, duration, and preparation) and timing of serologic follow-up were later refined through case series and experiences in civilian and military venereal disease clinics. It soon became apparent that *T. pallidum* was extremely sensitive to the action of penicillin and that this antibiotic could be used for treatment of patients with early and late syphilis, as well as infected pregnant patients and infants with congenital syphilis. In June 1944, the US Army adopted penicillin as the routine treatment for syphilis (Parascandola 2001). However, since penicillin had therapeutic value for treatment of war wounds, the limited supplies were mostly reserved for military use. In April 1945, as World War II was nearing an end, the US government lifted all restrictions on distribution of penicillin. Within a few years, the widespread use of penicillin for the treatment of all stages of syphilis, together with education and improved diagnostic tests, resulted in dramatic decreases in the US incidence of syphilis and its associated mortality. The Deputy Surgeon General of the US Public Health Service confidently forecast a "swift and final conquest of venereal disease by dosing every man, woman and child in the United States with penicillin" (Parascandola 2010). From 1944 to 1954, rates of reported cases of syphilis of any stage decreased by more than 75%, with the greatest declines seen in primary and secondary syphilis (Douglas 2009). Victory over syphilis was presumably at hand.

11.4 Syphilis Reemerges

In contrast to most bacteria that quickly acquired high-level resistance to penicillin, *T. pallidum* continued to remain sensitive to this antibiotic (Stamm 2001; Douglas 2009). However, despite the availability of an effective treatment, reported rates of primary and secondary syphilis in the USA increased in the early 1960s and the early to mid-1970s. Rates increased again in the early 1980s, mostly among men who have sex with men (MSM), and yet again in the late 1980s through 1990. The latter epidemic occurred primarily among heterosexuals and was strongly associated with crack cocaine use. More than 50,000 cases of primary and secondary syphilis were reported to the Centers for Disease Control and Prevention (CDC) in 1990, the highest number of cases since 1948. However, primary and secondary syphilis rates declined from 1991 to 2000, reaching their lowest rate since reporting began in 1941 (Centers for Disease Control and Prevention 2010). The historic low rates and geographic concentration of the majority of syphilis cases in the South and large urban areas led the CDC, in October 1999, to launch a *National Plan to Eliminate Syphilis*. Despite this action, syphilis rates began to increase in 2001, mostly among MSM, and have continued to do so. In 2009, nearly 14,000 cases of primary and secondary syphilis were reported to the CDC, representing a 3.7% increase from 2008 (Centers for Disease Control and Prevention 2010). Syphilis rates also increased in several other developed countries including Russia, where an explosive epidemic occurred in the 1990s (Tichonova et al. 1997; Simms et al. 2005; Velicko et al. 2008). After the near eradication of syphilis 50 years ago (Cohen et al. 1996), China experienced a precipitous increase in reported cases of adult and congenital syphilis, beginning in the late 1990s and continuing through 2009 (Chen et al. 2007; Tucker and Cohen 2011). Factors driving the reemergence of syphilis include: (1) demographic, behavioral, and social changes; (2) limited access to health care services; (3) reduced support for screening and control programs; and (4) treatment failure. Additionally, it has been proposed, based on computer modeling, that host immunity may be a factor in the perceived cyclical nature of syphilis epidemics (Grassly et al. 2007).

11.5 Macrolide-Resistant *T. pallidum*

In the absence of a vaccine, syphilis control is mainly dependent upon identification and treatment of infected individuals and their partners. The CDC Sexually Transmitted Diseases Treatment Guidelines 2010 recommend the use of penicillin G benzathine, administered intramuscularly (IM) as a single dose of 2.4 million units (MU), for the treatment of uncomplicated, early syphilis in adults (Centers for Disease Control and Prevention et al. 2010). This form of the drug provides treponemicidal levels of penicillin in the blood for weeks, although it does not efficiently cross the blood–brain barrier (Marra 2009; Mohr et al. 1976) (for further details on penicillin treatment of different stages of syphilis, see (Centers for Disease Control and Prevention et al. 2010)). There are no proven alternatives to penicillin for the treatment of infected pregnant women, thus those who are penicillin-allergic must be desensitized and then treated with penicillin G benzathine. The need for deep intramuscular injection of penicillin and the potential for severe allergic reactions prompted some clinicians to substitute second-line, oral antibiotics, including macrolides (e.g., erythromycin and azithromycin) and tetracyclines (e.g., tetracycline and doxycycline) as first-line drugs for the treatment of syphilis (Mabey 2009). This use of alternative antibiotics is inconsistent with CDC guidelines and is now further complicated by the emergence and spread of *T. pallidum* strains with clinically significant, high-level resistance to macrolides (Stamm 2010; Katz and Klausner 2008).

Macrolides are bacteriostatic antibiotics that inhibit bacterial protein synthesis by binding reversibly to 23S rRNA of the 50S ribosomal subunit (Vester and Douthwaite 2001). Shortly after introduction of the first macrolide, erythromycin, in the 1950s, resistance was observed in several bacterial pathogens (Roberts 2008). Failure of erythromycin treatment for syphilis was reported by South et al. (1964) and Fenton and Light (1976) in pregnant women who subsequently delivered infants with congenital syphilis. It is unknown if these treatment failures were due to erythromycin-resistant *T. pallidum* since erythromycin may not efficiently cross the placental barrier (Philipson et al. 1973). In 1977, *T. pallidum* Street strain 14 was isolated from a US patient with active lesions of secondary syphilis who failed long-term erythromycin therapy (Stapleton et al. 1985; Stamm et al. 1988). Using an in vitro assay to assess the effect of antibiotics on treponemal protein synthesis, Stamm and coworkers (Stamm et al. 1988; Stamm and Parrish 1990) showed that Street strain 14 is resistant to high levels of erythromycin and cross-resistant to azithromycin, a macrolide approved by the US Food and Drug Administration (FDA) in 1991. Studies with the rabbit model of syphilis confirmed that Street strain 14 is resistant to erythromycin and azithromycin (Lukehart et al. 2004) and that the macrolide-resistant phenotype is highly stable, despite multiple passages in rabbits in the absence of antibiotic pressure (Stamm 2010; Marra et al. 2006).

Macrolide resistance is usually due to mutation or methylation of the target site (i.e., the peptidyl transferase region in domain V of 23S rRNA) (Vester and Douthwaite 2001; Roberts 2008). Since *T. pallidum* lacks genetic elements (e.g., plasmids, bacteriophage, and transposons) that are usually associated with acquisition of antibiotic resistance (Fraser et al. 1998), it seemed probable that Street strain 14 macrolide resistance is due to a spontaneous, low frequency, chromosomal mutation in the 23S rRNA gene. Consistent with this hypothesis, Stamm and Bergen (Stamm and Bergen 2000) showed that an adenine (A) to guanine (G) transition, at the position cognate to A2058 in the *E. coli* 23S rRNA gene, is present in both copies of the Street strain 14 23S rRNA gene (Table 11.1) (Centurion-Lara et al. 1996). This mutation was not present in *T. pallidum* Nichols, the type strain, which was isolated in 1912 and is sensitive to macrolides. Sequencing of the Street strain 14 genome to high accuracy with oligonucleotide arrays confirmed the A to G transition in both 23S rRNA genes (Matejkova et al. 2008). Point mutations at position A2058 have been identified in several macrolide-resistant bacteria, including spirochetes such as *Brachyspira hyodysenteriae* and *B. pilosicoli*, the agents of swine diarrheal disease (Table 11.1) (Vester and Douthwaite 2001; Roberts 2008; Karlsson et al. 1999, 2004; Prapasarakul et al. 2003; Ross et al. 1997). Additionally, Lee et al. (2002) reported that an A2058G mutation, present in one or both copies of the *T. denticola* 23S rRNA gene, confers equivalent, high-level resistance to erythromycin (Table 11.1). Macrolide-resistant *T. denticola* isolates were obtained after in vitro exposure to erythromycin, suggesting that antibiotic pressure is sufficient to select mutants with a point mutation that confers high-level resistance.

Despite documentation of high-level erythromycin resistance in *T. pallidum* Street strain 14 (Stapleton et al. 1985; Stamm et al. 1988), there was considerable enthusiasm initially for the use of azithromycin for the treatment of syphilis. Clinical efficacy of azithromycin for the treatment of syphilis was demonstrated in nonrandomized studies (Katz and Klausner 2008; Kiddugavu et al. 2005; Vernon et al. 1994) and in randomized controlled trials in the USA (Hook et al. 1999, 2002, 2010), Africa (Riedner et al. 2005), China (Bai et al. 2008), and Madagascar (Hook et al. 2010; Van Damme et al. 2009) that compared cure rates for azithromycin and penicillin. Azithromycin was used for syphilis treatment in Uganda in the mid-1990s (Kiddugavu et al. 2005), in the USA (San Francisco, CA) in 1999 and 2000 (Mitchell et al. 2006), and in the USA (Los Angeles, CA) and Canada (Vancouver, B.C.) in 2000 (Chen et al. 2002; Rekart et al. 2003). However, treatment failures were observed in eight patients in San Francisco in 2002–2003 (Katz and Klausner 2008; Klausner et al. 2004). Molecular analysis of *T. pallidum* in clinical specimens from two of the patients revealed the presence of a 23S rRNA gene mutation identical to that of Street strain 14 (i.e., A2058G) (Lukehart et al. 2004). Retrospective analysis of clinical specimens from Baltimore, San Francisco, Seattle, and Vancouver showed that while prevalence of the A2058G mutation varied between sites, it increased significantly over time within those sites that were analyzed temporally (Katz and Klausner 2008; Lukehart et al. 2004; Morshed and Jones 2006). For example, the percentage of San Francisco specimens containing *T. pallidum* with the A2058G mutation increased from 4% in 2002 to 41% in 2003, 56% in 2004, and 76.5% in 2005 (Katz and Klausner 2008; Mitchell et al. 2006). Similarly, the percentage of Seattle specimens containing *T. pallidum* with the A2058G mutation increased from 9% in 2002 to 23% in 2003, 50% in 2004, and 56% in 2005 (Marra et al. 2006). Recent studies in Shanghai, China showed that the percentage of specimens containing *T. pallidum* with the A2058G mutation is nearly 100% (Table 11.2) (Zhou et al. 2010). While only a limited number of clinical specimens were analyzed in these studies, the rapid increase in prevalence of the A2058G mutation is disturbing, particularly since treatment failure extends the "window of opportunity" for an infectious patient to transmit syphilis to others.

Macrolide-resistant *T. pallidum* with the A2058G mutation is now present in several areas of the USA, Canada, Europe, and China (Table 11.2) (Lukehart et al.

Table 11.1 Spirochete macrolide resistance associated with 23S rRNA gene mutations

Spirochete	Host	Mutation position[a]	Resistance(s)[b]	Clinical/induced[c]	References[d]
T. pallidum	Human	A2058G	Ery, Azi	Clinical	Stamm and Bergen (2000)
		A2058G	Cla	Clinical	Woznicova et al. (2010)
		A2059G	Spi	Clinical	Matejkova et al. (2009)
T. denticola	Human	A2058G	Ery	Induced	Lee et al. (2002)
B. hyodysenteriae	Swine	A2058G	Ery, Cli, Tyl	Induced	Karlsson et al. (1999)
		A2058T	Ery, Cli, Tyl	Clinical	Karlsson et al. (1999)
B. pilosicoli	Swine	A2058G	Tyl	Induced	Prapasarakul et al. (2003)
		A2058T	Ery, Cli, Tyl	Clinical	Karlsson et al. (2004)
		A2059C	Ery, Cli, Tyl	Clinical	Karlsson et al. (2004)
		A2059G	Ery, Cli, Tyl	Clinical	Karlsson et al. (2004)
Brachyspira spp.	Canine	A2062C	Tyl	Induced	Prapasarakul et al. (2003)

[a]Based on the numbering of the *E. coli* 23S rRNA gene nucleotide sequence
[b]Based on the antibiotics tested, *Azi* azithromycin, *Cla* clarithromycin, *Cli* clindamycin, *Ery* erythromycin, *Spi* spiramycin, *Tyl* tylosin
[c]Macrolide resistance present in clinical/field isolate versus experimentally induced
[d]Initial description of mutation and/or resistance associated with mutation

2004; Marra et al. 2006; Mitchell et al. 2006; Morshed and Jones 2006; Zhou et al. 2010; Martin et al. 2009a, b, 2010; Matejkova et al. 2009; Katz et al. 2010; Woznicova et al. 2010). Several outbreaks appear to involve multiple strains in which macrolide resistance emerged independently, although there is some evidence for limited spread of macrolide resistant clones (Marra et al. 2006; Martin et al. 2009a, b; Martin et al. 2010; Katz et al. 2010). The findings of Marra et al. (2006) suggest that recent macrolide use for unrelated infections (e.g., oral, skin, genital, etc.) contributes to the increased prevalence of macrolide-resistant *T. pallidum*, presumably by providing a selective pressure. At this juncture, macrolide-resistant *T. pallidum* appears to be uncommon in some developing countries (e.g., Madagascar, Tanzania, Uganda) (Katz and Klausner 2008; Kiddugavu et al. 2005; Riedner et al. 2005; Van Damme et al. 2009) where macrolide use has been limited. However, because macrolide use is increasing in these areas, it is likely that macrolide-resistant *T. pallidum* will emerge or will be introduced from an outside source. Regardless of the setting, the use of macrolides for the treatment of syphilis must be undertaken with caution, particularly in areas where the prevalence of resistance is unknown. CDC 2010 treatment guidelines state that azithromycin should be used only when treatment with penicillin or doxycycline is not feasible and that azithromycin must not be used in MSM or pregnant women (Centers for Disease Control and Prevention et al. 2010). All syphilis patients who receive macrolides must be closely monitored both clinically and serologically. Specimens from patients who fail treatment should be subjected to molecular analysis for detection of mutations in the *T. pallidum* 23S rRNA gene. Though problematic in resource-limited settings, availability of a real-time PCR assay for detection of the A2058G mutation and, with modification, a new mutation (A2059G; see below) would enable molecular surveillance efforts for rapid identification of macrolide-resistant *T. pallidum* in areas where macrolide use has been limited and therapy might still be effective (Pandori et al. 2007). Monitoring for macrolide resistance is important to identify treatment failures and to efficiently target control programs that are based on the prevalence and geographical distribution of resistance.

Until recently, the A2058G mutation was the only documented mutation associated with *T. pallidum* macrolide resistance. In vitro studies by Stamm and coworkers (Stapleton et al. 1985; Stamm et al. 1988; Stamm and Parrish 1990) showed that this mutation confers resistance to 14-member (e.g., erythromycin, roxithryomycin, clarithromycin) and 15-member lactone ring macrolides (e.g., azithromycin) but not to 16-member lactone ring macrolides (e.g., spiramycin, midecamycin, tylosin), possibly due to the differential binding of these antibiotics to the 23S rRNA of Street strain 14. The presence of *T. pallidum* harboring the A2058G mutation in clinical specimens from patients who failed erythromycin, azithromycin, and clarithromycin therapy support in vitro observations of the resistance spectrum associated with this mutation (Table 11.2) (Stamm 2010; Lukehart et al. 2004; Mitchell et al. 2006; Morshed and Jones 2006; Zhou et al. 2010; Martin et al. 2009a, b, 2010; Matejkova et al. 2009; Katz et al. 2010; Woznicova et al. 2010). In 2009, Matejkova et al. (2009) reported identification of a new mutation, A2059G, in the 23S rRNA gene of *T. pallidum* in clinical specimens from a Czech Republic patient with secondary syphilis who failed spiramycin therapy (Tables 11.1 and 11.2). This mutation is associated with resistance to 14-, 15-, and 16-member lactone ring macrolides in a number of bacteria including *B. pilosicoli* and is likely responsible for spiramycin resistance in *T. pallidum* (Table 11.1) (Vester and Douthwaite 2001; Karlsson et al. 2004). The prevalence of the A2059G or the A2058G mutation among clinical specimens collected from syphilis patients in the Czech Republic is ~18% for either mutation (Matejkova et al. 2009). There was no direct epidemiological relationship between syphilis patients with the A2059G mutation, suggesting that they had not been infected with a single, macrolide-resistant strain. A recent report from London, UK of infections due to *T. pallidum* harboring the A2059G mutation, indicates that emergence of *T. pallidum* with the A2059G mutation is not unique to a specific geographical site (Tipple et al. 2011).

11.6 Potential for Development of Tetracycline Resistance

Currently, there are no documented reports of tetracycline resistance in *T. pallidum*. However, due to the emergence of clinically relevant macrolide resistance, it is important to consider how this spirochete might develop resistance to tetracycline, an alternative antibiotic currently approved by the CDC for the treatment of non-pregnant, penicillin-allergic adult patients with early syphilis. Awareness of potential

Table 11.2 Evaluation for *T. pallidum* 23S rRNA gene mutations in clinical specimens from syphilis patients at geographically diverse sites

Site	Mutation	Number with mutation/total amplifiable[a]	Number of strains[b]	References
US, San Francisco	A2058G	12/55	NR[c]	Lukehart et al. (2004)
San Francisco	A2058G	46/118	NR	Mitchell et al. (2006)
San Francisco	A2058G	42/62	3	Katz et al. (2010)
Seattle	A2058G	2/19	NR	Lukehart et al. (2004)
Seattle	A2058G	20/58	>2	Marra et al. (2006)
Baltimore	A2058G	3/23	NR	Lukehart et al. (2004)
Canada, Vancouver	A2058G	5/56	NR	Morshed and Jones (2006)
Alberta/NWT	A2058G	7/43	6	Martin et al. (2010)
Alberta	A2058G	4/14	NR	Martin et al. (2009b)
Ireland, Dublin	A2058G	15/17	NR	Lukehart et al. (2004)
Czech Republic	A2058G	4/22	NR	Matejkova et al. (2009)
	A2059G	4/22	NR	Matejkova et al. (2009)
	A2058G	1/1	1	Woznicova et al. (2010)
China, Shanghai	A2058G	126/132	NR	Zhou et al. (2010)
Shanghai	A2058G	4/4	NR	Zhou et al. (2010)
Shanghai	A2058G	38/38	5	Martin et al. (2009a)
Madagascar (3 sites)	None	0/141	NR	Van Damme et al. (2009)

[a]Only specimens with PCR-amplifiable *T. pallidum* DNA were tested for mutations
[b]Number of strains with a mutation based on molecular typing
[c]*NR* not reported

mechanisms of tetracycline resistance would enable a more focused approach to detecting mutations in *T. pallidum* that are known to be associated with bacterial tetracycline resistance. Identification of the genetic basis of resistance is a prerequisite for developing molecular methods (e.g., real-time PCR assays, oligonucleotide arrays, pyrosequencing) for rapid detection and surveillance of antibiotic-resistant *T. pallidum* in clinical specimens. Presumably, such information could aid clinicians in their choice of antibiotic for syphilis treatment.

Tetracyclines are bacteriostatic antibiotics that inhibit protein synthesis by binding reversibly to 16S rRNA of the 30S ribosomal subunit (Roberts 2005). Four small-scale studies demonstrated that doxycycline, a tetracycline derivative with better bioavailability, a more convenient dosing schedule, and fewer gastrointestinal side effects than tetracycline, has a success rate similar to that of penicillin for the treatment of early adult syphilis (Ghanem et al. 2006; Harshan and Jayskumar 1982; Onoda 1979; Wong et al. 2008). Although no treatment failures were reported for three of the four studies, one patient in the Onoda (Onoda 1979) study did not show serological evidence of response to treatment. While it is unclear why this patient, who was later lost to follow-up, was nonresponsive, treatment failure due to doxycycline resistance is a possibility. Tetracycline resistance, which confers cross-resistance to doxycycline, can be due to efflux systems, ribosomal protection proteins, mutation of certain ribosomal proteins, or enzymatic inactivation of the antibiotic (Roberts 2005). Additionally, single-point mutations in the 16S rRNA gene of *H. pylori* and *Propionibacterium* at residues cognate to positions 965–967 or 1,058, respectively, in the *E. coli* 16S rRNA gene, confer tetracycline or doxycycline resistance in these bacteria (Wu et al. 2005; Ross et al. 1998). Pringle et al. (Pringle et al. 2007) hypothesized that point mutations in the 16S rRNA gene of spirochetes, which have just one or two copies of this gene, could result in decreased susceptibility to doxycycline, as they observed with *B. hyodysenteriae* (i.e., G1058C). Thus, point mutations in one or both copies of the *T. pallidum* 16S rRNA gene (Centurion-Lara et al. 1996) are a potential mechanism for the development of doxycycline resistance. Due to the emergence of macrolide-resistant *T. pallidum*, it is likely that doxycycline will become more widely used as an alternative drug for syphilis. Since treatment for early adult syphilis requires 100 mg of doxycycline taken orally twice daily for 14 days, patient compliance is an issue (Centers for Disease Control and Prevention et al. 2010). If treatment failure occurs, analysis of the 16S rRNA gene of *T. pallidum* in clinical specimens from patients who have been compliant should be a high priority to determine if mutations linked to doxycycline resistance are present.

11.7 Conclusions and Future Directions

Over five centuries after its emergence in Europe, syphilis is still with us, appearing as sporadic outbreaks (e.g., in the USA, Canada, and Europe) and as widespread epidemics (e.g., in Russia and China) (Stamm 2010). Today, syphilis is of particular concern since it significantly increases the risk of acquiring and transmitting human immunodeficiency virus (HIV) infection, which can be fatal if untreated. Introduction of penicillin in the 1940s revolutionized syphilis therapy and led to predictions, albeit premature, of elimination or possibly eradication of syphilis. Unlike most bacteria, *T. pallidum* subsp. *pallidum*, the syphilis agent, has not developed resistance to penicillin, an antibiotic that has remained the first-line drug for the treatment of all stages of syphilis. Emergence of penicillin resistance, while not impossible, appears unlikely since it would require multistep mutational changes in a bacterium that lacks the ability to acquire genes that encode penicillin resistance (Stamm 2010). Nevertheless, *T. pallidum* has a capacity for genetic change as demonstrated by the emergence of high-level macrolide resistance due to a single-point mutation (i.e., A2058G or A2059G) in the 23S rRNA gene. The effect of this mutation on the prevalence of syphilis is currently unknown. However, as observed with other antibiotic-resistant bacteria, treatment failure permits the persistence of infectious organisms in host tissues and thus enables the spread of infection by extending the "window of opportunity" for transmission.

At this juncture, macrolide resistance is the only clinically significant resistance that has developed in *T. pallidum* (Stamm 2010; Stamm et al. 2001). It is unknown if resistance will also emerge to doxycycline, the only oral antibiotic approved by the CDC as an alternative to penicillin for the treatment of early adult syphilis in nonpregnant patients. Due to movement away from macrolides for syphilis treatment, use of doxycycline is likely to increase. Interestingly, Wilson et al. (Wilson et al. 2011) recently proposed the prophylactic use of doxycycline as a public health intervention for mitigating syphilis epidemics. They are developing a multi-country chemoprophylaxis trial among HIV-infected gay men who engage in high-risk behavior to evaluate the impact of doxycycline on syphilis incidence. Although the length of doxycycline prophylaxis was not stated, presumably it would be of longer duration than that of treatment for early syphilis (i.e. 14 days). Poor compliance to long-term chemoprophylaxis, or to short-term treatment, would result in subtherapeutic concentrations of doxycycline in tissues, potentially acting as a selective pressure for doxycycline resistant mutants. If doxycycline-resistant *T. pallidum* emerges and becomes widespread, it could affect syphilis control.

Addressing the reemergence of syphilis and the emergence of macrolide-resistant *T. pallidum* requires effort on several fronts. First, since penicillin administration can be problematic, especially in developing countries, there is a need for new, single-dose, oral antibiotics to ensure patient compliance during syphilis treatment. Novel macrolides are in development, which are active against bacteria that harbor 23S rRNA gene mutations at positions A2058 and A2059. These macrolides merit investigation in the rabbit model of syphilis since it is conceivable that they could circumvent the current problem of azithromycin-resistant *T. pallidum*. Second, the availability and reliability of rapid point-of-care (POC) diagnostic tests must be improved. Castro et al. (2010) described a novel POC immunochromatographic test for simultaneous detection of nontreponemal and treponemal antibodies in sera of patients with syphilis. This dual platform, which acts as both a screening and confirmatory test, is designed to be used in primary health care clinics and resource-limited settings with serum, plasma, or whole blood. Rapid POC tests have the potential to improve treatment rates, particularly for patients who may not return for lab-based test results, thus interrupting transmission (Tucker et al. 2010; Sena et al. 2010). Third, molecular surveillance for detection of antibiotic-resistant *T. pallidum* must be expanded to enable a better understanding of the prevalence and geographic distribution of this problem. Additionally, further development of typing methods for *T. pallidum* is needed to aid in tracking movement of resistant strains (Marra et al. 2010). Fourth, efforts to develop a vaccine to prevent syphilis and thereby limit dependence on antibiotics should be renewed. Enhanced knowledge of *T. pallidum* proteogenomics, coupled with availability of novel strategies to modulate host innate and adaptive immune responses, suggest that this objective has a greater likelihood of success than in the past (Fraser et al. 1998; Cullen and Cameron 2006). Advances on these fronts (i.e., treatment, diagnostics, surveillance, and vaccines) would build on those of early twentieth century syphilis research and, if combined with a rapid, multifaceted approach for outbreak containment, could greatly accelerate control efforts, thus paving the way for syphilis elimination.

References

Bai ZG, Yang KH, Liu YL, Tian JH, Ma B, Mi DH, Jiang L, Tan JY, Gai QY (2008) Azithromycin vs. benzathine penicillin G for early syphilis: a meta-analysis of randomized clinical trials. Int J STD AIDS 19:217–221

Baker BJ, Armelagos GJ (1988) The origin and antiquity of syphilis: paleopathological diagnosis and interpretation. Curr Anthropol 29:703–738

Bialynicki-Birula R (2008) The 100th anniversary of Wassermann-Neisser-Bruck reaction. Clin Dermatol 26:79–88

Bosch F, Rosich L (2008) The contributions of Paul Ehrlich to pharmacology: a tribute on the occasion of the centenary of his Nobel prize. Pharmacology 82:171–179

Castro AR, Esfandiari J, Kumar S, Ashton M, Kikkert SE, Park MM, Ballard RC (2010) Novel point-of-care test for simultaneous detection of nontreponemal and treponemal antibodies in patients with syphilis. J Clin Microbiol 48:4615–4619

Centers for Disease Control and Prevention (2010) Sexually transmitted disease surveillance, 2009. U.S. Dept. Health and Human Services, Atlanta, GA

Centers for Disease Control and Prevention, Workowski KA, Berman S (2010) Sexually transmitted diseases treatment guidelines. MMWR Recomm Rep 59(RR-12):1–110

Centurion-Lara A, Castro C, van Voorhis WC, Lukehart SA (1996) Two 16S-23S ribosomal DNA intergenic regions in different *Treponema pallidum* subspecies contain tRNA genes. FEMS Microbiol Lett 143:235–240

Chargin L, Leifer W (1956) Massive dose arsenotherapy of early syphilis by intravenous drip. Arch Dermatol 73:482–484

Chen JL, Callahan DB, Kerndt PR (2002) Syphilis control among incarcerated men who have sex with men: public health response to an outbreak. Am J Public Health 92:1473–1474

Chen ZQ, Zhang GC, Gong XD, Lin C, Gao X, Liang GJ, Yue XL, Chen XS, Cohen MS (2007) Syphilis in China: results of a national surveillance programme. Lancet 369:132–138

Cohen MS, Henderson GE, Aiello P, Zheng H (1996) Successful eradication of sexually transmitted diseases in the People's Republic of China: implications for the 21st century. J Infect Dis 174(suppl 2):S223–S229

Cullen PA, Cameron CE (2006) Progress towards an effective syphilis vaccine: the past, present and future. Expert Rev Vaccines 5:67–80

de Melo FL, de Melo JC, Fraga AM, Nunes K, Eggers S (2010) Syphilis at the crossroads of phylogenetics and paleopathology. PLoS Negl Trop Dis 4:e575

Douglas JM Jr (2009) Penicillin treatment of syphilis—clearing away the shadow on the land. JAMA 301:769–771

Fenton LJ, Light IJ (1976) Congenital syphilis after maternal treatment with erythromycin. Obstet Gynecol 47:492–494

Fraser CM, Norris SJ, Weinstock GM, White O, Sutton GG, Dodson R, Gwinn M, Hickey EK, Clayton R, Ketchum KA, Sodergren E, Hardham JM, McLeod MP, Salzberg S, Peterson J, Khalak H, Richardson D, Howell JK, Chidambaram M, Utterback T, McDonald L, Artiach P, Bowman C, Cotton MD, Fujii C, Garland S, Hatch B, Horst K, Roberts K, Sandusky M, Weidman J, Smith HO, Venter JC (1998) Complete genome sequence of *Treponema pallidum*, the syphilis spirochete. Science 281:375–388

Frazen C (2008) Syphilis in composers and musicians—Mozart, Beethoven, Paganini, Schubert, Schumann, Smetana. Eur J Clin Microbiol Infect Dis 27:1151–1157

Ghanem KG, Erbelding EJ, Cheng WW, Rompalo AM (2006) Doxycycline compared with benzathine penicillin for treatment of early syphilis. Clin Infect Dis 42:e45–e49

Grassly NC, Fraser C, Garnett GP (2007) Host immunity and synchronized epidemics of syphilis across the United States. Nature 433:417–421

Harshan V, Jayskumar W (1982) Doxycycline in early syphilis: a long term follow up. Indian J Dermatol 27:119–124

Hook EW III, Stephens J, Ennis DM (1999) Azithromycin compared with penicillin G benzathine for treatment of incubating syphilis. Ann Intern Med 131:434–437

Hook EW III, Martin DH, Stephens J, Smith BS, Smith K (2002) A randomized, comparative pilot study of azithromycin versus benzathine penicillin G for treatment of early syphilis. Sex Transm Dis 29:486–490

Hook EW III, Behets F, Van Damme K, Ravelomanana N, Leone P, Sena AC, Martin D, Langley C, McNeil L, Wolff M (2010) A phase III equivalence trial of azithromycin versus benzathine penicillin for treatment of early syphilis. J Infect Dis 201:1729–1735

Karlsson M, Fellstrom C, Heldtander MU, Johansson KE, Franklin A (1999) Genetic basis of macrolide and lincosamide resistance in *Brachyspira* (*Serpulina*) *hyodysenteriae*. FEMS Microbiol Lett 172:255–260

Karlsson M, Fellstrom C, Johansson KE, Franklin A (2004) Antimicrobial resistance in *Brachyspira pilosicoli*, with special reference to point mutations in 23S rRNA gene associated with macrolide and lincosamide resistance. Microb Drug Resist 10:204–208

Katz KA, Klausner JD (2008) Azithromycin resistance in *Treponema pallidum*. Curr Opin Infect Dis 21:83–91

Katz KA, Pillay A, Ahrens K, Kohn RP, Hermanstyne K, Bernstein KT, Ballard RC, Klausner JD (2010) Molecular epidemiology of syphilis—San Francisco, 2004–2007. Sex Transm Dis 37:660–663

Kiddugavu MG, Kiwanuka N, Wawer MJ, Serwadda D, Sewankambo NK, Wabwire-Mangen F, Makumbi F, Li X, Reynollds SJ, Quinn TC, Gray RH, the Rakai Study Group (2005) Effectiveness of syphilis treatment using azithromycin and/or benzathine penicillin in Rakai, Uganda. Sex Transm Dis 32:1–6

Klausner JD, Engelman J, Lukehart SA, Berman SM, Mitchell SJ (2004) Brief report: azithromycin treatment failures in syphilis infections—San Francisco, California, 2002–2003. MMWR 53:197–198

Knell RJ (2004) Syphilis in Renaissance Europe: rapid evolution of an introduced sexually transmitted disease. Proc R Soc Lond B 271:S174–S176

Kohler W (2007) Zentrablatt fur Bacteriologie—100 years ago: the detection of the causative agent of syphilis, *Treponema pallidum*—the year after. Int J Med Microbiol 297:127–132

Lee SY, Ning Y, Fenno JC (2002) 23S rRNA point mutation associated with erythromycin resistance in *Treponema denticola*. FEMS Microbiol Lett 207:39–42

Lukehart SA, Gordones C, Molini BJ, Sonnett P, Hopkins S, Mulcahy F, Engelman J, Mitchell SJ, Rompalo AM, Marra CM, Klausner JD (2004) Macrolide resistance in *Treponema pallidum* in the United States and Ireland. N Engl J Med 351:154–158

Mabey D (2009) Azithromycin resistance in *Treponema pallidum*. Sex Transm Dis 36:777–778

Mahoney JF, Arnold RC, Harris A (1943) Penicillin treatment of early syphilis. A preliminary report. Am J Public Health Nations Health 33:1387–1391

Marra CM (2009) Update on neurosyphilis. Curr Infect Dis Rep 11:127–134

Marra CM, Colina AP, Godornes C, Tantalo LC, Puray M, Centurion-Lara A, Lukehart SA (2006) Antibiotic selection may contribute to increases in macrolide-resistant *Treponema pallidum*. J Infect Dis 194:1771–1773

Marra CM, Sahi SK, Tantalo LC, Gordones C, Reid T, Behets F, Rompalo A, Klausner JD, Yin YP, Mulcahy F, Golden MR, Centurion-Lara A, Lukehart SA (2010) Enhanced molecular typing of *Treponema pallidum*: geographical distribution of strain types and association with neurosyphilis. J Infect Dis 202:1380–1388

Martin IE, Gu W, Yang Y, Tsang RS (2009a) Macrolide resistance and molecular types of *Treponema pallidum* causing primary syphilis in Shanghai, China. Clin Infect Dis 49: 515–521

Martin IE, Tang RS, Sutherland K, Tilley P, Read R, Anderson B, Roy C, Singh AE (2009b) Molecular characterization of syphilis in patients in Canada: azithromycin resistance and detection of *Treponema pallidum* DNA in whole-blood samples versus ulcerative swabs. J Clin Microbiol 47:1668–1673

Martin IE, Tang RSW, Sutherland K, Anderson B, Read R, Roy C, Yanow S, Fonseca K, White W, Kandola K, Kouadjo E, Singh AE (2010) Molecular typing of *Treponema pallidum* strains in Western Canada: predominance of 14d subtypes. Sex Transm Dis 37:544–548

Matejkova P, Strouhal M, Smajs D, Norris SJ, Palzkill T, Petrosino JF, Sodergren E, Norton JE, Singh J, Richmond TA, Molla MN, Albert TJ, Weinstock GM (2008) Complete genome sequence of *Treponema pallidum* ssp. *pallidum* strain SS14 determined with oligonucleotide arrays. BMC Microbiol 8(76)

Matejkova P, Flasarova M, Zakoucka H, Borek M, Kremenova S, Arenberger P, Woznicova V, Weinstock GM, Smajs D (2009) Macrolide treatment failure in a case of secondary syphilis: a novel A2059G mutation in 23S rRNA gene of *Treponema pallidum* subsp. *pallidum*. J Med Microbiol 58:832–836

Mitchell SJ, Engelman J, Kent CK, Lukehart SA, Gordones C, Klausner JD (2006) Azithromycin-resistant syphilis infection: San Francisco, California, 2000–2004. Clin Infect Dis 42: 337–345

Mohr JA, Griffths W, Jackson R, Saadah H, Bird P, Riddle J (1976) Neurosyphilis and penicillin levels in cerebrospinal fluid. JAMA 236:2208–2209

Morens DM, Folkers GK, Fauci AS (2008) Emerging infections: a perpetual challenge. Lancet Infect Dis 8:710–719

Morshed MG, Jones HD (2006) *Treponema pallidum* macrolide resistance in BC. CMAJ 174:349

O'Shea JG (1990) 'Two minutes with venus, two years with mercury'—mercury as an antisyphilitic chemotherapeutic agent. J R Soc Med 83:392–395

Onoda Y (1979) Therapeutic effect of oral doxycycline on syphilis. Br J Vener Dis 55:110–115

Pandori MW, Gordones C, Castro L, Engelman J, Siedner M, Lukehart S, Klausner J (2007) Detection of azithromycin resistance in *Treponema pallidum* by real-time PCR. Antimicrob Agents Chemother 51:3425–3430

Parascandola J (2001) John Mahoney and the introduction of penicillin to treat syphilis. Pharm Hist 43:3–13

Parascandola J (2009) From mercury to miracle drugs: syphilis therapy over the centuries. Pharm Hist 51:14–23

Parascandola J (2010) The fall and rise of syphilis in America since World War II. MD Advis 3:22–35

Philipson A, Sabath LD, Charles D (1973) Transplacental passage of erythromycin and clindamycin. N Engl J Med 288:1219–1221

Prapasarakul N, Ochi K, Adachi Y (2003) In vitro susceptibility and a new point mutation associated with tylosin-resistance in Japanese canine intestinal spirochetes. J Vet Sci 65:1275–1280

Pringle M, Fellstrom C, Johansson KE (2007) Decreased susceptibility to doxycycline associated with a 16S rRNA gene mutation in *Brachyspira hyodysenteriae*. Vet Microbiol 123:245–248

Rekart ML, Patrick DM, Chakraborty B, Maginley JJL, Jones HD, Bajdik CD, Pourbohloul B, Brunham RC (2003) Target mass treatment for syphilis with oral azithromycin. Lancet 361:313–314

Riedner G, Rusizoka M, Todd J, Maboko L, Hoelscher M, Mmbando D, Samky E, Lyamuya E, Mabey D, Grosskurth H, Hayes R (2005) Single-dose erythromycin versus penicillin G benzathine for the treatment of early syphilis. N Engl J Med 353:1236–1244

Roberts MC (2005) Update on acquired tetracycline resistance genes. FEMS Microbiol Lett 245:195–203

Roberts MC (2008) Update on macrolide-lincosamide-streptogramin, ketolides, and oxazolidinone resistance genes. FEMS Microbiol Lett 282:147–159

Ross JI, Eady EA, Cove JH, Jones CE, Ratyal AH, Miller YW, Vyakrnam S, Cunliffe WJ (1997) Clinical resistance to erythromycin and clindamycin in cutaneous propionibacteria isolated from acne patients is associated with mutations in 23S rRNA. Antimicrob Agents Chemother 41:1162–1165

Ross JI, Eady EA, Cove JH, Cunliffe WJ (1998) 16S rRNA mutation associated with tetracycline resistance in a Gram-positive bacterium. Antimicrob Agents Chemother 42:1702–1705

Rothschild BM (2005) History of syphilis. Clin Infect Dis 40:1454–1463

Sena AC, White BL, Sparling PF (2010) Novel *Treponema pallidum* serologic tests: a paradigm shift in syphilis screening for the 21st century. Clin Infect Dis 51:700–708

Simms I, Fenton KA, Ashton M, Turner KM, Crawley-Boevey EE, Gorton R, Thomas DR, Lynch A, Winter A, Fisher MJ, Lighton L, Maguire HC, Solomou M (2005) The re-emergence of syphilis in the United Kingdom: the new epidemic phases. Sex Transm Dis 32:220–226

South MA, Short DH, Knox JM (1964) Failure of erythromycin estolate therapy in in utero syphilis. JAMA 190:70–71

Stamm LV (2001) Chapter 85—*Treponema pallidum*. In: Sussman M (ed) Molecular medical microbiology, 1st edn. Academic press, London, UK, pp 1795–1808

Stamm LV (2010) Global challenge of antibiotic-resistant *Treponema pallidum*. Antimicrob Agents Chemother 54:583–589

Stamm LV, Bergen HL (2000) A point mutation associated with bacterial macrolide resistance is present in both 23S rRNA genes of an erythromycin-resistant *Treponema pallidum* clinical isolate. Antimicrob Agents Chemother 44:806–807

Stamm LV, Parrish EA (1990) In-vitro activity of azithromycin and CP-63,956 against *Treponema pallidum*. J Antimicrob Chemother 25(suppl A):S11–S14

Stamm LV, Stapleton JT, Bassford PJ Jr (1988) In vitro assay to demonstrate high-level erythromycin resistance in a clinical isolate of *Treponema pallidum*. Antimicrob Agents Chemother 32:164–169

Stamm LV, Bergen HL, Shangraw KA (2001) Natural rifampin resistance in *Treponema* spp. correlates with presence of N531 in RpoB rif cluster I. Antimicrob Agents Chemother 45:2973–2974

Stapleton JT, Stamm LV, Bassford PJ Jr (1985) Potential for development of antibiotic resistance in pathogenic treponemes. Rev Infect Dis 7(Suppl 2):S314–S317

Tichonova L, Borisenko K, Ward H, Meheus A, Gromyko A, Renton A (1997) Epidemics of syphilis in the Russian federation: trends, origins, and priorities for control. Lancet 350:210–213

Tipple C, McClure MO, Taylor GP (2011) High Prevalence of macrolide resistant *Treponema pallidum* strains in London centre. Sex Transm Infact 87:486–488

Tognotii E (2009) The rise and fall of syphilis in Renaissance Europe. J Med Humanit 30:99–113

Tucker JD, Cohen MS (2011) China's syphilis epidemic: epidemiology, proximate determinants of spread, and control responses. Curr Opin Infect Dis 24:50–55

Tucker JD, Biu J, Brown LB, Yin YP, Chen XS, Cohen M (2010) Accelerating worldwide syphilis screening through rapid testing: a systematic review. Lancet Infect Dis 10:381–386

Van Damme K, Behets F, Ravelomanana N, Godornes C, Khan M, Randrianasolo B, Rabenja NL, Lukehart S, Cohen M, Hook E (2009) Evaluation of azithromycin resistance in *Treponema pallidum* specimens from Madagascar. Sex Transm Dis 36:775–776

Velicko I, Arneborn M, Blaxhult A (2008) Syphilis epidemiology in Sweden: re-emergence since 2000 primarily due to spread among men who have sex with men. Euro Surveill 13(50):pii: 19063

Vernon MS, Handsfield HH, Johnson RB (1994) Pilot study of azithromycin for treatment of primary and secondary syphilis. Clin Infect Dis 19:486–488

Vester B, Douthwaite S (2001) Macrolide resistance conferred by base substitutions in 23S rRNA. Antimicrob Agents Chemother 45:1–12

Waugh MA (1982) Role played by Italy in the history of syphilis. Br J Vener Dis 58:92–95

Waugh M (2005) The centenary of *Treponema pallidum*: on the discovery of *Spirochaeta pallida*. Int J STD AIDS 16:594–595

Wilson DP, Prestage GP, Gray RJ, Hoare A, McCann P, Down I, Guy RJ, Drummond F, Klausner JD, Donovan B, Kaldor JM (2011) Chemopropylaxis is likely to be acceptable and could mitigate syphilis epidemics among populations of gay men. Sex Transm Dis 38(7):573–579

Wong T, Singh AE, De P (2008) Primary syphilis: serological treatment response to doxycycline/ tetracycline versus benzathine penicillin. Am J Med 121:903–908

Woznicova V, Matejkova P, Flasarova M, Zakoucka H, Dastychova E, Valisova Z, Smajs D (2010) Clarithromycin treatment failure due to macrolide resistance in *Treponema pallidum* in a patient with primary syphilis. Acta Derm Venereol 90:206–207

Wright AD (1971) Venereal disease and the great. Br J Vener Dis 47:295–306

Wu JY, Kim JJ, Reddy R, Wang WM, Graham DY, Kwon DH (2005) Tetracycline-resistant clinical *Helicobacter pylori* isolates with and without mutations in 16S rRNA-encoding genes. Antimicrob Agents Chemother 49:578–583

Zhou P, Li K, Lu H, Qian Y, Gu X, Gong W, Tucker JD, Cohen MS (2010) Azithromycin treatment failure among primary and secondary syphilis patients in Shanghai. Sex Transm Dis 37: 726–729

Chapter 12
Borrelia burgdorferi Persistence Post-antibiotic Treatment

Monica E. Embers and Stephen W. Barthold

12.1 Introduction

A basic precept of *Borrelia burgdorferi* biology is that persistent infection is the rule, not the exception, in immunocompetent hosts. Persistence occurs in many reservoir and incidental hosts of *B. burgdorferi*, and has been unequivocally proven experimentally in *Peromyscus leucopus* mice (Schwan et al. 1988), laboratory mice (Barthold et al. 1993a), rats (Moody et al. 1990), hamsters (Goodman et al. 1991), gerbils (Preac-Mursic et al. 1990), guinea pigs (Sonnesyn et al. 1993), dogs (Straubinger et al. 1997), and nonhuman primates (Roberts et al. 1995a). There is no reason to believe that humans are different, and indeed there are a number of documented reports confirming persistent infection in humans by both culture (Asbrink and Hovmark 1985; Kuiper et al. 1994; Maraspin et al. 2011; Miklossy et al. 2004; Snydman et al. 1986; Stanek et al. 1990; Strle et al. 1995) and polymerase chain reaction (PCR) (Bradley et al. 1994; Frazer et al. 1992; Moter et al. 1994; Nocton et al. 1994a; von Stedingk et al. 1995). *Borrelia burgdorferi* has evolved to persist in its varied hosts as a survival strategy for maintaining its natural 2-year life cycle through its hosts and vector tick stages. That life cycle requires survival of *B. burgdorferi* in reservoir hosts over winter, since there is no transovarial transmission from female ticks to larvae. Uninfected eggs are laid in the spring, and larvae emerge during the summer, when they feed upon persistently infected reservoir hosts to acquire infection and subsequently pass infection through the life stages (transstadial transmission) to nymphs and adults (Anderson and

M.E. Embers, Ph.D. (✉)
Division of Bacteriology and Parasitology, Tulane National Primate Research Center,
Tulane University Health Sciences, 18703 Three Rivers Rd., Covington, LA 70433
e-mail: members@tulane.edu

S.W. Barthold, D.V.M., Ph.D.
Department of Pathology, Microbiology and Immunology Center of Comparative Medicine,
School of Veterinary Medicine, University of California, Davis, CA 95616

M.E. Embers (ed.), *The Pathogenic Spirochetes: Strategies for Evasion of Host Immunity and Persistence*, DOI 10.1007/978-1-4614-5404-5_12,

Magnarelli 1993). Without the essential role of persistently infected reservoir hosts, the life cycle of *B. burgdorferi* would fail. Mechanisms of *B. burgdorferi* immune evasion, dormancy, and activation in hosts and vectors are among the least studied, but most critically needed areas for research.

Lyme borreliosis is a multisystem disease of human and other incidental hosts. Borrelia spirochetes utilize an array of adhesins that engage virtually every component of the extracellular matrix to facilitate their dissemination and persistence within connective tissue of hosts (Brissette et al. 2009; Cabello et al. 2007; Grab et al. 1998; Guo et al. 1995; Probert et al. 1995; Verma et al. 2009; Zambrano et al. 2004). Dissemination is also facilitated by bacteremia during early infection, which is generally cleared during the immune phase of infection and is intermittent thereafter. In contrast to small rodents, which develop generalized infections, humans experience localized infections. The clinical progression of disease may be divided into phases reflecting tissue colonization. The early-localized phase is manifested as focal erythema migrans (EM), which spreads centrifugally from the tick feeding site. The early disseminated phase results from multifocal infection of other tissues through bacteremia. Subjective symptoms such as fatigue, myalgia, and arthralgia may follow disseminated infection, but additional signs aid in clinical diagnosis. Early disseminated infection may result in pauciarticular arthritis, secondary erythema migrans, and cardiac conduction abnormalities (carditis), whereas late disseminated disease may induce peripheral neuropathy, meningitis, and other objective clinical signs.

These clinical signs of Lyme borreliosis in humans can be ephemeral, with "spontaneous" resolution (without antibiotic treatment) (Steere 2001). Studies in animal models have shown similar resolution of inflammation (arthritis and carditis) that is mediated by the acquired humoral immune response of the infected host. In the mouse model, passive transfer of immune serum from persistently infected, immunocompetent mice into immunodeficient severe combined immunodeficient (SCID) mice with ongoing infection and progressive disease results in generalized reduction (but not elimination) of spirochete burdens in all tissues and resolution of inflammatory lesions in joints and hearts (Barthold et al. 1996a). In studies using genetically engineered mice, it has been shown that these events can occur in the complete absence of T cells, underscoring the importance of antibody in mediating the host response (McKisic and Barthold 2000). Examination of tissues during the persistent phase of infection reveals spirochetes sequestered within bundles of collagen (Fig. 12.1), suggesting a unique mechanism of immune evasion (Barthold et al. 2006). Collagen sequestration during persistent infection is maintained by antibody. In SCID mice, decay of passively transferred immune serum results in resurgence of spirochetes, renewed bacteremia, and inflammation (SWB, unpublished). If there can be ascribed a purpose to *B. burgdorferi's* behavior during its sojourn within mammalian hosts, it is to persist within the highly collagenous dermis with minimal harm to the host and to await a vector to acquire infection upon feeding. Involvement of other tissues is simply an incidental event that favors neither the host nor the bacterium.

This chapter provides an overview of antibiotic treatment studies in animal models (dogs, rodents, and nonhuman primates). Studies that relied upon culture as the only index of treatment efficacy have given way to more sensitive polymerase chain reaction (PCR) assays for nucleic acid amplification to determine infection status. PCR technology has evolved with increasing sensitivity over the last two decades.

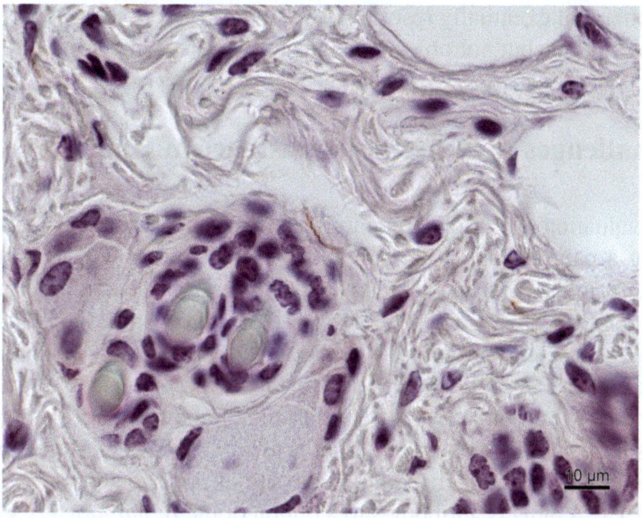

Fig. 12.1 Dermis of a mouse infected with *B. burgdorferi*. Note spirochetes sequestered within collagen bundles, in the absence of inflammation. This figure has not been published

With current detection technology, including real-time quantitative PCR, studies are revealing persistence of PCR-positive, noncultivable *B. burgdorferi* in tissues of animals for months following treatment with various classes of antibiotics. The medical significance of persisting noncultivable spirochetes following treatment, which occurs under experimental conditions in a wide range of animal hosts, compels the need for further investigation.

12.2 Posttreatment Lyme Disease Syndrome

The Infectious Diseases Society of America (IDSA) has issued guidelines for antimicrobial therapy of Lyme borreliosis (Wormser et al. 2006). The signs and symptoms of disease are ordinarily ameliorated with these accepted antimicrobial therapy regimens. However, some patients continue to have persistent subjective complaints (Asch et al. 1994; Shadick et al. 1994), while some patients fail outright to respond to antibiotic therapy, as made evident by signs of persistent infection (Wormser et al. 2006; Steere et al. 1994). The response to treatment in patients with late disseminated disease is often slower (Asch et al. 1994) and sometimes remains incomplete.

Posttreatment Lyme disease syndrome (PTLDS) is sometimes used to refer to the condition that occurs in some patients after treatment for Lyme borreliosis. The etiology of PTLDS is unknown but prolonged antibiotic therapy has not been experimentally proven to aid symptoms (Klempner et al. 2001). Several possible explanations for PTLDS may be put forth: (1) symptoms are caused by residual immunological responses, possibly autoimmune (Bolz and Weis 2004), despite the clearance of infection; (2) dead or attenuated organisms continue to trigger sporadic symptoms that will eventually resolve; or (3) active infection ensues at a low level, triggering symptoms but not objective signs of persistent infection.

12.3 Challenges to Studying Persistence in Humans

Objective evaluation of this phenomenon in humans is complicated by the difficulty in obtaining a patient population suitable for such studies. Clearly, only patients who have unequivocally been infected with *B. burgdorferi* could be included and those with confirmed Lyme borreliosis treated only post-dissemination are rare. Human study is further confounded by the vague, nonspecific symptoms with which patients present in PTLDS. The criteria for inclusion of patients into the PTLDS category are not well-defined; exposure duration and treatment regimens may vary tremendously. In addition, reliable measures to definitively determine that infection has been cleared from Lyme disease patients are difficult to attain.

The C6 ELISA detects antibodies to a region of the *B. burgdorferi* VlsE lipoprotein (located on plasmid 28-1) that is immunogenic in infected individuals and common to all infectious variants. Not only is the C6 test among the most reliable in terms of accuracy, but it has also been used as a predictor of treatment outcome (Philipp et al. 2005). In most patients with posttreatment Lyme disease syndrome (PTLDS), anti-C6 titers were found to generally persist at a low level compared to acute patient titers (Fleming et al. 2004). Thus, whether the C6 test indicates active infection in PTLDS patients remains to be explored.

In a recent review, it was stated that "in the treatment of other infections it is probably unrealistic to expect that antimicrobial therapy *per se* will eliminate every single microorganism from an infected host, and moreover, such an action would rarely if ever be required for a successful outcome…the role of antimicrobial therapy in vivo can be thought of in terms of "tipping the balance" in favor of the host's own defenses against a particular pathogen" (Wormser and Schwartz 2009). This may be true for "other infections" but when treating for *B. burgdorferi*, which persists in its fully immunocompetent hosts as the rule of its natural biology, "tipping the balance" may be problematic. The random, multifocal nature of human infection, the ephemeral clinical signs, the myriad of subjective symptoms, the impracticality of culture or PCR in clinical settings, the insensitivity of culture, and the retrospective nature of serology all make objective diagnostic criteria for testing outcome of antibiotic treatment in humans simply impossible. That is not the case with animal models.

12.4 Antibiotic Studies in Rodent Models Based Upon Culture

In studies based upon culture as the index of response, a number of *B. burgdorferi*-infected rodent models, including mice (Kazragis et al. 1996; Konishi and Nakao 1997; Malawista et al. 1994; Moody et al. 1994; Yrianainen et al. 2006), hamsters (Johnson et al. 1987; Johnson et al. 1990), and gerbils (Hansen et al. 1992; Preac-Mursic et al. 1987; Preac-Mursic et al. 1989), were utilized to evaluate the efficacy of antibiotic treatment, with outcome measured by culture of spirochetes from tissues at necropsy. Culture of tissues in artificial media was generally regarded as the gold-standard for testing outcome of antibiotic treatment, but it can only be so if surviving spirochetes are replicatively robust. During persistent infection, *B. burgdorferi* is no longer disseminating and is sequestered within collagen, suggesting that at least a subpopulation of spirochetes may be dormant and less apt to grow out in culture. In contrast to cultures taken during early infection, examination of cultures from tissues of untreated mice taken at 6–12 months of infection, for example, reveals the presence of sparse numbers of nonmotile spirochetes at the bottom of many of the culture tubes and only modest replicative activity in others. Similar rare nonmotile spirochetes have been noted in cultures from synovial fluid of human Lyme disease patients (Snydman et al. 1986). Culture, therefore, selects for those spirochetes that are capable of robust replication in artificial media.

Results derived from different animal model studies in different laboratories have varied somewhat, depending upon individual antimicrobial drugs, dose, treatment regimen, therapeutic drug levels achieved, pharmacodynamics of drugs in the species tested, stage of infection, culture sensitivity among different laboratories, and tissues sampled for culture. Generalities are possible, however, with the finding that beta-lactam antibiotics, particularly ceftriaxone, are generally the most effective at eliminating cultivable spirochetes from tissues of mice (Malawista et al. 1994; Moody et al. 1994) (Yrianainen et al. 2006; Pahl et al. 1999; Pavia et al. 2002), including infected SCID mice (Kazragis et al. 1996). Indeed, in one study, treatment with ceftriaxone for only 24 h was touted as being "100% effective in sterilizing tissue samples of C3H mice infected with… *B. burgdorferi*" (Pavia et al. 2002). Such over-confident conclusions based upon culture are now being challenged by more sensitive PCR assays, which detect persistence of viable, noncultivable spirochetes following antibiotic treatment in a wide variety of animal model species (dogs, mice, and nonhuman primates) treated with different classes of antibiotics.

12.5 Antibiotic Studies in the Dog Model, Based Upon Culture and PCR

The seminal study that began to challenge prevailing Lyme disease dogma regarding effectiveness of eliminating *B. burgdorferi* infection with antibiotics, utilized laboratory beagle dogs infected by exposure to multiple adult *Ixodes scapularis* ticks that were naturally infected with *B. burgdorferi*. The study demonstrated for the first time persistence of spirochetal DNA (presumably noncultivable *B. burgdorferi*) following 30 days of antibiotic treatment (Straubinger et al. 1997). Groups of five dogs were treated with doxycycline (10 mg/kg twice daily per os), amoxicillin (20 mg/kg three times per day or twice daily per os), or no treatment. Dogs were confirmed positive by culture of skin biopsies prior to treatment (one amoxicillin-treated dog was uninfected and remained negative throughout the study), which commenced at 2 months after tick exposure. Skin biopsies were cultured at 12, 40, 54, 68, and 97 days after treatment of each dog. Spirochetes were cultured from only one skin biopsy of a single doxycycline-treated dog at 97 days, whereas all other biopsies were culture-negative. Multiple tissues were cultured at necropsy 2 and 6 months following treatment, and only 1 site (regional lymph node near the tick feeding sites) was positive in one doxycycline-treated dog and one amoxicillin-treated dog. All dogs that received no treatment were culture-positive. In striking contrast to culture results, skin biopsies taken from all antibiotic-treated dogs were PCR-positive (*23srRNA* and *ospA*) at one or more intervals of treatment, and at necropsy, multiple tissues from 3/4 doxycycline-treated dogs and 2/3 amoxicillin-treated dogs were PCR-positive, as were tissues of dogs that received no treatment. Minimal inflammatory changes were found in joints of treated dogs at necropsy.

A similar study was performed in groups of 4 dogs treated for 30 days with azithromycin (25 mg/kg once daily per os), ceftriaxone (25 mg/kg once daily intravenously), doxycycline (10 mg/kg twice daily per os), or no treatment at 120 days after tick exposure (Straubinger 2000; Straubinger et al. 2000). Dogs were euthanized up to 455 days after completion of treatment. Skin punch biopsies were collected at 4 week intervals and cultured or analyzed by *ospA* PCR. Spirochetes were cultured only from biopsies of untreated dogs, but none of the 13 biopsies obtained from treated dogs after onset of treatment were culture-positive. In contrast, one or more biopsies were PCR-positive in 3/4 azithromycin-treated dogs, 2/4 ceftriaxone-treated dogs, 3/4 doxycycline-treated dogs, and 4/4 untreated dogs. At necropsy, 25 different tissues were analyzed from each dog, with none being culture-positive in treated dogs, whereas 3/4 azithromycin-treated dogs, 0/4 ceftriaxone-treated dogs, 2/4 doxycycline-treated dogs, and 4/4 untreated dogs were *ospA* PCR-positive in one or more tissues. In addition, treated dogs had minimal or no inflammatory changes in tissues, whereas all control dogs had mild polyarthritis.

In both of the aforementioned studies, serum antibody levels to *B. burgdorferi* rapidly declined, in some cases to undetectable levels, following antibiotic treatment, despite persistence of noncultivable spirochetes. Sera from dogs in the second study were also tested for C6 antibody titers, which waned rapidly after treatment (Philipp et al. 2001).

In summary, both of these dog studies demonstrated persistence of spirochetal DNA (presumably noncultivable spirochetes) in dogs for months following azithromycin, ceftriaxone, or doxycycline treatment, with decline in antibody titers that did not correlate with cure. Declining antibody following treatment suggested that noncultivable spirochetes remained occult to the immune system, despite their persistence. In addition, some of the treated dogs experienced rises in antibody at later intervals (6 months) after treatment, suggesting a response to resurgent populations of spirochetes. Inflammatory disease was minimal or absent in treated dogs, despite persistence of noncultivable spirochetes.

12.6 Antibiotic Studies in the Mouse Model, Based Upon Culture and PCR

The first PCR-based study to evaluate antibiotic response in an animal model involved C3H/HeN mice inoculated intradermally (dorsal thoracic midline) by syringe with cultured *B. burgdorferi* N40 (10^3), then treated with ceftriaxone (16 mg/kg twice daily subcutaneously) for 5 days, starting on day 15 after inoculation (Malawista et al. 1994). Urinary bladder and ear tissue were collected from 5 antibiotic-treated, 5 sham-treated, and 2 uninfected-treated mice on days 20, 30, and 60 after inoculation. Tissues were cultured and probed by standard PCR for *ospA* and *ospB* targets. There was excellent concordance between culture and PCR results derived from both tissue types. Two of five urinary bladders from antibiotic-treated mouse at 60 days were culture-positive and one was PCR-positive, but all other samples were both culture- and PCR-negative in antibiotic treated mice at all intervals tested. Thus, results suggested resurgence of spirochetes in incompletely treated mice, but did not detect noncultivable (PCR-positive) spirochetes in tissues of treated mice using standard PCR amplification.

Another PCR-based study evaluated response of mice treated with penicillin G, comparing sensitivity of nested PCR with real-time quantitative PCR, targeting *flagellin* (Pahl et al. 1999). C3H/HeJ mice were infected by syringe with cultured *B. burgdorferi* N40 (5×10^5) in the right footpad, and then treated subcutaneously with penicillin G (10,000 IU, twice daily for 14 days) commencing at 8 days of infection. Spirochetal DNA was detected at significantly reduced copy levels in treated mice compared to controls in all tissues (inoculation site, lymph node, kidney, and spleen) for up to 8 weeks after treatment. Arthritis (joint swelling) also improved in treated mice compared to controls. Borreliae could be cultured from tissues of treated mice, but data (tissues cultured) were not shown. Thus, treatment significantly reduced spirochete burdens in tissues and improved arthritis, but failed to clear tissues of cultivable spirochetes. This study missed the mark by using already documented sub-effective levels of penicillin G in the mouse model (Moody et al. 1994), but demonstrated the equal sensitivity (and greater specificity) of real-time PCR compared to nested PCR, with the advantage of quantification of spirochetes in tissues.

A comprehensive antibiotic mouse study was performed in mice that utilized real-time PCR to detect spirochetes in tissues of C3H/HeJ mice infected via *B. burgdorferi*-infected nymphal ticks (five ticks per mouse, placed in the ear canals). At 30 days of infection, groups of ten mice were treated with 16 mg/kg ceftriaxone twice daily subcutaneously for 5 days, then once daily for an additional 25 days; sham-treated; or treated with 50 mg/kg doxycycline twice daily by gavage for 30 days (Bockenstedt et al. 2002a). Outcome of treatment was evaluated by histopathology of heart base and joints of a rear leg; culture of blood, ear, heart, spleen, and urinary bladder; real-time PCR (*flagellin* and *16S rRNA*) of ear, urinary bladder, tibiotarsal joint, heart, and spleen; and xenodiagnosis, by feeding uninfected nymphal ticks on each mouse at 1, 3, 6, and 9 months after the last antibiotic dose.

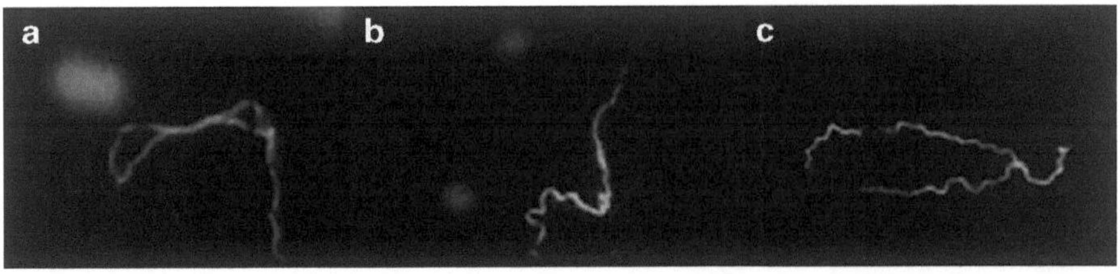

Fig. 12.2 Figure reproduced from Bockenstedt et al. (2002a) with permission from Oxford University Press. Direct immunoflourescent labeling of spirochetes within the midguts of xenodiagnosis ticks from mice treated with saline (**a**), ceftriaxone (**b**), or doxycycline (**c**)

In addition, at the 3-, 6-, and 9-month intervals, 100 larval ticks were fed upon 1 mouse per group, allowed to molt and harden into nymphs, and then 10 nymphs were fed upon C3H-*scid* (SCID) mice to assess transmissibility. Xenodiagnosis ticks were tested by conventional *ospA, BBE21.1* (lp-25 of B31), and *arp* (lp28-1 of B31) PCR, and direct immunofluorescence with anti-*B. burgdorferi* IgG. Serum *B. burgdorferi* IgG titers were determined by ELISA.

At the end of the aforementioned study, tissues from ceftriaxone- and doxycycline- treated mice were all culture-negative, but very low copy numbers of spirochetal DNA were detectable in urinary bladder tissue of 2 ceftriaxone-treated mice, and multiple tissues of doxycycline-treated mice. In contrast, spirochetes were cultured and spirochetal DNA was detectable in multiple tissues of saline-treated mice. Xenodiagnosis ticks were able to acquire spirochetes from all infected control mice at 1 and 3 months after completion of sham treatment. At 3 months, xenodiagnosis detected spirochetal DNA in ticks from 2/5 mice in each antibiotic treatment group. Spirochetes were visualized by immunofluorescence in most of the ticks that fed upon saline-treated mice, with fewer evident at the 3-month interval. In contrast, rare spirochetes were visualized in a small proportion of ticks that fed upon mice treated with ceftriaxone or doxycycline at 1 and 3 months (Fig. 12.2). Ticks were negative among all treatment groups thereafter. Serum *B. burgdorferi* IgG titers declined over the course of 9 months after treatment in all groups, including saline-treated mice, but were significantly lower in both antibiotic treatment groups compared to saline-treated mice.

Attempts were made to transmit noncultivable spirochetes from treated mice into naïve mice by tissue transplants and tick feeding. Transmission by transplantation of ear tissue from PCR-positive antibiotic-treated mice into SCID mice failed (due no doubt to the PCR-negative status of ear tissue following treatment), but transmission was documented with ear tissue from infected saline-treated control mice. Larval ticks were then fed upon infected saline- or ceftriaxone-treated mice. Samples of ticks from both groups tested positive by *ospA* PCR. Ticks from those cohorts were allowed to molt into nymphs, and subsequently fed upon naïve C3H mice (10 ticks per mouse). Only the mice that were infested with ticks from saline-treated mice seroconverted, suggesting lack of transmission from PCR-positive ticks from ceftriaxone-treated mice. Further analysis was performed on small numbers of ticks from ceftriaxone-, doxycycline-, or saline-treated mice for the presence of gene targets on 2 plasmids, lp25 (BBE21.1) and lp28-1 (*arp, BBF01*), in addition to *ospA*. All targets were amplified from ticks that fed upon saline-treated mice, whereas plasmid targets were variably amplified in ticks from antibiotic-treated mice. Based upon the inability to transmit infection by ear tissue transplant into SCID mice, inability to transmit infection from PCR-positive ticks to naïve C3H mice, and the variable presence of plasmid DNA in *ospA*-positive ticks from antibiotic-treated mice, it was concluded that persisting spirochetes in antibiotic-treated mice were genetically attenuated and noninfectious.

The noncultivable and noninfectious state of spirochetes that persist following antibiotic treatment is clearly an important phenomenon that needs further exploration. Another mouse study found that treatment of mice with ceftriaxone plus anti-tumor necrosis factor (TNF) antiserum, but not ceftriaxone alone, resulted in successful culture of spirochetes from multiple tissues of 20–40% of mice (Yrjanainen et al. 2007). The study involved C3H/HeN mice infected by intracutaneous syringe inoculation with 10^6 *B. burgdorferi* N40 or *B. garinii.* In addition to five-uninfected and five-infected control mice, groups of ten-infected mice were treated at 2 weeks of infection with ceftriaxone (50 mg/kg once daily intraperitoneally for 5 days); ceftriaxone plus anti-TNF (rat antiserum) at the onset of antibiotic treatment; ceftriaxone plus anti-TNF at 4 weeks after antibiotic; or anti-TNF without antibiotic. Ear, urinary bladder, and tibiotarsal joints were cultured, and nested *flaB* PCR was performed on urinary bladder tissue. At 14 weeks of infection, spirochetes were cultured from tissues of all infected untreated mice, whereas all ceftriaxone-treated mice were culture negative. In contrast, some of the antibiotic-treated mice infected with either species of Borrelia were culture-positive if they were also treated with anti-TNF antiserum. *Borrelia burgdorferi* isolates were probed by PCR for 16 plasmid targets, verified by sequence analysis, and found to be genetically intact, and also fully infectious in naïve C3H mice. These findings were certainly intriguing, but group sizes were small and statistically marginal.

In the study noted above, PCR analysis of urinary bladders revealed only one positive sample (*B. garinii*, ceftriaxone plus anti-TNF) among the 20 culture-negative samples. This result was counter to the study performed by Bockenstedt et al. (Bockenstedt et al. 2002a), which yielded positive PCR results from nearly half of ceftriaxone or doxycycline-treated mice. There are many possible reasons for this difference, including mouse substrain, interval of infection, treatment regimen, and other experimental variables. In a subsequent study, the presence of persisting spirochetes was assessed in C3H/HeN mice by culture and *flaB* PCR (Yrianainen et al. 2006). Mice were inoculated intradermally by syringe with 10^6 *B. burgdorferi* N40, and then treated at 2 or 18 weeks of infection with ceftriaxone (50 mg/kg once daily intraperitoneally for 5 days or 18 days), in addition to infected and uninfected controls. Mice were killed at 10, 21, or 30 weeks, depending upon treatment group. All tissues (ear, joint, urinary bladder, heart, brain, kidney, lymph node, spleen, liver, and eye) were culture-negative among all antibiotic-treated mice. In contrast, a significant number of samples, particularly joint tissue (30–100%), were PCR-positive. These results confirmed that analysis of outcome from antibiotic treatment requires assessment of appropriate tissue samples, particularly joint tissue. Furthermore, the large number of mice in this study and the extensive analysis of tissues confirmed that a short treatment regimen of only 5 days was equally effective at eliminating cultivable spirochetes as a longer 18-day treatment regimen, as was treatment during early (2 weeks) or late (18 weeks) of infection.

The issue of differential susceptibility to antibiotic treatment, depending upon stage of infection, is relevant to human Lyme disease, in that patients treated during early infection tend to have a more favorable response to treatment than those treated during late infection (Steere 2001; Wormser et al. 2006). Persistence of noncultivable spirochetes following ceftriaxone treatment was examined in C3H/HeN mice treated during the early (1 week) or late (4 months) stages of infection, and results suggested a more favorable outcome when treatment was commenced during early infection (Hodzic et al. 2008). In that study, mice were inoculated subdermally on the dorsal thoracic midline with 10^4 *B. burgdorferi* N40 spirochetes. Groups of mice were treated with ceftriaxone (16 mg/kg twice daily intraperitoneally for 5 days, then once daily for 25 additional days) or saline, starting at 1 week or 4 months of infection. Infection status was evaluated at 1 or 3 months after completion of treatment by culture (inoculation site, urinary bladder, and quadriceps muscle) and real-time *flaB* PCR (inoculation site, heart base, ventricular muscle, tibiotarsal joint, quadriceps muscle). Mice were also subjected to xenodiagnosis, by feeding approximately 40 larval ticks on each mouse several days prior to necropsy. Ticks were allowed to molt and harden into nymphs, and then samples of each cohort were cultured or tested by *flaB* PCR for spirochetal DNA. In addition, allografts of ear tissue from each antibiotic- or saline-treated mouse were transplanted subcutaneously into recipient SCID mice to determine infectivity of tissues.

Tissues from ceftriaxone-treated mice, regardless when treatment commenced or interval after treatment, were consistently culture-negative, whereas persistence of low copy numbers of spirochetal DNA (compared to saline-treated controls) were detected in tissues, particularly tibiotarsus and heart base, in many of the treated mice. Treatment during early infection appeared to be more effective at eliminating persisting spirochetal DNA. As previously published (Bockenstedt et al. 2002a), serum antibody titers to *B. burgdorferi* declined markedly following treatment, regardless of persisting DNA (spirochetes).

Immunohistochemistry was performed on heart and tibiotarsus of each mouse. Spirochetes were identified in heart and tibiotarsus tissues of nearly all of the saline-treated mice, and in several of the antibiotic-treated mice. Spirochetes were located within collagen-rich areas and appeared morphologically intact (Fig. 12.3). Xenodiagnosis also revealed a low rate of acquisition of spirochetal DNA, whereas ear allograft transmission to SCID mice was consistently negative. PCR-positive ticks that fed upon antibiotic-treated mice contained significantly lower *flaB* copy numbers (mean 120) compared to ticks that fed upon saline-treated mice (mean 36,000), indicating reduced replicative ability following acquisition. Fifteen to twenty nymphal ticks from each of nine PCR-positive tick cohorts were fed upon each of nine SCID mice. Infection status of the SCID mice was assessed by culture of urinary bladder, ear, and tick feeding site, and by *flaB* PCR of ear, tick feeding site, heart base, and

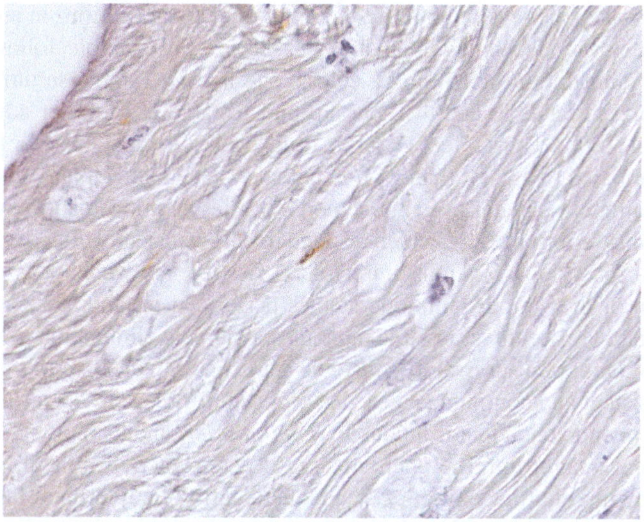

Fig. 12.3 Immunohistochemical labeling of a spirochete in tibiotarsal ligament of a mouse at 1 month following completion of treatment with ceftriaxone

tibiotarsal tissue at 3 weeks after tick feeding. With the exception of one ear culture, which contained a single nonmotile spirochete, all other samples were culture-negative. In contrast, all of the SCID mice had one or more PCR-positive tissues, particularly heart base and tibiotarsus. Thus, larval ticks were able to acquire spirochetal DNA, spirochetal DNA survived the molt of ticks from larvae to nymphs, nymphal ticks could transmit spirochetal DNA to SCID mice, and the spirochetal DNA disseminated in the SCID mice. Spirochetal DNA did not diminish with each step, but rather appeared to modestly increase. These are not features of nonviable spirochetes or residual DNA, but rather slowly replicating, noncultivable spirochetes, confirmed by visualization of spirochetes by immunohistochemistry in tissues of treated mice.

The results of the above study extended results and challenged conclusions of previous studies by showing the infectivity of noncultivable spirochetes by tick feeding upon naïve hosts, identified the presence of antigenically reactive and morphologically intact spirochetes in tissues of antibiotic-treated mice, and verified the predilection of persisting spirochetes for the joint tissue, as noted by others (Yrjanainen et al. 2010) as well as heart base tissue. In contrast to the previous study (Bockenstedt et al. 2002a) that failed to demonstrate tick transmission of noncultivable spirochetes, the present study unequivocally demonstrated that noncultivable spirochetes were indeed infectious following tick feeding on SCID mice. The paucity (or absence) of spirochetal DNA in ear tissue following antibiotic treatment was the likely explanation for the failure of ear allograft transmission noted in the present and the previous studies (Bockenstedt et al. 2002a). Furthermore, tissues and ticks from antibiotic-treated mice were assessed for gene targets on lp28-1 and lp25, and found to be intact, rather than genetically attenuated, as surmised by Bockenstedt, et al. (Bockenstedt et al. 2002a).

The pharmacokinetics of antibiotics are important considerations in animal model studies, as they vary depending upon drug and species. For example, following intraperitoneal administration of 16 mg/kg ceftriaxone in mice, ample serum MIC/MBC levels are present at 1, 2, and 4 h, but are undetectable at 8 h after administration (Hodzic et al. 2008). Thus, effective serum drug levels are maintained for less than 8 h in mice, compared to greater than 24 h in humans (Patel and Kaplan 1984). The ceftriaxone treatment regimens used in the mouse studies mentioned above involved one or two administrations per day, which resulted in only transiently effective serum drug levels, albeit for various durations of treatment. Tissue levels were not assessed. Upon examination of the various antibiotic studies performed in laboratory animals, it is apparent that there are three possible outcomes: complete elimination of spirochetes, indicative of cure; rare survival of cultivable spirochetes, indicative of treatment failure; or consistent and common persistence of noncultivable spirochetes, indicative of some other phenomenon that is unique to *B. burgdorferi* and its biology.

Additional insight into the phenomenon of persistence of noncultivable spirochetes following pharmacokinetically adequate antibiotic treatment has been provided in a study in which C3H/HeN mice were treated with tigecycline compared to ceftriaxone at different intervals of infection (Barthold et al. 2010a). Tigecycline, a new first-in class antibiotic, is highly effective in vitro against multiple strains of *B. burgdorferi*, has a serum half-life in mice of approximately 12 h, and a serum maximum 100 times greater than the *B. burgdorferi* N40 tigecycline MBC and 70 times greater than the *B. burgdorferi* N40 ceftriaxone MBC (Koomanachai et al. 2009; Sauve et al. 1996; Wang et al. 2005). Groups of 8 C3H/HeN mice were inoculated subdermally on the dorsal thoracic midline with 10^4 *B. burgdorferi* N40, then treated with 12.5 or 50 mg/kg tigecycline subcutaneously once daily for 10 days; 16 mg/kg ceftriaxone intraperitoneally twice daily for 5 days, and then once daily for 25 days (for comparison to previous studies); or saline (controls). Treatments commenced at three different intervals of infection: 1 week, 3 weeks, or 4 months, and mice were necropsied 3 months following treatment. At necropsy, inoculation site and urinary bladder were cultured, and real-time *ospA* PCR was performed on inoculation site, heart base, ventricular muscle, tibiotarsus, and quadriceps muscle.

As expected, none of the mice treated with antibiotics were culture-positive, whereas saline-treated controls were culture-positive. Among the mice treated at 1 week of infection, approximately half were PCR-positive in one or more tissues, regardless of antibiotic group, and all saline-treated controls were PCR-positive in multiple tissues. Mice treated at 3 weeks of infection were all PCR-positive in one or more tissues. All tissue types were PCR-positive among saline-treated mice, but nearly all PCR-positive tissues among antibiotic-treated mice were heart base and tibiotarsus. Mice treated at 4 months of infection were also all PCR-positive, but the great majority of PCR-positive tissues in antibiotic-treated mouse groups were tibiotarsal tissue, and only rarely heart base tissue, and less commonly other tissue types. In addition to confirming consistent persistence of noncultivable spirochetes (DNA) following both types of antibiotics, results revealed strikingly different patterns of PCR reactivity among tissues, depending upon the stage of infection when treatment was commenced. These tissue patterns suggested that persisting spirochetes (DNA) were not changing their tissue locations following treatment and their differential location represented their location at the time mice were treated. Furthermore, results confirmed that treatment was most effective when commenced during early (1 week) infection.

The high rate of PCR-positive heart base tissue in mice treated at 3 weeks of infection allowed assessment of tissues for RNA transcription of selected *B. burgdorferi* N40 genes, using nested RT-PCR for cDNA of *flaB*, *arp*, *ospA*, *dbpA*, *ospC*, *oppA-2*, *bbk32*, and mouse *beta-actin*. Remarkably, none of the samples from antibiotic-treated mice indicated transcription of *flaB* (compared to all saline-treated mice), and most other targets also appeared to be downregulated, with the exception of *oppA-2*. Sample sizes were too small to make absolute conclusions regarding patterns of RNA transcription, but they clearly confirmed that DNA PCR-positive tissues were also RNA PCR-positive, further confirming viability of persisting spirochetes.

This study also afforded the opportunity to further examine tick acquisition (xenodiagnosis) and transmission of persisting spirochetes. Xenodiagnosis was performed on the high-tigecycline and saline-treated mouse groups at the 4-month interval of infection. Larval ticks were fed to repletion on each mouse, allowed to molt into nymphs, and 15–23 nymphs from mice in each treatment group were pooled an allowed to feed upon SCID mice. SCID mice infested with ticks from saline-treated mice were culture-positive and PCR-positive, but SCID mice infested with ticks from tigecycline-treated mice were culture-negative and PCR-positive, confirming previous observations with ceftriaxone (Hodzic et al. 2008). Inflammatory lesions (arthritis and carditis) were present in the SCID mice infested with ticks from saline-treated but not tigecycline-treated mice, and *ospA* copy numbers were markedly lower in the tigecycline/tick mice compared to the saline/tick group. In addition, larval ticks were fed upon ceftriaxone-treated mice from the 3-week treatment group. Samples of tick cohorts from each mouse were evaluated by ospA PCR. Tick cohorts with positive samples were allowed to molt into nymphs, and nymphs were fed upon SCID mice, again confirming transmission of noncultivable spirochetes. Nymphs that fed upon the SCID mice were then allowed to molt into adults, and adults were tested by *ospA* PCR. Some of the ticks were PCR-positive, but culture-negative, demonstrating that noncultivable spirochetes survived molting through all stages of the tick.

Because previous studies (Bockenstedt et al. 2002a; Hodzic et al. 2008) were unable to demonstrate transmission of infection by transplant of ear tissue from PCR-positive antibiotic-treated mice, tibiotarsal and heart tissue were utilized for allograft transplants into groups of 9 or 10 SCID mice. These tissue sites were the most common sites for *B. burgdorferi* PCR-positivity. Tissues from saline-, ceftriaxone-, and tigecycline-treated mice from the 4-month treatment groups were transplanted into 9 or 10 SCID mice, which were necropsied at 4 weeks after transplant. SCID mice receiving transplants from saline-treated mice were all culture- (blood, inoculation site, and/or urinary bladder) and PCR-positive (transplant site, heart base, ventricular muscle, quadriceps muscle, and tibiotarsal tissue).

In contrast, none of the mice receiving transplants from antibiotic-treated mice were culture-positive, but 4/9 mice receiving transplants from tigecycline-treated mice and 4/10 mice receiving transplants from ceftriaxone-treated mice were PCR-positive in one or more tissues. Therefore, in contrast to previous conclusions of noninfectivity of tissues based upon ear transplants, transplantation of tissues with a high rate of spirochetal DNA clearly demonstrated that transplantation of tissues from PCR-positive mice transmits noncultivable spirochetes which disseminate in recipient SCID mice.

Nearly all of the mouse studies examining persistence of noncultivable spirochetes following antibiotic treatment have utilized C3H/He mice infected with *B. burgdorferi* N40 (Bockenstedt et al. 2002a) (Yrjanainen et al. 2007; Hodzic et al. 2008), with the exception of one study that also utilized *B. garinii* A218 (Yrjanainen et al. 2007). In unpublished studies (Barthold, et al.), C3H/HeN mice were infected with *B. burgdorferi* N40, B31, 72a, 118a, or JD-1, treated with ceftriaxone (16 mg/kg for 5 days) or saline, and then necropsied 1 month later. All were culture-negative but PCR-positive in multiple tissues, whereas saline-treated mice were both culture- and PCR-positive. Thus, the phenomenon of persistence of noncultivable spirochetes following antibiotic treatment in mice is universal among *B. burgdorferi* s.s. strains and likely *B. burgdorferi* s.l. as well. It is also apparently universal among a range of other species, including dogs (detailed above) and nonhuman primates (detailed below).

12.7 Studies in Nonhuman Primates

12.7.1 The Nonhuman Primate Model of Lyme Disease

Several animal species can be used to study Lyme disease, including mice, guinea pigs, dogs, rabbits, and monkeys (Barthold 1995). Of those studied, rhesus macaques most closely mimic the multi-organ character of the of human disease (Pachner et al. 1995). Unlike other animal models, disease hallmarks such as erythema migrans, carditis, arthritis, and neuropathy of the peripheral and central nervous systems are all observed in macaques. In mice, the reservoir host for *B. burgdorferi*, disease varies by mouse strain and age (Barthold et al. 1990), while the early- and late-disseminated manifestations are uncommon (Barthold et al. 1993b). Rodents, rabbits, and canines all fail to exhibit neurological disease from *B. burgdorferi* infection (Philipp and Johnson 1994). Importantly, macaques exhibit signs that are characteristic of all three phases of Lyme borreliosis, namely, early-localized, early-disseminated, and late-stage Lyme disease (Roberts et al. 1995b, 1998; Philipp et al. 1993). Erythema migrans (EM) is thought to occur in 70–80% of human cases (Steere and Sikand 2003), and in a similar proportion of rhesus macaques (Pachner et al. 1995; Philipp et al. 1993). In macaques as in humans, spirochetes disseminate from the site of inoculation to multiple organs. Spirochetal DNA has been detected in skeletal muscles, heart, bladder, peripheral nerves, and in the central nervous system (cerebrum, brainstem and cerebellum, spinal cord, and dura mater) (Cadavid et al. 2000). Arthritis and carditis, which are signs of disseminated disease, may both be studied in rhesus macaques (Roberts et al. 1995b; Cadavid et al. 2004). In a study of five animals that were tick-infected with the JD1 strain of *B. burgdorferi* sensu stricto, the joints most commonly affected were, as with humans (Steere and Sikand 2003), the knee and elbow joints (Roberts et al. 1995b). Observed changes included hyperplasia of the synovial membrane, periarticular fibrosis, and even severe erosion of articular cartilage. Mononuclear cell infiltrates were evident in at least one joint of four of the five infected animals in this study (Roberts et al. 1995b). In a study that involved 14 immune competent and 11 immune suppressed animals that received a needle inoculum with *B. burgdorferi* of the N40 strain, all but one of the animals showed carditis postmortem (Cadavid et al. 2004); this sign also was evident in three of five monkeys that were infected by tick with *B. burgdorferi* spirochetes of the JD1 strain (Roberts et al. 1995b). Collectively, these studies convincingly demonstrate that the macaque model most closely resembles human borreliosis and provides a relevant model for attempting to validate or dispel the notion that pathogenic spirochetes are retained posttreatment and are responsible for PTLDS.

A disadvantage for studying infection in primates (both human and nonhuman) is that the cultivability of *B. burgdorferi* appears to be inversely related to the duration of host adaptation, even in the absence of antibiotic treatment (Pachner et al. 1998; Nowakowski et al. 2009). The most sensitive method of detection for disseminated, low-level *B. burgdorferi* infection may therefore be the feeding of uninfected ticks for detection by uptake of organisms by those ticks, a procedure called xenodiagnosis (Bockenstedt et al. 2002b). Spirochete burden in the mouse appears to be much higher than it would be in a primate host, given that the mouse is a reservoir host and ample growth in this mammal is necessary for the transmission cycle. This is evident from previous studies (Pachner et al. 1998; Persing et al. 1994; Nocton et al. 1994b). Mice also generate different immune responses to *B. burgdorferi*, so the immune-mediated selection and clearance of spirochetes following infection may not be comparable to what occurs in human Lyme disease (Pachner et al. 2002; Barthold et al. 1996b). Thus, the phenotype of spirochetes selected before and after antibiotic treatment is likely not identical to those selected in the murine host. The importance of utilizing both mouse and primate hosts for studies of antibiotic efficacy can therefore not be underestimated.

12.7.2 Results of Nonhuman Primate Antibiotic-Treatment Experiments

As mentioned in Sect. 12.2, putative antibiotic treatment failure of in patients with PTLDS may be attributed to several causes, including (1) spirochetes that persist in the tissues, inaccessible or impervious to antibiotic; (2) inflammatory responses to lingering antigens from dead organisms; or (3) autoimmune responses, possibly

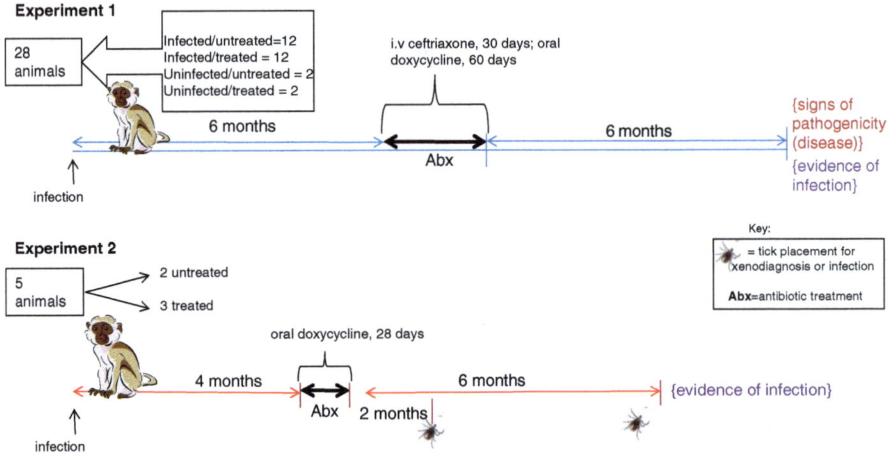

Fig. 12.4 Experimental design of two separate studies in nonhuman primates aimed at evaluating antibiotic efficacy. Both experiments involved infection, treatment post-dissemination, serology, and detection of spirochetes in tissues. The two varied by the number of animals, *B. burgdorferi* strain used, duration prior to treatment, antibiotic treatment regimen, and detection methods

elicited by antigenic mimicry (Bolz and Weis 2004). In an effort to better understand the feasibility of these possible explanations, experiments (Fig. 12.4) that assessed the efficacy of two regimens of ceftriaxone and/or doxycycline treatment in rhesus macaques were conducted (Embers et al. 2012). Here, treatment was withheld until early or late disseminated disease, commencing at 4–7 months after primary infection with *B. burgdorferi* (Fig. 12.4). In Experiment 1, treatment of late-disseminated phase infection (6 months p.i.) was followed by assessment of animals for signs of pathology and evidence of infection using histology, immunohistochemistry, and molecular detection. In Experiment 2, treatment was given at the early disseminated phase (4 months p.i.) and xenodiagnosis was employed.

In Experiment 1, an aggressive treatment regimen of intravenous ceftriaxone for 30 days followed by 60 days of oral doxycycline was employed. Serum samples were collected longitudinally for C6 serology, and infection status postmortem was assessed by in vitro culture, DNA PCR, RT-PCR, and immunohistochemistry (IHC) at 6 months after the last treatment. Only 3 of the 11 (one died during anesthesia) treated animals were negative by all criteria for the presence of the spirochetes; 2 of 12 untreated animals were negative by all criteria. Table 12.1 shows a summary of results for *B. burgdorferi*-infected macaques that received antibiotics for late-stage infection (Treated) versus those that did not (Untreated). Figure 12.5 shows an inflammatory lesion (A) and positive staining for *B. burgdorferi* (B) in the heart tissue from a treated animal. Anti-C6 antibody levels were quantified longitudinally and declined to below baseline (pre-immune) levels in all treated animals. Not surprisingly, none of the treated animals yielded a positive culture and only one of the untreated animals was culture positive (lung tissue).

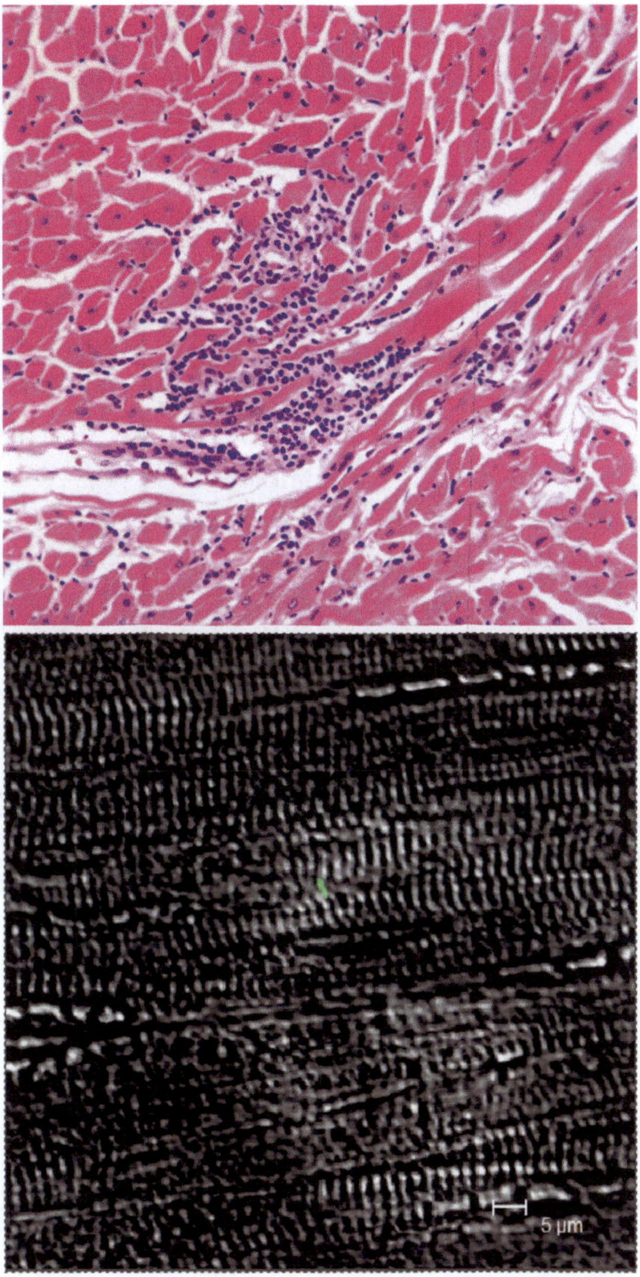

Fig. 12.5 Images of tissue inflammation and *B. burgdorferi* antigen in heart of one infected animal. For antigen detection, samples of tissue were stained for fluorescent detection (IFA) with anti-*B. burgdorferi* antibody. *Top*: H&E stain showing monocytic and lymphocytic infiltrate in a heart section (×20) from a treated animal. *Bottom*: Image of positive staining from the heart tissue of the same animal (modified from Embers et al. 2012)

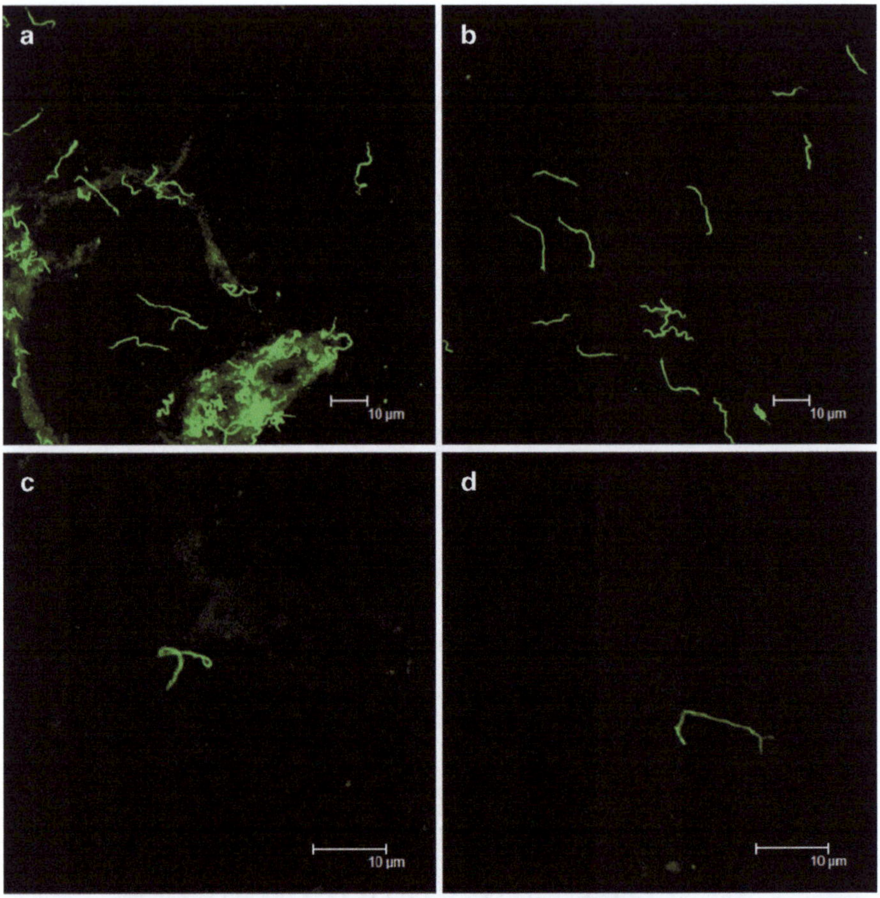

Fig. 12.6 Spirochetes recovered from *B. burgdorferi*-infected, antibiotic-treated animals 2 months after the cessation of treatment (Experiment 2). Shown are images from direct fluorescent staining of *B. burgdorferi* spirochetes found in xenodiagnostic tick midgut culture (**a** and **b**) direct tick midgut preparation (**c** and **d**) from treated animals GA59 and GB56, respectively (modified from Embers et al. 2012)

Table 12.1 Number of animals that tested positive/total number by the indicated method

	PCR	IHC	C6	RT-PCR
Treated	1/11	7/11	0/11	3/11
Untreated	2/12	9/12	5/12	2/12

No gross lesions were observed in any of the animals. Fragments of heart and meninges were collected postmortem, and fixed, sectioned, and stained for histology and immunofluorescence. Three animals, all of them treated, had moderate to severe inflammatory lesions in the heart (Table 12.1, Fig. 12.5). Numerous

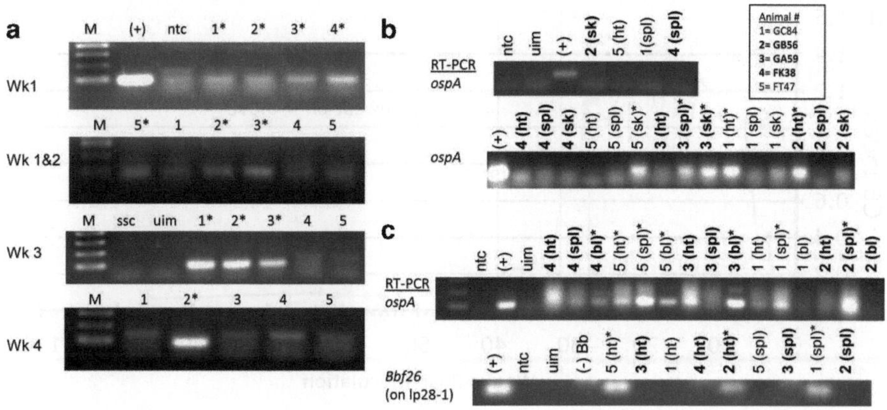

Fig. 12.7 Detection of *B. burgdorferi*-specific nucleic acids (Experiment 2). Detection by DNA PCR or RT-PCR using primers for the indicated genes from: (**a**) skin biopsy samples; (**b**) organ tissue culture pellets; and (**c**) directly from tissues known to harbor the spirochetes. Animal numbers in *bold* are of those that were treated. *Asterisks* indicate clear positive amplimers. *sk* skin, *ht* heart, *bl* bladder, *spl* spleen. Labels include: M = marker (100 bp ladder); (+) = *B. burgdorferi*, strain B31 DNA/RNA; ssc = spirochete strain control; *Bb* strain JD1 DNA; ntc = no template control; uim = uninfected monkey DNA/RNA (modified from Embers et al. 2012)

immunofluorescence-positive specimens were found both in the untreated and the treated groups (Table 12.1, Fig. 12.5).

In Experiment 2, at 7 and 11 months postinfection, all monkeys were fed upon by naive *I. scapularis* nymphs for xenodiagnosis. The number of nymphs that fed to repletion varied considerably between animals, so spirochetes were not recovered from all animals. However, intact spirochetes were detected from the cultured midgut contents or directly from tick midgut smears (Fig. 12.6) of two antibiotic-treated animals at 7 months postinfection. Several *B. burgdorferi* genes were used both for detection of spirochetes (by standard DNA PCR) and for detection of metabolically active spirochetes (by RT-PCR). The detection of genes throughout the course of infection, in tissues, by standard PCR/RT-PCR is shown in Fig. 12.7. A few spirochetes grew in cultures of organ tissues from each animal after >9 weeks, but spirochetes could not be subcultured from either treated or untreated animals due to their slow growth. Those culture samples were therefore pelleted to confirm their identity and test their viability by DNA/RNA analysis. Transcription was detected in culture pellets from the tissues of treated animals, indicating that the bacteria were metabolically active (Fig. 12.7b). *OspA* transcription was also detected directly in tissues harvested from treated and untreated animals (Fig. 12.7c). We chose to amplify the *ospA* transcript because it has been shown to be transcribed by host-adapted *B. burgdorferi* (Crowley and Huber 2003), in disseminated infection (Narasimhan et al. 2003), and due to the induction of an anti-OspA response in patients (Kalish et al. 1993) post-dissemination.

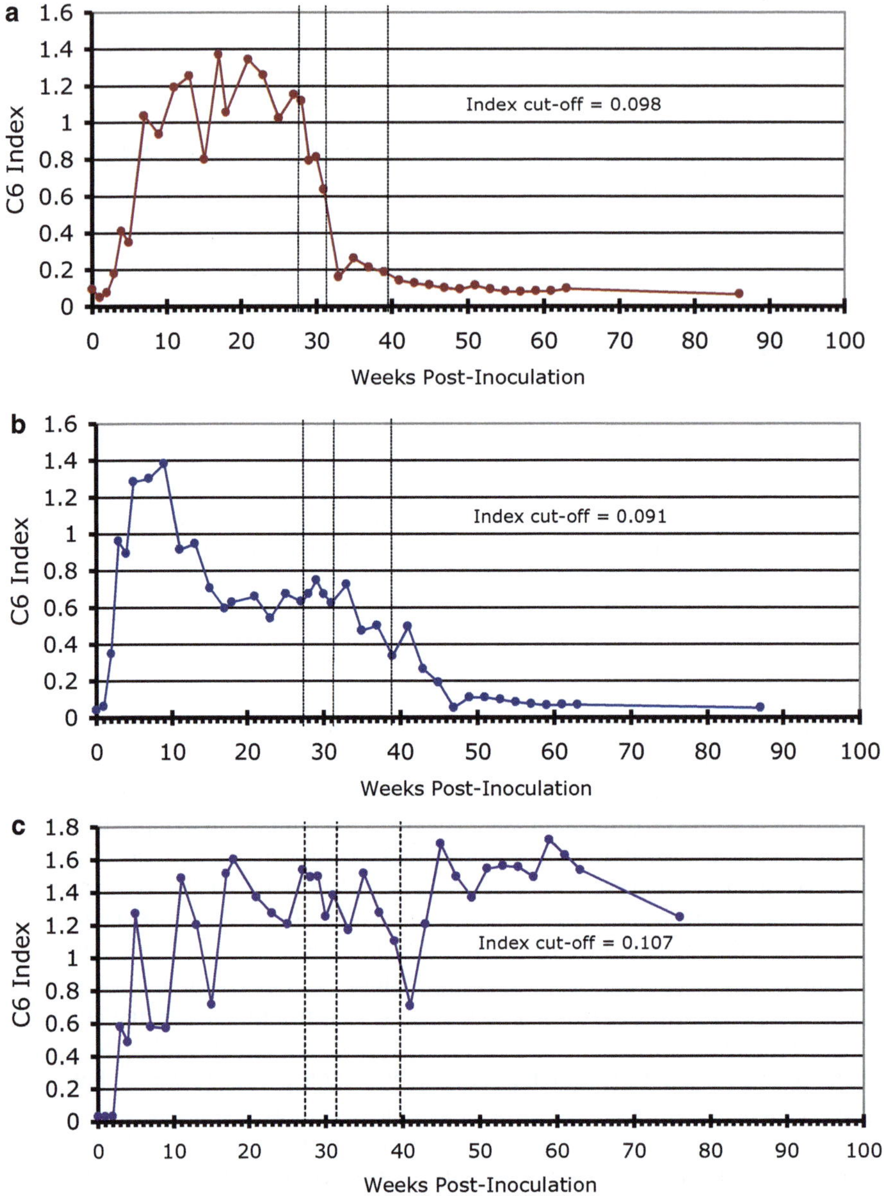

Fig. 12.8 Patterns of antibody response to the C6 peptide in infected animals. The C6 index measured as a function of time PI with *B. burgdorferi* followed three distinct patterns (Experiment 1). Each graph is the longitudinal antibody response to C6 from one representative animal for each pattern. In treated ($n=12$) animals (**a**) the response declined sharply, almost to background levels, within the antibiotic administration periods (*solid vertical lines*, weeks 27–31 for ceftriaxone, weeks 31–39 for doxycycline). In sham-treated animals (**b**, **c**) the response either declined, but independently from the sham treatment intervals, to eventually reach background level (**b**, $n=7$), or oscillated throughout the study period without ever declining steadily (**c**, $n=5$). The pattern in (**c**) corresponds to the C6-positive animals, indicated in Table 10.1. The *dotted lines* in (**b**) and (**c**) indicate the sham-treatment intervals (modified from Embers et al. 2012)

Table 12.2 Comparison of titers (reciprocal end point dilution) of anti-C6 antibodies between antibiotic-treated (*bold*) and untreated animals in Experiment 2 (modified from Embers 2012)

Animal designation	Week 16[a]	Week 32	Week 47
GA59	>2,560	640	320
FK38	>2,560	40	160
GB56	>2,560	160	320
GC84	>2,560	>2,560	>2,560
FT47	>2,560	>2,560	>2,560

[a]Treatment was given between week 16 and 20

12.7.3 The Antibody Response to C6

The antibody response to C6 was measured in serially collected serum specimens. In all of the infected animals, the C6 antibody index rose steeply within the first 5–8 weeks PI. Thereafter, the responses fit into three patterns (Fig. 12.8), depending on whether the animals were or were not treated with antibiotics. In the treated group, the response declined steadily during the treatment period and reached background levels at the endpoint in all animals (Fig. 12.8a, Table 12.1). In contrast, the responses of the untreated group remained either largely unchanged (in 5 of 12 animals, Fig. 12.8c, scored positive (+) in Table 12.1) or returned to background levels (7 of 12 animals, scored negative (–) in Table 12.1) but not in parallel with the kinetics of the treated group's decline in specific antibody (Fig. 12.8b). The vertical lines in the figures denote the treatment intervals with both ceftriaxone and doxycycline, or sham. The decline in C6 titers for the untreated animals would appear to indicate that these animals self-cured. However, based on the immunofluorescence and PCR data, spirochetes remained detectable. One possible explanation is that those present have lost the lp28-1 plasmid that encodes the protein to which anti-C6 antibodies are produced. Studies indicate that lp28-1(-) spirochetes lack infectivity (Embers et al. 2007; Labandeira-Rey et al. 2003; Labandeira-Rey and Skare 2001), but it is possible that they may persist in a permissive host.

In Experiment 2, we also measured the responses to C6 longitudinally to include before, during, and after treatment. The C6-specific antibody levels declined in all animals treated with antibiotic. In untreated animals, the anti-C6 responses either leveled off or slowly declined. To determine if antibody titers declined to baseline with treatment, the endpoint dilution titers for each of the five animals in Experiment 2 were determined for sera collected before treatment (16 weeks PI), shortly after treatment (32 weeks PI), and at the end point (47 weeks PI). For all untreated animals the titers remained high and the titers from the treated animals declined markedly, and remained at low level even at 27 weeks after the cessation of treatment (Table 12.2).

We hypothesized that antibiotic-persistent spirochetes may also lose linear plasmid 28-1 (lp28-1), which encodes the VlsE antigen bound by the anti-C6 antibody. Transcription of a lp28-1 gene (*bbf26*) was verified in organ tissue from both untreated animals and one treated animal (Fig. 12.7c). This would indicate that, while this plasmid may be lost by some spirochetes, it is not lost by all of those present in the tissue samples tested.

12.8 Conclusions and Future Directions

As briefly discussed in the introduction to this phenomenon of PTLDS, patients who have been diagnosed with Lyme disease can present with signs and symptoms that can vary from person to person. Multiple variables exist within the population of humans diagnosed with Lyme, such as the duration between the tick bite or onset of symptoms to medical care, the multiple *B. burgdorferi* types present in nature (Liveris et al. 1999; Wormser et al. 2008), the possible coinfections, the criteria used for diagnosis, and the antibiotic treatment regimen prescribed. Here, we have used nonhuman primates as a model to test the ability of spirochetes to persist in a defined, experimental system. We controlled the spirochete strain and dose, and knew the exact point of infection. For humans, the cases of PTLDS are not clearly defined. While patients may experience symptoms after the recommended treatment, the assertion that further antibiotic treatment is warranted and effective in these cases is a matter of contention and considerable debate (Auwaerter 2007; Stricker 2007; Feder et al. 2007; Mayer and Merz 2008). The studies presented herein do not offer an answer to this question nor do the experiments aim to do so. Our results rather offer proof of the principle that disseminated spirochetes of two different *B. burgdorferi* strains can persist in the primate host following high dose, or long-lasting antibiotic therapy.

The pathogenic capability of these persistent organisms, however, is an open question of considerable importance. In terms of disease, objective signs of macroscopic or microscopic disease post-therapy can be measured in an animal model, and only macroscopic disease (such as rash or joint swelling) may be seen in humans. The symptoms (fatigue, arthralgia, cognitive impairment) that are usually associated with PTLDS in humans are largely subjective and not easily measured. Specifically with regard to the nonhuman primate studies, we did not find macroscopic signs of disease postmortem, but we did identify heart sections with inflammatory foci in three of the treated animals. In addition, several animals both treated and untreated, exhibited heart and meningeal sections that were positive by immunofluorescence for *B. burgdorferi.* At the molecular level, detection of *B. burgdorferi* DNA could indicate the presence of organisms, live or dead. The detection of mRNA (transcript), however, more clearly indicates that those present are metabolically active and thus, alive. In nonhuman primate Experiment 1, *B. burgdorferi* DNA and RNA were detected in the tissues of only a few animals, independent of treatment. This may reflect the paucity of spirochetes in this host, lack of *flaB* transcription (Barthold et al. 2010b), and/or seclusion in untested tissues.

In nonhuman primate Experiment 2, we were able to recover spirochetes by xenodiagnosis in two of the three treated animals. The inability to recover spirochetes from all five macaques could be a reflection of the number of ticks that fed, which was significantly fewer for control animals. A few slow-growing organisms were recovered by culture from each animal, but these grew too slowly and were too sparse to try to subculture or store. Importantly, we detected the *ospA* transcript in culture pellets from tissues of four animals and directly from at least one tissue from each animal. These data compel further inquiry into the phenotype of spirochetes that persist through the antibiotic environment of the treated host.

Antibiotic tolerance has been demonstrated in vitro with several bacterial species, both gram negative (*E. coli*) and gram positive (*Staphylococcus* ssp.). The ability of bacteria to endure in the presence of antibiotics, such as penicillin and cephalosporins (ceftriaxone), that interfere with cell wall synthesis, appears to stem from their ability to enter a dormant, nondividing state (Lewis 2007, 2010). Here, they may avert the need for cell wall synthesis to continue growth. The tetracycline class of antibiotics (to which doxycycline belongs) interferes with translation at the ribosome; therefore, a global reduction in translation by dormant organisms would leave them less affected. The Borreliae have not been explored specifically for their ability to enter into a "persister" phenotype. Specific transcriptomic changes associated with host adaptation and microenvironmental adaptation must be responsible for the recalcitrance made evident by the mouse and nonhuman primate studies outlined in this chapter. As more genetic tools become available, the ability to identify these changes becomes more and more likely.

Clearly, there is much work to be done on this important area of research. The etiology of chronic conditions, such as PTLDS continues to intrigue scientists at many levels. Studies that focus on the pathogen and those that explore the host response in multiple model systems will no doubt, in combination shed light on the unique capabilities of infectious agents to evade host immunity and establish persistence.

12.9 Summary of Animal Model Studies

Collectively, studies performed in different laboratories have demonstrated the following.

- Antibiotic treatment failure results in recovery of cultivable spirochetes
- Treatment with a variety of antibiotic classes can result in persistence of noncultivable spirochetes
- Noncultivable spirochetes have been documented following antibiotic treatment in a variety of animal models, including dogs, mice, and nonhuman primates
- Treatment with antibiotics is more effective when administered during early infection
- Noncultivable spirochetes in tissues of mice and monkeys treated with antibiotics can be identified as morphologically intact spirochetes by immunohistochemistry and are therefore expressing antigen

- Noncultivable spirochetes in tissues of mice and monkeys treated with antibiotics transcribe RNA, thereby confirming their viability
- Persisting noncultivable spirochetes in mice have a predilection for collagenous tissue such as tibiotarsal joints
- Ticks feeding upon antibiotic-treated mice and nonhuman primates can acquire morphologically intact, antigen-expressing, noncultivable spirochetes
- Noncultivable spirochetes in ticks are transmissible to new hosts
- Tick-transmitted noncultivable spirochetes disseminate in new hosts
- Noncultivable spirochetes survive through different stages of the tick
- Noncultivable spirochetes are infectious and disseminate when tissues from antibiotic-treated mice are transplanted into recipient hosts
- Noncultivable spirochetes occur following antibiotic treatment of mice and nonhuman primates infected with a variety of different *B. burgdorferi* strains

Major questions remain regarding the biological and medical significance of persisting noncultivable spirochetes, including

- Are noncultivable spirochetes genetically attenuated as an explanation for their reduced ability to replicate?
- Do noncultivable persisting spirochetes eventually die out in tissues following treatment?
- Can anti-TNF antiserum or other immunomodulatory treatments activate noncultivable spirochetes into a replicative and cultivable state?
- Does this phenomenon occur in human Lyme disease patients, and is it a possible explanation for post-Lyme disease syndrome or other symptoms?

References

Anderson JF, Magnarelli LA (1993) Epizootiology of Lyme disease-causing Borreliae. Clin Dermatol 11:339–351

Asbrink E, Hovmark A (1985) Successful cultivation of spirochetes from skin lesions of patients with erythema chronica migrans afzelius and acrodermatitis chonica atrophicans. Acta Pathol Microbiol Immunol Scand 93:161–163

Asch ES, Bujak DI, Weiss M, Peterson MG, Weinstein A (1994) Lyme disease: an infectious and postinfectious syndrome. J Rheumatol 21(3):454–461

Auwaerter P (2007) Point: antibiotic therapy is not the answer for patients with persisting symptoms attributable to Lyme disease. Clin Infect Dis 45(2):143–148

Barthold SW (1995) Animal models for Lyme disease. Lab Invest 72(2):127–130

Barthold SW, Beck DS, Hansen GM, Terwilliger GA, Moody KD (1990) Lyme borreliosis in selected strains and ages of laboratory mice. J Infect Dis 162(1):133–138

Barthold SW, de Souza MS, Janotka JL, Smith AL, Persing DH (1993a) Chronic Lyme borreliosis in the laboratory mouse. Am J Pathol 143(3):951–971

Barthold SW, de Souza MS, Janotka JL, Smith AL, Persing DH (1993b) Chronic Lyme borreliosis in the laboratory mouse. Am J Pathol 143(3):959–971

Barthold SW, de Souza M, Feng S (1996) Serum-mediated resolution of Lyme arthritis in mice. Lab Invest 74:57–67

Barthold SW, Hodzic E, Tunev S, Feng S (2006) Antibody-mediated disease remission in the mouse model of Lyme borreliosis. Infect Immun 74:4817–4825

Barthold SW, Hodzic E, Imai D, Feng S, Yang X, Luft BJ (2010) Ineffectiveness of tigecycline against persistent *Borrelia burgdorferi*. Antimicrob Agents Chemother 54:643–651

Bockenstedt LK, Mao J, Hodzic E, Barthold SW, Fish D (2002) Detection of attenuated, noninfectious spirochetes after antibiotic treatment of *Borrelia burgdorferi*-infected mice. J Infect Dis 186:1430–1437

Bolz DD, Weis JJ (2004) Molecular mimicry to Borrelia burgdorferi: pathway to autoimmunity? Autoimmunity 37(5):387–392

Bradley JF, Johnson RC, Goodman JL (1994) The persistence of spirochetal nucleic acids in active Lyme arthritis. Ann Intern Med 120:487–489

Brissette CA, Verma A, Bowman A, Cooley AE, Stevenson B (2009) The *Borrelia burgdorferi* outer-surface protein ErpX binds mammalian laminin. Microbiology 155:863–872

Cabello FC, Godfrey HP, Newman SA (2007) Hidden in plain sight: Borrelia burgdorferi and the extracellular matrix. Trends Microbiol 15:350–354

Cadavid D, O'Neill T, Schaefer H, Pachner AR (2000) Localization of Borrelia burgdorferi in the nervous system and other organs in a nonhuman primate model of lyme disease. Lab Invest 80(7):1043–1054

Cadavid D, Bai Y, Hodzic E, Narayan K, Barthold SW, Pachner AR (2004) Cardiac involvement in non-human primates infected with the Lyme disease spirochete Borrelia burgdorferi. Lab Invest 84(11):1439–1450

Crowley H, Huber BT (2003) Host-adapted Borrelia burgdorferi in mice expresses OspA during inflammation. Infect Immun 71(7):4003–4010

Embers ME, Wormser GP, Schwartz I, Martin DS, Philipp MT (2007) Borrelia burgdorferi spirochetes that harbor only a portion of the lp28-1 plasmid elicit antibody responses detectable with the C6 test for Lyme disease. CVI 14(1):90–93

Embers ME, Barthold SW, Borda JT, Bowers L, Doyle L, Hodzic E, Jacobs MB, Hasenkampf NR, Martin DS, Narasimhan S et al (2012) Persistence of *Borrelia burgdorferi* in Rhesus Macaques following Antibiotic Treatment of Disseminated Infection. PLoS One 7(1):e29914

Feder HM Jr, Johnson BJ, O'Connell S, Shapiro ED, Steere AC, Wormser GP, Ad Hoc International Lyme Disease G (2007) A critical appraisal of "chronic Lyme disease". N Engl J Med 357(14):1422–1430

Fleming RV, Marques AR, Klempner MS, Schmid CH, Dally LG, Martin DS, Philipp MT (2004) Pre-treatment and post-treatment assessment of the C(6) test in patients with persistent symptoms and a history of Lyme borreliosis. Eur J Clin Microbiol Infect Dis 23(8):615–618

Frazer DD, King LI, Miller FW (1992) Molecular detection of persistent *Borrelia burgdorferi* in a man with dermatomyositis. Clin Exp Rheumatol 10:387–390

Goodman JL, Jurkovich P, Kodner C, Johnson RC (1991) Persistent cardiac and urinary tract infections with *Borrelia burgdorferi* in experimentally infected Syrian hamsters. J Clin Microbiol 29:894–896

Grab DJ, Givens C, Kennedy R (1998) Fibronectin-binding activity in Borrelia burgdorferi. Biochim Biophys Acta 1407:135–145

Guo BP, Norris SJ, Rosenberg LC, Hook M (1995) Adherence of *Borrelia burgdorferi* to the proteoglycan decorin. Infect Immun 63:3467–3472

Hansen K, Hovmark A, Lebech A-M, Lebech K, Olsson I, Halkier-Sorenson L, Olsson E, Asbrink E (1992) Roxithromycin in Lyme borreliosis: discrepant results of an in vitro and in vivo animal susceptibility study and a clinical trial in patients with erythema migrans. Acta Derm Venereol 72:297–300

Hodzic E, Feng S, Holden K, Freet KJ, Barthold SW (2008) Persistence of *Borrelia burgdorferi* following antibiotic treatment in mice. Antimicrob Agents Chemother 52:1728–1736

Johnson RC, Kodner C, Russell M (1987) In vitro and in vivo susceptibility of the Lyme disease spirochete, *Borrelia burgdorferi*, to four antimicrobial agents. Antimicrob Agents Chemother 31:164–167

Johnson RC, Kodner C, Russell M, Girard D (1990) In vitro and in vivo susceptibility of *Borrelia burgdorferi* to azithromycin. J Antimicrob Chemother 25(suppl A):33–38

Kalish RA, Leong JM, Steere AC (1993) Association of treatment-resistant chronic Lyme arthritis with HLA-DR4 and antibody reactivity to OspA and OspB of Borrelia burgdorferi. Infect Immun 61(7):2774–2779

Kazragis RJ, Dever LL, Jorgensen JH, Barbour AG (1996) In vivo activities of ceftriaxone and vancomycin against *Borrelia spp.* in the mouse brain and other sites. Antimicrob Agents Chemother 40:2632–2636

Klempner MS, Hu LT, Evans J, Schmid CH, Johnson GM, Trevino RP, Norton D, Levy L, Wall D, McCall J et al (2001) Two controlled trials of antibiotic treatment in patients with persistent symptoms and a history of Lyme disease. N Engl J Med 345(2):85–92

Konishi T, Nakao M (1997) Antibiotic treatment in mice infected with Japanese Borrelia garinii: efficacy of ceftriaxone for eradicating the infection induced by Ixodes persulcatus ticks. Microbiol Immunol 41:165–168

Koomanachai P, Kim A, Nicolau DP (2009) Pharmacodynamic evaluation of tigecycline against *Acinetobacter baumannii* in a murine pneumonia model. J Antimicrob Chemother 63: 982–987

Kuiper H, van Dam AP, Spanjaard L, de Jongh BM, Widjojokusumo A, Ramselaar TCP, Cairo I, Vos K, Dankert J (1994) Isolation of *Borrelia burgdorferi* from biopsy specimens taken from healthy-looking skin of patients with Lyme borreliosis. J Clin Microbiol 32:715–720

Labandeira-Rey M, Skare JT (2001) Decreased infectivity in *Borrelia burgdorferi* strain B31 is associated with loss of linear plasmid 25 or 28-1. Infect Immun 69(1):446–455

Labandeira-Rey M, Seshu J, Skare JT (2003) The absence of linear plasmid 25 or 28-1 of Borrelia burgdorferi dramatically alters the kinetics of experimental infection via distinct mechanisms. Infect Immun 71(8):4608–4613

Lewis K (2007) Persister cells, dormancy and infectious disease. Nat Rev Microbiol 5(1):48–56

Lewis K (2010) Persister cells. Annu Rev Microbiol 64(1):357–372

Liveris D, Varde S, Iyer R, Koenig S, Bittker S, Cooper D, McKenna D, Nowakowski J, Nadelman RB, Wormser GP et al (1999) Genetic diversity of Borrelia burgdorferi in lyme disease patients as determined by culture versus direct PCR with clinical specimens. J Clin Microbiol 37(3):565–569

Malawista SM, Barthold SW, Persing DH (1994) Fate of *Borrelia burgdorferi* DNA in tissues of infected mice after antibiotic treatment. J Infect Dis 170:1312–1316

Maraspin V, Ogrinc K, Ruzic-Sabljic E, Lotric-Furlan S, Strle F (2011) Isolation of *Borrelia burgdorferi* sensu lato from blood of adult patients with borrelial lymphocytoma, Lyme neuroborreliosis, Lyme arthritis and acrodermatitis chronica atrophicans. Infection 39:35–40

Mayer L, Merz S (2008) An appraisal of "chronic Lyme disease". [comment]. N Engl J Med 358(4):428, author reply 430–421

McKisic MD, Barthold SW (2000) T-cell-independent responses to *Borrelia burgdorferi* are critical for protective immunity and resolution of Lyme disease. Infect Immun 68:5190–5197

Miklossy J, Khalli K, Gern L, Ericson RL, Darekar P, Bolle L, Hurlimann J, Paster BJ (2004) *Borrelia burgdorferi* pesists in the brain in chronic Lyme neuroborreliosis and may be associated with Alzheimer disease. J Alzheimers Dis 6:639–649

Moody KD, Barthold SW, Terwilliger GA, Beck DS, Hansen GM, Jacoby RO (1990) Experimental chronic Lyme borreliosis in Lewis rats. Am J Trop Med Hyg 42:65–74

Moody KD, Adams RL, Barthold SW (1994) Effectiveness of antimicrobial treatment against *Borrelia burgdorferi* infection in mice. Antimicrob Agents Chemother 38:1567–1572

Moter SE, Hofmann H, Wallich R, Simon MM, Kramer MD (1994) Detection of *Borrelia burgdorferi* sensu lato in lesional skin of patients with erythema migrans and acrodermatitits chronica atrophicans by ospA-specific PCR. J Clin Microbiol 32:2980–2988

Narasimhan S, Caimano MJ, Liang FT, Santiago F, Laskowski M, Philipp MT, Pachner AR, Radolf JD, Fikrig E (2003) Borrelia burgdorferi transcriptome in the central nervous system of non-human primates. [erratum appears in Proc Natl Acad Sci USA. 2004 Feb 3;101(5):1426 Note: Camaino MJ [corrected to Caimano MJ]]. Proc Natl Acad Sci USA 100(26):15953–15958

Nocton JJ, Dressler F, Rutledge RJ, Rys PN, Persing DH, Steere AC (1994) Detection of *Borrelia burgdorferi* DNA by polymerase chain reaction in synovial fluid from patients with Lyme arthritis. N Engl J Med 330:229–234

Nowakowski J, McKenna D, Nadelman RB, Bittker S, Cooper D, Pavia C, Holmgren D, Visintainer P, Wormser GP (2009) Blood cultures for patients with extracutaneous manifestations of Lyme disease in the United States. Clin Infect Dis 49(11):1733–1735

Pachner AR, Delaney E, O'Neill T, Major E (1995) Inoculation of nonhuman primates with the N40 strain of Borrelia burgdorferi leads to a model of Lyme neuroborreliosis faithful to the human disease. Neurology 45(1):165–172

Pachner AR, Zhang WF, Schaefer H, Schaefer S, O'Neill T (1998) Detection of active infection in nonhuman primates with Lyme neuroborreliosis: comparison of PCR, culture, and a bioassay. J Clin Microbiol 36(11):3243–3247

Pachner AR, Dail D, Li L, Gurey L, Feng S, Hodzic E, Barthold S (2002) Humoral immune response associated with lyme borreliosis in nonhuman primates: analysis by immunoblotting and enzyme-linked immunosorbent assay with sonicates or recombinant proteins. Clin Diagn Lab Immunol 9(6):1348–1355

Pahl A, Kuhlbrandt U, Brune K, Rollinghoff M, Gessner A (1999) Quantitative detection of Borrelia burgdorferi by real-time PCR. J Clin Microbiol 37:1958–1963

Patel IH, Kaplan SA (1984) Pharmacokinetics profile of ceftriaxone in man. Am J Med 77:17–25

Pavia C, Inchiosa MA, Wormser GP (2002) Efficacy of short-course ceftriaxone therapy for Borrelia burgdorferi infection in C3H mice. Antimicrob Agents Chemother 46:132–134

Persing DH, Mathiesen D, Podzorski D, Barthold SW (1994) Genetic stability of Borrelia burgdorferi recovered from chronically infected immunocompetent mice. Infect Immun 62(8):3521–3527

Philipp MT, Johnson BJ (1994) Animal models of Lyme disease: pathogenesis and immunoprophylaxis. Trends Microbiol 2(11):431–437

Philipp MT, Aydintug MK, Bohm RP Jr, Cogswell FB, Dennis VA, Lanners HN, Lowrie RC Jr, Roberts ED, Conway MD, Karacorlu M (1993) Early and early disseminated phases of Lyme disease in the rhesus monkey: a model for infection in humans. Infect Immun 61(7):3047–3059

Philipp MT, Bowers LC, Fawcett PT, Jacobs MB, Liang FT, Marques AR, Mitchell PD, Purcell JE, Ratterree MS, Straubinger RK (2001) Antibody response to IR6, a conserved immunodominant region of the VlsE lipoprotein, wanes rapidly after antibiotic treatment of *Borrelia burgdorferi* infection in experimental animals and humans. J Infect Dis 184:870–878

Philipp MT, Wormser GP, Marques AR, Bittker S, Martin DS, Nowakowski J, Dally LG (2005) A decline in C6 antibody titer occurs in successfully treated patients with culture-confirmed early localized or early disseminated Lyme Borreliosis. Clin Diagn Lab Immunol 12(9):1069–1074

Preac-Mursic V, Wilske B, Schierz G, Holmburger M, Sub E (1987) In vitro and in vivo susceptibility of *Borrelia burgdorferi*. Eur J Clin Microbiol 6:424–426

Preac-Mursic V, Wilske B, Schierz G, Suss E, Gross B (1989) Comparative antimicrobial activity of the macrolides against *Borrelia burgdorferi*. Eur J Clin Microbiol Infect Dis 8:651–653

Preac-Mursic V, Patsouris E, Wilske B, Reinhardt S, Gos B, Mehraein P (1990) Persistence of *Borrelia burgdorferi* and histopathological alterations in experimentally infected animals; comparison with histopathological findings in human Lyme disease. Infection 18:332–341

Probert WS, Allsup KM, LeFebvre RB (1995) Identification and characterization of a surface-exposed, 66-kilodalton protein from *Borrelia burgdorferi*. Infect Immun 63:1933–1939

Roberts ED, BohmJr RP, Cogswell FB, Lanners HN, LowrieJr RC, Povinelli L, Piesman J, Philipp MT (1995a) Chronic Lyme disease in the rhesus monkey. Lab Invest 72:146–160

Roberts ED, Bohm RP Jr, Cogswell FB, Lanners HN, Lowrie RC Jr, Povinelli L, Piesman J, Philipp MT (1995b) Chronic lyme disease in the rhesus monkey.[comment]. Lab Invest 72(2):146–160

Roberts ED, Bohm RP Jr, Lowrie RC Jr, Habicht G, Katona L, Piesman J, Philipp MT (1998) Pathogenesis of Lyme neuroborreliosis in the rhesus monkey: the early disseminated and chronic phases of disease in the peripheral nervous system. J Infect Dis 178(3):722–732

Sauve C, Azoulay-Dupuis E, Moine P, Darras-Joly C, Rieux V, Carbon C, Bedos JP (1996) Efficacies of cefotaxime and ceftriaxone in a mouse model of pneumonia induced by two penicillin- and cephalosponin-resistant strains of *Streptococcus pneumoniae*. Antimicrob Agents Chemother 40:2829–2834

Schwan TG, Burgdorfer W, Schrumpf ME, Karstens RH (1988) The urinary bladder: a consistent source of *Borrelia burgdorferi* in experimentally infected white-footed mice (Peromyscus leucopus). J Clin Microbiol 26:893–895

Shadick NA, Phillips CB, Logigian EL, Steere AC, Kaplan RF, Berardi VP, Duray PH, Larson MG, Wright EA, Ginsburg KS et al (1994) The long-term clinical outcomes of Lyme disease. A population-based retrospective cohort study. Ann Intern Med 121(8):560–567

Snydman DR, Schenkein DP, Berardi VP, Lastavica CC, Pariser KM (1986) *Borrelia burgdorferi* in joint fluid in chronic Lyme arthritis. Ann Intern Med 104:798–800

Sonnesyn SW, Manivel JC, Johnson RC, Goodman JL (1993) A guinea pig model for Lyme disease. Infect Immun 61:4777–4784

Stanek G, Klein J, Bittner R, Glogar D (1990) Isolation of *Borrelia burgdorferi* from the myocardium of a patient with longstanding cardiomyopathy. N Engl J Med 322:249–252

Steere AC (2001) Lyme disease. N Engl J Med 345:115–125

Steere AC, Sikand VK (2003) The presenting manifestations of Lyme disease and the outcomes of treatment. N Engl J Med 348(24):2472–2474

Steere AC, Levin RE, Molloy PJ, Kalish RA, Abraham JH 3rd, Liu NY, Schmid CH (1994) Treatment of Lyme arthritis. Arthritis Rheum 37(6):878–888

Straubinger RK (2000) PCR-based quantification of *Borrelia burgdorferi* organisms in canine tissue over a 500-day postinfection period. J Clin Microbiol 38:2191–2199

Straubinger RK, Summers BA, Chang YF, Appel MJG (1997) Persistence of *Borrelia burgdorferi* in experimentally infected dogs after antibiotic treatment. J Clin Microbiol 35:111–116

Straubinger RK, Straubinger AF, Summers BA, Jacobson RH (2000) Status of *Borrelia burgdorferi* infection after antibiotic treatment and the effects of corticosteroids: an experimental study. J Infect Dis 181:1069–1081

Stricker R (2007) Counterpoint: long-term antibiotic therapy improves persistent symptoms associated with Lyme disease. Clin Infect Dis 45(2):149–157

Strle F, Cheng Y, Cimperman J, Maraspin V, LotricFurlan S, Nelson JA, Picken MM, RuzicSabljic E, Picken RN (1995) Persistence of *Borrelia burgdorferi* sensu lato in resolved erythema migrans lesions. Clin Infect Dis 21:380–389

Verma A, Brissette CA, Bowman A, Stevenson B (2009) *Borrelia burgdorferi* BmpA is a laminin-binding protein. Infect Immun 77:4940–4946

von Stedingk LT, Olsson I, Hanson HS, Asbrink E, Hovmark A (1995) Polymerase chain reaction for detection of *Borrelia burgdorferi* DNA in skin lesions of early and late Lyme borreliosis. Eur J Clin Microbiol Infect Dis 14:1–5

Wang E, Bergeron Y, Bergeron MG (2005) Ceftriaxone pharmacokinetics in interleukin-10-treated murine pneumococcal pneumonia. J Antimicrob Chemother 55:721–726

Wormser Gary P, Dattwyler Raymond J, ShapiroEugene D, Halperin John J, Steere A, Klempner Mark S, Krause P, Bakken J, Strle F, Stanek G et al (2006) The clinical assessment, treatment, and prevention of lyme disease, human granulocytic anaplasmosis, and babesiosis: clinical practice guidelines by the Infectious Diseases Society of America. Clin Infect Dis 43(9):1089–1134

Wormser GP, Schwartz I (2009) Antibiotic treatment of animals infected with *Borrelia burgdorferi*. Clin Microbiol Rev 22:387–395

Wormser GP, Dattwyler RJ, Shapiro ED, Halperin JJ, Steere AC, Klempner MS, Krause PJ, Bakken JS, Strle F, Stanek G et al (2006) The clinical assessment, treatment, and prevention of Lyme disease, human granulocytic anaplasmosis, and babesiosis: clinical practice guidelines by the Infectious Diseases Society of America. Erratum in: Clin Infect Dis 45:941, 2007. Clin Infect Dis 43:1089–1134

Wormser GP, Brisson D, Liveris D, Hanincová K, Sandigursky S, Nowakowski J, Nadelman RB, Ludin S, Schwartz I (2008) Borrelia burgdorferi genotype predicts the capacity for hematogenous dissemination during early Lyme disease. J Infect Dis 198(9):1358–1364

Yrianainen H, Hytonen J, Soderstrom KO, Oksi J, Hartiala K, Viljanen MK (2006) Persistent joint swelling and Borrelia-specific antibodies in *Borrelia garinii*-infected mice after eradication of vegetative spirochetes with antibiotic treatment. Microbes Infect 8:2044–2051

Yrjanainen H, Hyotenen J, Song SR, Oksi J, Hartiala K, Viljanen MK (2007) Anti-tumor necrosis factor-alpha treatment activates *Borrelia burgdorferi* spirochetes in 4 weeks after ceftriaxone treatment in C3H/He mice. J Infect Dis 195:1489–1496

Yrjanainen H, Hytonen J, Hartiala P, Oksi J, Vijanen MK (2010) Persistence of borrelial DNA in the joints of *Borrelia burgdorferi*-infected mice after ceftriaxone treatment. APMIS 118(9): 665–673

Zambrano MC, Beklemisheva AA, Bryksin AV, Newman SA, Cabello FC (2004) Borrelia burgdorferi binds to, invades, and colonizes native type I collagen lattices. Infect Immun 72: 3138–3146

Index

M. Embers (ed.), *The Pathogenic Spirochetes*, DOI 10.1007/978-1-4614-5404-5,

MIX
Papier aus verantwortungsvollen Quellen
Paper from responsible sources
FSC® C105338

If you have any concerns about our products,
you can contact us on
ProductSafety@springernature.com

In case Publisher is established outside the EU,
the EU authorized representative is:
Springer Nature Customer Service Center GmbH
Europaplatz 3, 69115 Heidelberg, Germany

Printed by Libri Plureos GmbH
in Hamburg, Germany